THE LOST ORCHID

THE LOST ORCHID

A STORY *of* VICTORIAN
PLUNDER *and* OBSESSION

SARAH BILSTON

HARVARD UNIVERSITY PRESS

Cambridge, Massachusetts | *London, England* 2025

Copyright © 2025 by Sarah Bilston
All rights reserved
Printed in the United States of America

First printing

Library of Congress Cataloging-in-Publication Data

Names: Bilston, Sarah, author.
Title: The lost orchid : a story of Victorian plunder and obsession / Sarah Bilston.
Description: Cambridge, Massachusetts ; London, England : Harvard University Press, 2025. | Includes bibliographical references and index.
Identifiers: LCCN 2024035466 | ISBN 9780674272606 (cloth) |
    ISBN 9780674300408 (epub) | ISBN 9780674300415 (pdf)
Subjects: LCSH: Orchids—History—19th century. | Orchid industry—England—History—19th century. | Orchid industry—America—History—19th century. | Orchid industry—Environmental aspects—History—19th century. | Orchids in literature. | Plant collectors—History—19th century. | Cattleyas—History—19th century.
Classification: LCC SB409.48 .B55 2025 | DDC 635.9/34720942—dc23/eng/20240904
LC record available at https://lccn.loc.gov/2024035466

CONTENTS

Color plates follow page 102.

    Prologue   1

PART I ~ THE CASE OPENS

1. Origins   *17*
2. A Naturalist Arrives   *25*
3. Trouble Brewing   *40*
4. Swainson Vanishes   *46*
5. "Old Antiquity"   *53*
6. The Russia Connection   *62*

PART II ~ EXOTIC

7. First Sight   *73*
8. False Trail?   *86*
9. Orchid Varieties   *97*
10. "Orchids for the Million"   *103*
11. Battles at the Gate   *115*

PART III ~ THE DARKLING PLAIN

12. A New Dynasty   *127*
13. Hatching a Plan   *135*

14	Doctoring Orchids	*141*
15	Darwin's Crisis	*151*
16	"I Much Want a Cattleyea"	*160*

PART IV ~ SURVIVAL OF THE FITTEST

17	"The Whole World's Gone Cattleya Crazy"	*173*
18	Fakes	*182*
19	Damage	*195*
20	Entering the Lists	*202*
21	Ericsson and Bungeroth	*214*
22	The Great Discovery	*226*
23	The Second Great Discovery	*239*
24	Errors	*247*
25	"Some Brazilian"	*258*
26	King of the Orchids	*268*
	Epilogue: Fall—and Rise	*287*

NOTES	*295*
ACKNOWLEDGMENTS	*365*
ILLUSTRATION CREDITS	*368*
INDEX	*370*

Few Orchids have so interesting a history . . . but it is extremely probable that we shall never hear the whole of the facts connected with this beautiful Orchid, for many of the links in the chain of evidence have been lost. . . . It is certain, however, that for many years every effort made to discover the native habitat . . . failed completely; collectors were repeatedly dispatched by the leading firms, and thousands of miles have been explored without the slightest success.

—Lewis Castle, 1891

Nature loves to hide. —Heraclitus, c. 500 BCE

THE LOST ORCHID

PROLOGUE

ON JUNE 1, 1891, a Swedish plant hunter named Claes Ericsson set sail from Barbados on a steamer bound for Pernambuco in the northeast of Brazil. He was in pursuit of the "lost orchid," a possibly mythical plant that had been desired and debated for seven decades. He was also seething.

Ericsson's employer was the internationally famous businessman Frederick Sander, a German-born forty-four-year-old with expansive premises outside London and New York. Bearded and imposing, Sander was approaching the zenith of a glittering career: his company specialized in orchids at a time when passion for the plants—"orchidomania" or "orchidelirium"—was raging around the world. Official "Purveyor of Orchids" to Queen Victoria, supplier to many of the world's crowned heads of state, Sander was also the kind of showman the era applauded. Commissioned to prepare the Queen's bouquet for her Golden Jubilee, Sander arrived at Buckingham Palace with what the startled London *Times* termed a "structure" fully seven feet in diameter, five feet high, and bursting with orchids culled from Britain's many dominions. The letters VRI, or *Victoria Regina Imperatrix*, were picked out in red blooms.[1]

Sander's staggeringly lavish nursery outside St. Albans, meanwhile, was one of the wonders of Europe. The shimmering pools and immense greenhouses overflowed with hundreds of thousands of orchids of every size, shape, and color, rendering dazzled visitors speechless. Sander's second branch, recently opened in Summit, New Jersey, aimed to capitalize on a growing enthusiasm for orchids in the United States, and Sander would soon open a third in Bruges, Belgium, in an effort to dominate the continental market as well.

Operating behind the scenes of glitz and glamor, meanwhile, were plant hunters like Ericsson. Sander & Co. was in the throes of battle with other large nurseries in Europe and North America for market dominance. Competition was intense, and Sander kept his dozens of agents on a painfully short financial leash. When interviewed by the press, Sander reported spending lavishly on their maintenance, up to £3000 (in today's money, £405,000 or $520,000) per hunter per year—but in private he fought with them over every last penny.[2] Plant hunters were often migrants, socially peripheral figures—rootless, working-class, ill-educated, and sometimes managing behavioral or physical disabilities. Gustav Wallis was born deaf and learned to speak at age six; Benedict Roezl, Sander's most successful collector, lost his hand in Cuba and collected orchids with a prosthetic hook. Claes Ericsson's chronic health issues included obesity and possible alcoholism. He worried constantly about debts and his elderly mother (when he remembered her). He, like many of his colleagues, fought almost every day with his employer and the Sander & Co. business manager, Joseph Godseff. Frustrated by Sander's penny-pinching and insistent demands, Ericsson gambled life, health, and sanity for a chance at money and fame.

The dream, for Ericsson and his fellow collectors, was the discovery of a new species or variety that would thrill the newspapers, sell for large sums, and ultimately bear the name of the plant hunter who retrieved it. The task was physically, mentally, financially, and practically exhausting. As Ericsson headed to Brazil, he knew that fellow hunter William Digance had died of fever at age twenty-eight just weeks earlier while hunting the "lost orchid" in the mountains outside Rio de Janeiro. Digance was the son of an impoverished agricultural laborer; desperate to send back something remarkable, Digance wrote to Sander that he was willing, if necessary, to "die in the attempt."[3] His trip was a disaster from start to finish. The inexperienced young man was tricked by other collectors and struggled with the language. He was baffled by the distances he had to cover, stunned by the amount of money everything cost, stranded by rising water and thunderstorms in rainy season, and he fell repeatedly ill. In his final letters, Digance protested he'd been badly prepared by Godseff, who assured him the journeying would be easy. The business manager could only make that claim, Digance wrote bitterly, "because he has not been here." It was easy to travel in Rio or along a railway, but the situation was quite different once one headed into the Brazilian interior "& everything has to be done with animals they are not English roads— it is a series of climbing up one mountain & down another through rivers & woods where one has to sometimes stop & cut a road. it is not very easy I can assure you." To add to his misery, Digance was unable to find much of anything—the region was "played out," as he put it, stripped of its botanical riches. "I dare say you begin to think I am a failure," he wrote hopelessly, seeing his one slight window of opportunity closing, "but I shall make up for it before I return to england."[4]

Ericsson set sail to continue where Digance left off but, he warned Sander: "I'm getting sick of this life." Still, he had no real intention of quitting. Ericsson hated northern Europe. He kept his visits home, to the "beastly weather" of Stockholm, to a minimum.[5] He also, unlike Digance, knew the job backward and forward; Ericsson had been working for Sander for four years.

Plant hunting required cunning, resilience, and a seemingly unshakeable sense of entitlement to the resources of other nations. Locating plants new to Europe, removing and transporting them across continents, required hunters to journey vast distances, spy on competitors, identify (and harass) locals with knowledge and expertise, hire work teams, hack apart trees, pack up living plants for arduous journeys, and interface with networks of bankers, merchants, and shipping agents. Ericsson was not concerned about the environmental consequences of stripping plants from their habitat, nor about the rights and connections of local communities to the land. Ericsson viewed plant hunting as a competition for resources between him and rival hunters, and on this occasion he was competing against a particularly storied and mysterious adversary, a German named Erich Bungeroth.

It was Bungeroth whom the Sander team suspected of knowing the location of the fabulous "lost orchid." After decades of fruitless searching, after the recovery of numerous plants whose likeness to the *vera* or "true" orchid tested the boundaries of scientific knowledge and the understanding of species to a breaking point, Sander's office believed they finally had a credible lead.

Erich Bungeroth was born in Prussia. In the 1880s he worked in a nursery outside Liverpool, where he lodged with a railwayman. He now hunted for Sander's most bitter rival in the business, L'Horticulture Internationale, operated by the Linden family, based in Brussels; Ericsson irritably termed them "Bung. and the Belgies" in his letters. Bungeroth was skilled at sourcing plants while leaving no trail: hunters used code names and pseudonyms when registering with local authorities to make it harder for competitors to learn their whereabouts. Bungeroth was, a baffled Ericsson wrote to his employer, a "mysterious person nobody seems to know." When Ericsson tried to shake information about Bungeroth from other German speakers he encountered on the road, the Swede discovered they seemed "spellbound" by his competitor.[6] But this was only the more reason to find him. Ericsson's first task in Pernambuco, then, was to track down the elusive Bungeroth, find the local teams he was using to source plants, and bribe them to work for Sander instead. His second task was to use Bungeroth's men to help him find the habitat of the lost orchid, strip all the plants of the population from the region, and send back crates of them as quickly as possible, by steamship, to Sander.

Where exactly this fabulously rare orchid grew, nobody knew. It had been hunted all over South America, most recently in Colombia, Venezuela, and in the mountains outside Rio. Large, with a particularly lustrous purple-and-crimson bloom, it was often termed the "Queen of the Orchids." Even today, it remains celebrated as the "pinnacle of orchid eminence, the epitome of floral beauty."[7] Many believed it no longer even existed in the wild. As James Veitch & Sons, another rival business to Sander and Co., stated in 1889, the orchid "is believed to have been exterminated many years ago, a belief unfortunately strengthened by the fact that no plants of it have been known to have been imported for upwards of forty years, although it has been diligently sought for by collectors sent out by English and other horticultural firms."[8] Highly desired, highly valued, it was the orchid after which a genus was named; it was also widely thought to be, and continues to be credited as, the orchid that incited the craze that would sweep across both sides of the Atlantic.[9]

Sander & Co. and many other companies had been on the hunt for the lost orchid for years. A single rare orchid could raise as much as 650 guineas at auction, around £95,000 or $121,000 today.[10] In April, two months before setting sail for Brazil, Ericsson learned Bungeroth had been sighted in Bolivia, then on the Rio Meta in Colombia, within the past year. "I hope I'm on the right track this time," he scribbled to Sander, reporting that he had feelers out with contacts to learn more.[11] Sander, meanwhile, was using his own information networks, managing his agents like a war general with troops on the ground. In the third week of May, Sander sent a telegram to Ericsson commanding him to leave Barbados for Pernambuco at once.

Ericsson landed in Brazil and cabled his safe arrival to Sander, following up with a letter to explain the latest: "I thought you would be anxious to hear from me." Bungeroth had indeed been collecting in the area, Ericsson wrote, and the Swede sent men off into the interior of the state in pursuit of more information. "I do not think B. has been further than Garanhus [a city in the region], I think in that neighborhood as far as I can make out at present." He continued, somewhat incoherently, "I can get to Garanhus from here in about a day so soon I arrive will write you full particulars." "I'm very glad to see a finish on this bussiness," he concluded; "it has taken a very long time."[12]

Indeed, the hunt for this particular orchid had taken nearly seventy-five years. Ericsson set off again after Bungeroth; Sander and his team in London anxiously awaited telegrams and letters from Brazil. The hunt entered the final chapters of a narrative that had electrified Europe and North America for decades.

THIS IS THE STORY of a decades-long international quest for an orchid. It is also about the story making that helped power the mania for orchids—and the stories left off the printed page.

It picks up events at what was, effectively, a pivotal moment in the history of globalization, of trade and communication between and among nations. The first transatlantic cable, patented in 1839, stretched across the Atlantic by 1866; thereafter the telegraph connected people and business interests across the globe, while fast-growing railway and shipping networks moved goods across continents, print media advertised to and informed an ever-expanding reading and purchasing public, and migrants and immigrants moved to take up new lives. Roughly 12 million people did so without choice. Enslaved Africans, compelled to leave their homes, traveled across oceans to labor in the economic service of others. For them, connections with home and loved ones were brutally severed. For the estimated 40 million Europeans who pursued new lives in North and South America, on the other hand, ties between old and new countries could plausibly be maintained—through the trade in familiar goods as well as letters, cables, and journeys home.[13] Colonizing powers strove to impose their institutions and values on the land they conquered and the people they subjugated; Britain especially pitched its language, religion, culture, literature, science, economy, and commerce as models to unite the world. Other nations, even those outside the British Empire, experienced huge pressure to adapt, to accommodate the values and demands of an empire so charismatically confident in its supposed superiority.

The orchid business was powered by the ships and railways that moved plants across continents; by cables and mail that connected networks of employers, collectors, bankers, and shippers; by scientific discoveries that shone new light on plant biology and ecology and that demanded, in turn, more and more plants for analysis; by advances in print technologies that blazoned fresh discoveries to the world; by free trade policies that aimed to ease the flow of goods and capital between nations but that exploited Black and Brown people and the poor; and by the extraction of the resources of other lands, including plants. As many scholars have shown, imperial expansion was a driver of global ecological change, with a goal of supposedly "improving" nature and a thirst for the knowledge of new plants to achieve that end. That drive was not limited to countries that the colonists directly controlled, either, for imperialism's energy was to sweep the whole world into its networks. The parameters of the British Empire in particular offered—to its supporters—both a set of possibilities and constraints they were determined to resist.[14] The boundary lines separating *ours* from *not-ours* were subjected to constant pressure: a growing infrastructure of roads, railways, ships, telegraph cables, trade routes, and banking systems were deployed to extend Britain's cultural and economic reach ever farther around the globe, producing what historians have called an "informal empire." British-employed botanists and plant hunters who traveled beyond the politically and militarily controlled borders of empire stabilized these networks as they looked for resources to fuel the interests of the nation for which they labored. While seeking their own fame and fortune, botanists and plant hunters also advanced their nation's economic, commercial, and political interests by pinpointing resources for extraction. "It is no accident," remark the authors of a recent study on eco-cultural networks, "that disciplines such as botany and ecology emerged in concert with economic concerns of finding more efficient ways of exploiting natural resources."[15]

Understanding orchidomania, in other words, begins with tracing the networks of a globalizing world that advantaged some and exploited much and many. This book follows one particular variety of orchid along and through those networks, from habitat to packing case to ship to customs house to herbarium and nursery, from advertisement to auction room to greenhouse, from horticultural competition to naturalist's table, botanical illustration to news article, imperial adventure romance, gothic horror story, and even, eventually, West End and Broadway musical. I examine some of the many stories told about orchids in the nineteenth century through letters, bills of lading, memos, catalogs, botanical slips, advertisements, newspaper columns, illustrations, and more, and ask: What words, what terms, were used to identify them? What stories were told, what narrative arcs were employed? How is the reader situated, positioned, by the text? How did each narrative attempt to achieve its purpose? What was *not* said, and why? Whose words, whose labor, is concealed?[16]

It was the gaps in one particular narrative, in fact, that first launched this project. A tale of a hauntingly lovely "lost orchid" appears in numerous nineteenth-century books and periodicals and reappears in modern histories of botany and horticulture. Yet the most basic facts were and remain the subject of intense disagreement. Even though many of the events were written about soon after they'd taken place, I found wildly differing accounts of where the orchid came from, who received it, who first flowered it, where and when it was named, and by whom: so many confusions crept in that even a much-cited source like Merle A. Reinikka's classic *A History of the Orchid* made claims about provenance that were directly at odds with first-person accounts written by the people involved.[17] The business of the orchid's supposed disappearance and rediscovery was even more puzzling. Some argued that the lost orchid was found in the 1830s; many disputed this. Some said a few rare specimens appeared over the decades in St. Petersburg, in aristocratic collections in

Britain, in those of wealthy men in France, Germany, Italy, and the United States; others did not.[18] Then it was rediscovered at a ball in Paris, in a lady's corsage, historian of science Jim Endersby explains (with wry amusement at the obvious ridiculousness of the tale), where "an orchid enthusiast attached to the British legation saw it. He looked once, twice (in fact as often as etiquette would permit a gentleman to do). Convinced that this was the lost orchid, he somehow obtained the flower and an expert confirmed his guess as to its identity."[19]

Others claimed it was rediscovered when a collector met with rubber planters in the jungle, who happened to mention the whereabouts of a gorgeous orchid.[20] Some Victorian historians suggested it was found when a nurseryman's agent spied it on a drawing in the British Museum; others pointed to a painting of the orchid's close relative at the 1889 Paris World's Fair; still others suggested it was rediscovered when an entomologist named Moreau sent a collector to Brazil. The collector sent back fifty orchids along with his beetles, and "By an unbelievable coincidence, Frederick Sander stopped to visit Moreau when the plants were just starting to flower" and realized it was the lost orchid.[21] I found it strange that an orchid so desirable, the object of so many periodical and newspaper column inches in the nineteenth century and of historical interest today, should be the subject of so much disagreement, to the point that scholars past and present regularly concede their "unbelievable" stories strain the limits of credulity.

My search for more information was advanced by a remarkable discovery. An extensive archive of documents between Sander and his plant hunters lay almost forgotten at the Royal Botanic Gardens, Kew. These documents reveal that key figures, in the rediscovery phase especially, had much to conceal. Letters, cables, memos, and bills of lading issued by plant hunters, botanists, collectors, and businessmen are not necessarily authentic statements of fact. Yet the existence of private papers that agreed with one another but conflicted with public statements issued at the same time, by the same people, suggest that what was published was constructed for a purpose other than the expression of known facts. What was that purpose? Why did some even participate in the distribution of what we might today call "fake news"? This book aims to tease out some of the answers to these questions, to explore the changing stories that structured and sustained orchidomania in print over decades, even as it also works to recover the narratives of those whose work and expertise rarely made it onto the published page.

The Lost Orchid, then, brings a range of texts into conversation with one another, from letters and other unpublished documents to botanical illustrations and text in printed works of botany and natural history, horticultural magazines aimed at the new middle classes (the "million," as the Victorians called them), novels, short stories, and nonfiction memoirs. Like many scholars before me, I begin from the premise that nineteenth-century science and literature were intimately intertwined. Indeed there was, as George Levine puts it, almost "one culture" in the period, influence flowing powerfully in both directions. As Gillian Beer observes, in her pioneering study of literature's influence on Darwin and Darwin's influence on literature, "The traffic . . . was two-way."[22] Naturalists and novelists struggled together to make sense of a vast globe, that is, to grasp the power of kinship ties in an era of increasing mobility, to make sense of strangers, to comprehend the rise and fall of populations, to understand growth, transformation, and metamorphosis in an era of rapid change, to make sense of history while unlocking the potential of the present and future, and to identify an overarching structure or system that might give meaning in a world of happenstance. Novels answered these many challenges by organizing events around plots in which, for instance, families are remarkably reunited, dangerous strangers are appropriately identified, hidden secrets from the past are progressively uncovered, virtue and diligence are rewarded. Works of natural history, printed texts about plants and the environment, not only explored the same issues but often drew on similar plots, characters, discursive strategies, and techniques.[23]

Letters by hunters like Ericsson speak to similar concerns. But the stories they tell in those letters typically *narrativize* orchids and plant hunting rather differently. "Storytelling, it must be recognized from the start, is always associated with the exercise, in one sense or another, of power, of control," Michael Hanne remarks, in *The Power of the Story*.[24] Orchid hunting in Victorian published works typically presented hunters' courage and derring-do as leading to seemingly inexorable successes, to accumulation and the acquisition of scientific knowledge. In the letters, however, diligence is often *not* rewarded; collaborations dissolve into conflict; the reaches of empire and European knowledge meet firm resistance. Thus even as many of the letters express, as Lynn Voskuil notes, pro-imperialist attitudes and "a blatant disregard for the effects of large-scale plunder," they simultaneously evoke plant hunting as an endeavor that forced them to confront, time and again, their own and the empire's limitations. This narrative—of uncertainty and failure, of circling rather than advance, of loss rather than accumulation, of growing bafflement in place of certainty—complements and complicates the more familiar tales of imperialist progress, European superiority, and even human exceptionalism that, Voskuil notes, so often underpin published horticulturalist and botanical literatures.[25] From freight to scientific specimen, from transformative spectacle to commodity to "friend" and Grail, orchids played many different roles in a broad array of published and unpublished texts.[26]

To reintegrate lesser-known narratives into the larger story of orchidomania, I approach the unpublished letter as itself a literary form. I follow in the footsteps of literary historians like Christopher Hager, who adopt "a capacious definition of literature that embraces the richness of image, figure, and narrative in all written expression" and who combine "methods of reading inspired by history, composition studies, and literary scholarship" in their interpretative approach. "There is more to the letters than the information they contain," Hager argues in his treatment of Civil War correspondence, encouraging us to think beyond the words and what they literally say.[27]

In the case of orchids, letters between hunters and their mentors, patrons, editors, and employers often consisted of lists of plants and places visited, with formulaic statements acknowledging receipt and well-wishes to an addressee's family. On first glance, these details may seem, at best, to offer mere data points for a historian to sweep into larger narratives culled from better-known published texts.

But there are ways for us to recover the "richness of image, figure, and narrative" in these documents—to retrieve other, less familiar narratives beyond data points. Who were these writers? I aim to recover as much information as possible about who was writing to whom, where they lived, and what else was happening at the time of writing, paying particular attention to family lives, children, partners, elderly parents, illnesses, holidays, housing worries, debts, and money. I follow and try to identify figures mentioned by the letter writer, even just in passing: who were they? Where did they come from, what were their jobs, what parts did they play in facilitating the writer's journey, the movement of plants, people, and capital? Then, too, we can learn much from *how* the writers write, the ways they respond to the challenges of form. Letters were "an established genre that long had been the domain of the upper and middle classes," Hager points out, and only slowly became "adapted . . . to the needs of common people."[28] Writers' literacy issues and / or struggles with the epistolary form may be read not as evidence of ineptitude or failure but as versions of battles with other, larger social, political, and infrastructural systems. I have found this especially helpful when reading letters by semiliterate plant hunters, like Ericsson, in a language to which he was not native, whose challenges on a torn page of foolscap point to those on the steamer, at the post office, in the bank.

Plant hunters wrote because they had to. Young, working- and middle-class botanists earlier in the century needed to account for their time with wealthy mentors and patrons. Commercial hunters, later in the century, were required by employers to give detailed accounts of their activities almost minute-by-minute: money for reimbursements and advance payments for expenses were all dependent on receipt of satisfactory reports. Frederick Sander demanded information on the area, plants found and their condition, the number of boxes shipped and their routes, local people participating in plant extraction, and also detailed spying notes on rival hunters in the region. The resulting letters were obviously not spontaneous efforts to communicate feelings, like the kinds of documents one imagines hunters penned to families and friends, nor were the authors of these texts aspiring to produce high-class epistles for publication. Writers do deploy epistolary conventions, but the bulk of each letter is shaped around a form that was worked out on-the-go by them, with their employer. Through the back and forth of sending and receipt, hunters learned the features of the letter-form their employer or, earlier in the century, patron / mentor, demanded: "write me longer letters," Sander scolded Ericsson.[29] Thereafter, like a serial author conforming to the demands of an editor, letter writers worked to reproduce the required form week in, week out, in all conditions, on little money, from all over the world. At times, however, for all that the hunters' survival depended on meeting the employer's expectations, authors *disrupted* the form of the letter in ways that suggest an almost overwhelming urge to disrupt the systems in which they operated. Dismay over lack of money, anger about the driving demands placed on them, the physical shakes of fever all led writers to put down their pens. Letters and census data indicate that many of the men doing the hunting were drawn to the work because they lacked more stable means of support; what they commonly express, on page after damaged page, in the margins, in postscripts, and sometimes in the middle of a paragraph, is the feeling of being unjustly treated.[30]

This is part of the story—or, more properly, stories—*The Lost Orchid* aspires to uncover: the words emerging on thin, damaged paper, alongside other stories in the press of triumphant, spreading orchidomania, of increasing access to orchids and the world's most precious resources for Britain's fast-growing middle classes. Exploring how letters and diaries as well as ledgers, memos, telegrams, botanical illustrations, catalogs, magazines, how-to horticultural texts, fictions, and drama narrativize orchids and plant hunting, I set scenes of *Boy's Own* high drama beside the exploitation of working-class migrants in Europe. Through census data, baptismal registers, local newspapers, and even a remarkable oil painting by a Black Brazilian artist, I point to the lives, labors, and insights of peoples in South America—Brazil especially—left off the printed page. (I am deeply conscious that much important work, in this area especially, remains to be done.) Moving between published accounts and archival documents and ephemeral media week by week, month by month, I show the practices of erasure at work as, particularly later in the century, European businessmen and journalists deployed the tropes and themes of contemporary imperial romance fiction to stoke consumer appetite for orchids. This book uncovers some of the many complex stories that were actively concealed, the voices of the socially marginal and colonially oppressed that were silenced in the process, shining fresh light too on the mechanics of plant hunting, how plants got from point a to point b. Setting a wide array of texts side-by-side helps us identify more clearly the construction of public-facing stories that sought to inflame consumer ardor and incentivize further resource extraction, more colonialist violence; I hope this approach will help us look again, a little more closely, at the mechanisms by which a powerful imperial machine swept along on its self-justifying, violent course.

At a time when the Victorians' loud accounts of themselves have been mobilized for expressly political purposes, as when Brexiteer Jacob Rees-

Mogg writes of the Victorian "heroes of old who possessed belief and patriotism, a sense of duty, a confidence in progress and knowledge of civilization," I am particularly keen to make accessible some of the many stories the nineteenth century strove, so assiduously, to conceal.[31] In order to do this, my book makes judicious recourse to story itself—foregrounding the feelings, ambitions, fears, and everyday struggles of its protagonists. Meanwhile, the scholarship that grounds this book is largely detailed in the endnotes. Readers interested in learning more about my archival research and the scholarship with which the book engages—on, for instance, the culture of Victorian plant collecting, the rise of middle-class consumer culture, the flourishing of print media, the human and environmental consequences of British imperial expansion, the expanding boundaries of scientific knowledge—will find sources and broader debates documented there. In this way, I hope that the endnotes will also serve as a guide to further reading.

I'm not unconscious of the risks of my narrative approach, however. A literary history must confront the limits of what we can know about the past and acknowledge the potential seductions of storytelling and narrative invention. Hayden White famously pointed to the dangers of storytelling in historical analysis: we will likely edit out disorder and incoherence, he warns us, in our urge to find, recover, and make sense.[32] Yet it is obviously impossible to write a narrative and avoid narrativization of one sort of another; choices must be made, facts, events, texts placed in some sort of order. To explore, as this book aspires to do, the diverse ways in which orchids were narrated, the many characters they played, is not to show that story is inexorably dangerous but to remind us that story inexorably shapes our lives.[33] Since writers can never be absolved of responsibility for our choices, we must subject them carefully, constantly to question.

With this imperative in mind, I'm influenced by the arguments of intellectual historians like Marci Shore, who argue that the writer's job is "to evoke what was at stake—and for whom—at different moments." In order to do this, historians must help readers *feel*. "Our mandate," Shore explains, "is a tactile one"—to help others "touch the contours of the burden." Shore urges historians to try to capture how ideas were experienced by people like us, living complicated lives, in a way that encourages "an imaginative leap into another time and another place." "Only in this way," Shore explains, "do we have a chance to make the impact of ideas—their historical heaviness—felt. The risk is that, otherwise, the weight of the ideas is lost."[34]

ORCHIDOMANIA, IN ITS VICTORIAN FORM, seemed over by the time the Great War began, yet the market in orchids continues to be a multimillion-dollar industry, and books about orchid hunting—like Susan Orlean's *The Orchid Thief*—remind us that there are still fanatics willing to pay a fortune to feed their obsession. Orchid hunting was, and is, about taking; and Brazil, where Ericsson quested for the lost orchid, is a nation whose deforestation and exploitation impacts the well-being of its peoples and environment, and of our globe. As we navigate climate catastrophe, as deforestation threatens the most biodiverse country in the world, the story of the lost orchid reminds us to reflect on how, in a postcolonial, postindustrial world, we treat the environment we share.

But for Claes Ericsson—about to plunge into the Brazilian interior in July 1891—questions about the orchid's meanings or the hunt's impact on the environment or the rights and struggles of Brazilian peoples were not uppermost. Other issues plagued him as he sweated and stumbled along narrow paths on his mule: Would he find his rival, Erich Bungeroth? Had Bungeroth found the precious orchid they both sought? Could Bungeroth's local agents be bribed on the shoestring budget Sander allowed him? Could he definitively identify the long-lost orchid, given uncertain documentation and perplexing questions about the slippery nature of species? Was there a population of plants left to be collected, or had they all been destroyed? Could he survive the many threats—fevers, falls, encounters with wild animals—that killed off plant hunters around the world? Would he live to achieve lasting fame as the Sir Galahad of the hunt—or would his trip, like that of Digance and many others, end in ignominious death?

"Few orchids have so interesting a history," marveled the writer Lewis Castle in 1891, as "the lost orchid," "but it is extremely probable that we shall never hear the whole of the facts connected with this beautiful Orchid, for many of the links in the chain of evidence have been lost."[35]

This book represents my own hunt to recover those links at long last.

PART I

THE CASE OPENS

The firearm overcame my arrow,
My nudity became scandalous,
My language was kept in anonymity,
They changed my life, destroyed my land.

"TERRITÓRIO ANCESTRAL,"
MÁRCIA WAYNA KAMBEBA (2013)

1

ORIGINS

EUROPEAN PLANT HUNTERS like William Digance and Claes Ericsson figured themselves as brave discoverers of "new" plants, but of course there were orchids growing in what we call *Brazil* long before Europeans arrived. Long even before Theophrastus began to organize and systemize botanical knowledge in *Enquiry into Plants,* in 370–285 BCE, designating the "orkhis" in a literary text, people lived and interacted with orchids.[1]

The northeast of Brazil today comprises nine states, including Pernambuco and, bordering it, Paraíba, Alagoas, Ceará, Piauí, and Bahia. The region is rich in archeological sites offering tantalizing glimpses of human occupation stretching back at least 30,000 years. Rock art at Serra da Capivara National Park in Piauí, for instance, shows, in bold red and ochre drawings, animals and people—men, women, even pregnant women—interacting with the landscape as they dance, walk, and process in formation across the undulations of rock. Figures move around trees, touching them, perhaps worshiping them with uplifted arms. Deer commonly figure in these scenes, as do armadillos, jaguars, and rhea, a now-extinct kind of ostrich.[2]

Numerous tribes had inhabited the northeast for thousands of years before Portuguese colonists arrived at the turn of the sixteenth century. The larger of the tribes in the region included, along the coastline, the

Tabajara, whose approach was cautiously cooperative with the new arrivals, and the Caeté, who chose more active resistance. These and many other tribes bewildered the invaders with their different languages, cultural practices, and intricate social organizations. The new arrivals tended to simplify these complexities rhetorically in texts that smoothed their confusing differences into, as historian Hal Langfur puts it, a simple binary of potential "convert" or "cannibal"—that is, the manageable or the dangerous brute savage. Both figures seemingly required the "civilizing" influence of Europeans—that is, both limited and limiting alternatives, forged in European texts, supported the colonizers' ways of understanding the world and facilitated the expansion of their power.[3]

Written descriptions of Indigenous peoples from this era, shaped by European stereotypes and agendas, are structured to tell a story of need and lack that European arrival and colonization apparently supplied. Such texts have little to tell us about fact, about what "really happened," of course. But they are opportunities to identify in action the kinds of narrative strategies Europeans used to manage such encounters in and through text and thereby justify (at least to themselves) exploitative and colonial practices. Glimpses of complex tribal practices and beliefs may emerge in the silences, the margins, in what is left out or noted as absences.[4]

The first written account of the Tupinambá people, for example, who lived to the south of the Caeté, was penned by Pero Vaz de Caminha, a secretary and government official accompanying Pedro Álvares Cabral, commander of a fleet of thirteen ships, in 1500. Caminha stressed, in a formal letter to King Manuel I, the ignorance, simplicity, and baffling (to him) lack of clothing of the people they met when they came ashore. "They were dark brown and naked, and had no covering for their private parts," Caminha explained, making the point of their nakedness often, in ways that suggest he was both fascinated by their attitude to their bodies and keen to emphasize their difference, their lack of "civilization," hence their need for Christian teachings.

He was confident the people could be easily converted, he wrote, given their willingness to honor the wooden cross the Portuguese set up in the country they decided to call, as soon as they saw it, the Land of the True Cross. Almost the first thing the new arrivals did, in fact, was gather wood for the cross: tribe members walked with the new arrivals to set the cross in the ground, and "When the Gospel came and we all stood with uplifted hands, they arose with us, lifted their hands, and stayed like that till it was ended. After which they again sat, as we did"; only they "were so silent that I assure Your Majesty it much increased our devotion."

It clearly never crossed the writer's mind that the people of the tribe might be drawing on their own devotional practices as they raised their arms to the wood. Such practices were invisible to him. Caminha wrote that the people he met were lacking in spirituality, reading them as a community of "no worship." They were blanks, empty vessels, he informed the king: "Any stamp we wish may be easily printed on them." The complexity of their existing spiritual, social, and cultural customs and traditions appears only briefly in his text, and in passing—Caminha reports tribal people made a red paint from berries for a skin dye, for instance, and painted their foreheads with a thick black line; they pierced their lips, shaved their eyelids and eyelashes, and forged necklaces of "spiky green seed-shells off some tree." Each of these practices is listed with little commentary, for Caminha is focused only on what they looked like (the fact that the black line looked like a ribbon, for instance) to *him*.[5] Reading through his belief system, his own frames of reference, Caminha asked no questions, either of himself or the tribe members, about what their customs signified to *them*.

The next explorers sent by King Manuel, in 1501, met people from the Potiguara tribe, to the north of the Tabajara. That encounter unfolded rather differently, if the Italian merchant Amerigo Vespucci (after whom America is named) is to be believed. The tribe members he said he encountered on his Third Voyage were considerably less receptive to newcomers. Indeed, in Vespucci's telling, in a letter to the head of the Florentine Republic, they were dangerous, bloodthirsty, duplicitous monsters. Sirenlike women gazed in seeming admiration at one of the first crew members to advance, touching the young man lovingly—even as another woman advanced on him with a club and knocked him down dead. While the men of the tribe rushed to attack the rest of the shocked crew with bows and arrows, the women tore apart the body, then roasted and ate it, showing pieces with glee to the invaders. "What shocked us much," Vespucci wrote, "was seeing with our eyes the cruelty with which they treated the dead, which was an intolerable insult to all of us."[6]

Vespucci often embroidered the stories he addressed to people of power; he assured Lorenzo di Medici, in another letter, that tribal people "live for 150 years, and are rarely sick."[7] Still, some tribes may well have practiced cannibalism as a means of symbolically ingesting the strength of the enemy, though scholars have argued that the amount of cannibalism actually performed was almost certainly exaggerated in order to emphasize the animality of the tribespeople; to (as Langfur puts it) "condemn them as savages, to legitimate their slaughter, to justify their enslavement and the seizure of territory."[8] These first, simplistic stereotypes were repeated in stories by Vespucci and others to further justify the culture brought by the European arrivals by claiming the barbarism of those who were oppressed and exterminated.

Certainly, tribespeople seem to have treated the *land* they inhabited with dramatically less violence than those who would follow them. What struck almost every European at the turn of the sixteenth century was the richness and teeming diversity of the landscape they encountered in the "New World"—Caminha, gazing at the astonishingly large numbers of birds, noted that "the trees are very tall and thick and of an infinite variety."[9] Vespucci, too, stressed "the resources and fertility of the land" in his letter to di Medici.[10] Over the next three centuries new arrivals continued to gasp at the environment, routinely referring to the forests and jungles of Brazil as "untouched" and "virgin." Because the land was not mined nor the forests threaded with roads, because animals and plants thrived in rich diversity, and because natural resources were not being used up faster than they could be replaced or repaired, the colonists viewed Brazil as "untouched."

Terms like "untouched" imply and presume inaction and ignorance on the part of the Indigenous tribes. Such language points to the colonizers' efforts to reframe the act of violently appropriating the land of others as somehow moral and just by indicating that the people living on it were ignorant and incapable of realizing its potential. But the land was obviously not "untouched." Between five and thirteen million people subsisted on and with it: the more than one thousand tribes took a range of approaches to the landscape, depending on the kinds of environments they inhabited (forest, savannah, coastal, mountainous, scrubby, grassy, wet, dry, hot, cold, and so on), as well as their spiritual and cultural practices. Whether as hunter gatherers or semisedentary agriculturalists, the land

and its resources fed, housed, warmed, clothed, healed, inspired, solaced, and connected them.[11] Tribal approaches to the environment typically involved interacting with it in ways that were invisible only to newcomers acclimated to a far more destructive, extractive approach.

Archaeologists, anthropologists, and historical ecologists are still learning to appreciate the complexities of precolonial tribal communities and their interactions with the land. Using a range of scholarly techniques, they aim to help make visible what was ignored, dismissed, and then near-exterminated, to understand afresh complex systems of communication that were rarely recorded on a page, in ink, or in print. The ways in which Indigenous communities transformed the forests emerge in satellite photographs: for instance, scholars have discovered traces of settlements in the Amazon in the Upper Xingu region connected by a network of pathways and bridges from around 1200–1600 CE. There is also evidence of agriculture and artificially created ponds, dams, and trenches. Studies of plant life, meanwhile, indicate cultivation practices along tracking and hunting trails oriented around food production and medicines, while soil analyses reveal the use of fires in cooking and agriculture—the alkalinity of soil in certain areas indicates the presence of charcoal and ash, while traces of ceramic, bone, and kitchen waste in the soil reveal further evidence of occupation dating back millennia.[12]

Complex, detailed vocabularies and classification systems for plants, animals, and insects, still used by some Indigenous peoples today, express an immensely subtle, mindful relationship with the environment that has been passed down orally through generations. (Traditional knowledge is also reinforced, Tracy Guzmán points out, through videos, newspapers, e-books, and websites that help sustain and nurture traditions among more urbanist Indigenous communities.)[13] In the Ka'apor language, for instance, a range of nuanced terms for the verb "to plant" depend on how the seed enters the soil. The Kayapó in the Amazon recognize fifty-six species of stingless bee divided into fifteen families, depending on how the bees behave, where they visit, their flight patterns, sounds, nest structure, secretions, shape, size, smell, the quality and quantity of their honey, pollen, wax, and more. The Kayapó distinguish eight types of forest, eight types of Cerrado, or tropical savannah, with named transitions between, while the Matses people of Amazonian Peru identify fully 178 habitats in their area. This fine-grained approach to the natural world is shared by other tribes too—the Mēbêngôkre distinguish multiple ecosystems along the spectrum from forest to savanna, with names for up to nine different types of savanna alone.[14] Meanwhile Indigenous scholars, writers, artists, and activists take key roles in the critically important business of protecting the environment and rethinking land use, "grounded in the notion of the natural world [as] sacred."[15] The Portuguese, at the start of the sixteenth century, saw little of what was and would remain for centuries a steadfast, sophisticated, discriminating relationship between the tribes and the landscape. Focused on what they understood and could use, they recorded only blankness and absence, what was *not* there, "empty vessels."

In Caminha's written account, two very different approaches to natural resources and the environment meet. Almost as soon as the Portuguese fleet arrived, the crew began cutting timber. The men carried some wood to their fleet to store it for the future, and they built their cross, using an iron tool the tribespeople examined with interest. Caminha reported that the tribe cut timber with less-efficient stones fashioned into wedges. The Portuguese approach—extracting natural resources for both immediate and future needs—was in contrast to that of the tribespeople, according to Caminha, for they took to fulfill short-term requirements. They did not practice farming, Caminha claimed: they "do not plough or breed cattle. There are no oxen here, nor goats, sheep, fowls, nor any other animal accustomed to live with man. They only eat this *inhame* [yam], which is very plentiful here, and those seeds and fruits that the earth and the trees give of themselves." Caminha concluded, with some wryness, of the tribespeople: "Nevertheless, they are of a finer, sturdier, and sleeker condition than we are for all the wheat and vegetables we eat."[16] He recognized, in other words, that this approach did not harm and perhaps even benefited humans. But he did not suggest learning from this approach nor express the urge to understand it further, nor did he question the premises of extraction. Extractive practices, by the time Claes Ericsson arrived some four hundred years later, would be turned squarely on the immense botanical riches of Brazil, both those Caminha noticed when he arrived and those he evidently did not.[17]

Orchids seem to have been invisible to Caminha. He made no mention of them, at least: "Look as we would, we could see nothing but land and woods, and the land seemed very extensive," he reported, unconsciously foregrounding his inability to see what he did not understand. "Till now we have been unable to learn if there is gold or silver or any other kind of metal or iron," he went on, focused on the riches that could be gained through mining. Still, he mused, "There is a great plenty, an infinitude of waters. The country is so well-favored that if it were rightly cultivated, it would yield everything."[18]

There was indeed "great plenty." Brazil bursts with a diverse range of flora and fauna, to this day home to around 13 percent of all the world's known species. Orchids were surely among them on the day Caminha's gaze ranged across the forest, for they have inhabited our Earth in some form for millennia. The oldest known orchid fossil is a pollinia, or mass of pollen grains, on the back of a bee trapped in amber that dates to the Late Cretaceous period (76–84 million years ago).[19] But orchids were not the focus of the colonizers' extraction practices, which over the next three hundred years focused especially on trees (particularly Brazilwood, logged almost to extinction), sugar, coffee, and, from 1692, gold. Forests were destroyed and transformed into plantations, mines, and cattle ranches. Railways and roads were thrown down across the country to move goods from the interior to ports. And human beings were dragged by the million from Africa and forced to labor in the harvesting of sugar and coffee while the Indigenous population was decimated by around 90 percent. Millions died, either of new diseases or violently, in conflict and by forced labor.[20]

Tribespeople likely interacted with orchids, if contemporary practices are anything to go by: some species of orchid are used by the Kayapó in the Amazon for contraception, for instance. Tribespeople also remove orchids from the forest and relocate them nearer to hand to facilitate medicinal applications.[21] The orchids are made accessible, in other words, though with an eye to their organic growth habits; many species of Brazilian orchids grow on, in, and around trees and shrubs. They exist in a complex relationship with other plants, insects, and the environment in ways that botanists and horticulturalists in Europe were just beginning to understand by the late eighteenth century, a period of intense and rising engagement with plants and ecology.

Orchids were becoming a focus of scholarly investigation across Latin America, too, at that time, among intellectuals, naturalists, and colonial bureaucrats. In Rio, naturalist and Russian consul Georg Heinrich (or

Grigory) von Langsdorff, for instance, was a key point person for new arrivals in Brazil.[22] Access to an engaged, resident community of thinkers and intellectuals in South America was transformative for incoming European arrivals like Prussian Alexander von Humboldt, Englishman Joseph Banks, and other, later generations of explorers, scholars have shown. Indeed, it is revealing, Jorge Cañizares-Esguerra suggests, that so many European naturalists would be celebrated as innovators for the research they published *after* they journeyed to the continent. It was not just the flora and fauna that fostered intellectual flourishing in the careers of so many who voyaged there, though European texts downplay, even erase, indebtedness to the intellectual communities new arrivals encountered. Cañizares-Esguerra, for instance, building on the scholarship of Catalan geographer Pablo Vila, identifies close relationships between Humboldt's work on the geographical distribution of plants in the Andes and the work of Colombian naturalist Francisco José de Caldas and, more broadly, established traditions of writing in South America about the Andes as "a providentially designed space."[23]

What would become known as the "lost orchid" was not, through the first decade of the nineteenth century, distinguished by scholars or naturalist explorers for particular notice. *Epiphytic* in habit—meaning that it grows on other plants but is not parasitic—it loves clambering in tall canopies; it enjoys a range of possible habitats, including the silk-cotton tree, the floss-silk tree, a fig tree, and a tree from the mimosa family. It can also be *lithophytic*, growing on rocks, boulders, and cliffs in particularly hard-to-reach places. It co-exists with other orchids, bromeliads, and lichens and prefers high altitudes (between 500 and 1,000 meters, or 1,600–3,300 feet), with sunshine, high humidity, and nightly dews.[24] It typically begins its new growth in the late Brazilian spring—September / October—and flowers starting in the summer, November / December. Flowering extends into the autumn, deep into March, after which the orchid enters a period of dormancy; its seasons vary a little with the rains. The orchid's white roots extend on either side of it, thrusting into the humid air to find nutrients.[25]

But the fortunes of this particular variety of orchid were about to change dramatically; for one day in late 1816, some seven decades before Bungeroth and Ericsson, British naturalist William Swainson arrived in Brazil animated by a passionate, chaotic, destructive urge to "discover."

2

A NATURALIST ARRIVES

WILLIAM SWAINSON LANDED in Recife, on the coast of Pernambuco, in late December 1816.[1] Rather like Ericsson, Swainson was in difficult financial circumstances. But while the Swedish plant-hunter's battles, at the end of the century, were with a close-fisted boss, Swainson was incensed by the lack of support he'd received from the British government to fund a specimen-hunting expedition in pursuit of "new" species.

Swainson was the son of a customs collector, and he did not have the gift of charisma. He often rubbed people the wrong way, and throughout his career he struggled to receive the recognition he felt he deserved. Jealous and stubborn, Swainson had big reproachful eyes and crinkly black hair that, as a biographer wryly observed, "early deserted his front and crown." He also had a speech impediment and possibly dyslexia: certainly, his teachers despaired of him. He was forced out of formal schooling at fourteen, judged as possessing "not the least aptitude for the ordinary acquirements" of education.[2] He became a clerk in the Liverpool Customs office and struggled with that too, but the Napoleonic Wars soon opened doors to opportunity. Swainson's father managed to arrange a job for the young man in the Commissariat, and he sailed at eighteen to Malta, Sicily, and Greece, where, like an early Gerald Durrell, he was fascinated by the wildlife he encountered. A man who struggled to get out a word, who reported lifelong difficulties with languages, Swainson found that whistling and humming were all it took to arrest a Sicilian lizard. Entranced by the sound, the little creatures stood and examined him; Swainson sat for hours and watched them in return, taking careful notes.

Throughout the nineteenth century, travelers and explorers faced epidemics and disease, such as malaria, yellow fever, scarlet fever, dysentery, tuberculosis, and cholera. Swainson's collections in Malta in 1814 were halted by an outbreak of the plague, which forced him into lockdown for two months. At least it allowed him to get some writing done: "I was almost sorry, on my own account, when our street was released from quarantine," he reported, though he also admitted he spent a lot of time carrying stones up and down a staircase to pass the long hours.

Then in Palermo in 1815, after lockdown was lifted, he discovered "some new or little known species of Orchidae: natives of this island."[3] Swainson was intrigued: could the curious roots perhaps be packed up and shipped safely to England? Imported orchids were still rare in Britain: the few that were collected (by officers in the navy and merchant service, typically) tended to die on the sea-journeys home, poorly packed and exposed to salt air and water. Those that made it over generally failed to thrive in cool northern climes. For this reason, many regarded their cultivation at that time, according to the journal *Botanical Register,* as "hopeless."[4]

Undeterred, Swainson wrote to James Shepherd at the Liverpool Botanic Garden to ask for advice on packing up orchids. Shepherd's eccentrically spelled reply, preserved in the library of the Linnean Society, London, exudes uncertainty:

> Dear Sir,
>
> In reply to your note, I should think the Plant should be carefully taken out from the rest, root as well as the other parts are generly Figured of any of the Orchis Tribe & raped up in a little soft Papper & then Packed in a small Tin Box with soft Moss, which will keep it alive & prevent the Plant from Moving in the box. I should imagine that by this Method it May be convey'd to Its Place of destination in safety -
> I am yrs Sir,
> Yours very
>
> John Shepherd[5]

Perhaps Swainson's equipment looked a little like this, from George Graves's *The Naturalist's Pocket-Book, or Tourist's Companion* (London: Longman, 1817). 1. "A Box of Plants packed and shut down in inclement weather"; 2. "A Cask prepared for transporting Plants."

Shepherd's evident doubts ("I should *think*," "should *imagine*") did not dissuade Swainson. He collected up the roots "in a small Box" and sent them with an awkward, excited letter introducing himself to Sir James Edward Smith, the eminent botanist and Linnean Society president. And so, quietly, an important new step was taken in the history of orchids in Europe.[6]

Remarkably, the Sicilian orchids survived. Smith responded in January 1816 with gratifying warmth, beginning a relationship that most likely helped Swainson win election as a full fellow of the Society in December of that year; Smith would later write job references on Swainson's behalf. A species of the Sicilian orchid, "Orchis Longicornu," was soon written up in the *Botanical Register* in 1817 and explicitly attributed to Swainson.[7]

But Swainson wanted more. When he returned to his homeland, he found himself dull, at a loose end. "I sighed for my Sicilian cottage," he later wrote, "I longed again to ramble over mountains clothed with luxurious plants—to sketch delightful scenery—to rise with the sun, gallop on the sands, climb precipices, and swim in the sea. In place of this," he added disgustedly, "I had to join dinner parties." Then, that very year, naturalist Henry Koster published a well-received book, *Travels in Brazil*, in which

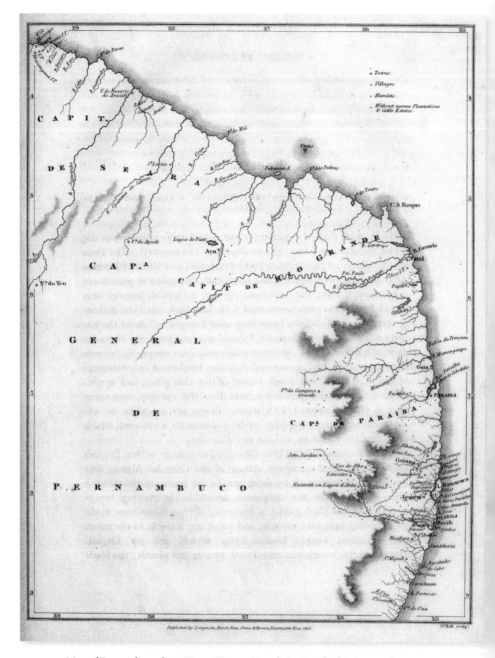

Map of Pernambuco from Henry Koster, *Travels in Brazil*. This image of near-empty space was taken from an 1814 map by Aaron Arrowsmith, an English cartographer.

he described the rich botanical possibilities of the country in thrilling, colorful detail, accompanied by maps of (apparently) blank, near-unoccupied space.[8] Swainson, electrified, resolved to follow suit.

Brazil remained under Portuguese rule, but Britain was working hard to grow its presence in the nation. Britain had guaranteed to protect and maintain Portuguese authority in its territories in a series of treaties in the seventeenth century: by the eighteenth, British merchants based in Portugal had gained a privileged position in the trade to and from Brazil. That close trading relationship, together with the fact that Portugal allowed the British to use ports in Lisbon against the French, vexed Napoleon; after the emperor invaded Portugal in 1807, igniting the Peninsular War, the British assisted the Portuguese prince regent, many of the leading families, and much of the apparatus of government to move to Brazil.[9] Thereafter Britain pledged to continue to preserve Portugal and its empire, including the new royal court: the reward for support was a pliant government and a flourishing, preferential trade relationship. British communities began to form in cities, including in Rio de Janeiro and Recife in the northeast; in and among these communities, knowledge, ideas, and expertise were exchanged.[10]

It was at this point that Swainson wrote to the British government, proposing a trip like Koster's to Brazil for the benefit of "our national museums and gardens." All he asked for in return, he suggested, was a little funding and "patronage as naturalist to the British Government."[11]

The government politely declined, and Swainson was intensely aggrieved. From earliest childhood, he declared, "sleeping or waking, my thoughts were constantly bent on how I could get abroad, and revel in the zoology of the tropics."[12] Swainson determined to go anyway, even if "entirely left to myself and my own resources." And indeed, he rather embraced the image of himself as a lone rogue in his published work: in a "Sketch" of the trip, published after his return, he effectively wrote Koster out of the picture: "This province had never been visited by any modern naturalist."[13] His diaries reveal that he traveled out with Koster himself on November 22, 1816, and stayed with him in the Capunga neighborhood for weeks. But Swainson was determined to prove himself, to show the British government it had made a mistake by refusing to fund his research. In order to garner the world's admiration, he was determined—like many Europeans before and after him—to call himself *first*; first, in this case, to find "new" species in Brazil at a time when the study of natural history was fast becoming a national, even global obsession.

Interest in natural history had intensified throughout the eighteenth and early nineteenth centuries in Britain and Europe more broadly; the journal *Critical Review* declared, in 1763, that "Natural History is now, by a kind of national establishment, become the favourite study of the time."[14] But quite what natural history *was*—the limits of the field— remained unclear, encompassing geology, entomology, ornithology, conchology, mycology, botany, and more. There was little agreement among those working in all these areas about core principles or practices. As historian of science Paul Farber puts it: "Natural history was ramifying into separate branches, later to become independent scientific disciplines. The nature of its institutions and audience was in the process of metamorphosis; the growth of its empirical base was accelerating; and its theoretical foundations were in flux."[15] What the naturalist was, should do, and how they should go about doing it, was commensurately unclear.

The job of *scientist* did not exactly exist, either; no one described themselves as a "professional scientist," still less drew a salary for it, at a time when the pursuit of knowledge still tended to be imagined, culturally and socially, as the preserve of the gentleman—meaning, to be undertaken for its own sake, not for "mere" money. The nineteenth century was increasingly to push back against the idea that the status of gentleman was bestowed only by birth and to embrace the conviction that, theoretically, anyone could be gentlemanly—even if they were the son of a Customs Collector, a gardener, or nurseryman. In this way the pursuit of knowledge was gradually becoming open to all—to all, that is, who could read and write (and who were born male, although that too was changing). Information about natural history was also becoming more widely available in a growing number of periodicals, journals, and books, like William Withering's 1776 *The Botanical Arrangement of all the Vegetables Naturally Growing in Great Britain,* and in groups and societies that fostered research and the acquisition of knowledge while giving members a sense of group identity and access to supportive networks and mentor figures.[16]

One of the greatest of these groups, for British men interested in natural history, was the Linnean Society of London. But becoming a fellow in that group, though an important first step, was not enough.

A Naturalist Arrives

When Swainson arrived in Recife, on the coast of Pernambuco, in late 1816, he must have known he needed to find species that would make his name and build a reputation with those who counted, those who could help him garner funding from wealthy individuals or even obtain one of the tiny handful of salaried university positions. His literary self-expression, then, was part of a larger project of presenting himself as worthy to elite readers. He depicted himself as a pioneer and foregrounded his devotion to his subject as disinterested, motivated by a passion for knowledge. "The motives of my travels," he assured readers in his "Sketch," were "only individual gratification and improvement." This statement is designed to communicate, not that Swainson was selfish or a dilettante, but that he was journeying out of deep, abiding, thoroughly gentlemanly commitment.[17]

ONCE HIS BAGS WERE UNPACKED and his earliest collections arranged, Swainson set off, paddling on a canoe into the local forests to see what new species he could recover and send back to England. But he had walked into a political powder keg, for Recife and its environs were on the edge of terrible, bloody revolt.

Pernambuco's coastal forests, edged by azure waters, had seemed paradisical at first. Recife was a "delightful" locale, Swainson pronounced. British-born and affiliated residents warmly accepted him into their community; he thoroughly enjoyed their hospitality and assistance as he set off in pursuit of fish, birds, termites, and plants in the forests and neighborhoods near his residence.[18] But what Swainson's diaries fail to record is that Recife was roiling.[19]

Almost at sea level, the city was framed by a "bold reef of rocks ... with the surf dashing violently against and over it" (*recife* means "reef").[20] A wealthy locale of five-story white-washed houses, it seemed to glitter in the sun, visitors reported. What also struck visitors was the diversity of the population. In the 1810s about a sixth was white; around a half, free people of color, and fully a third was enslaved. Ships arrived in the port regularly with more traumatized enslaved people, half-murdered on the voyage, even as "white ladies rode to church in elegant sedan chairs carried on the shoulders of black men."[21] Over seven thousand enslaved people disembarked from Africa in Pernambuco in 1817, and nearly nine thousand in 1818. Africans had been settled by force in the region since at least the 1570s, when the arrival of three slave ships was recorded. Almost a million people reached Pernambuco over the course of the following three centuries, and tens of thousands more perished on the month-long journey across the ocean.[22]

Recife's by-now-extensive enslaved population, compelled to suffer in labor camps devoted to sugarcane production and cotton, eyed the masters' wealth with agonized fury. Slavery had been technically abolished in Brazil in 1810, in the Peace and Friendship Treaty with Britain, but entrenched business interests in Brazil prevailed: in practice, little had changed.[23]

There were other causes of dissent and rebellion fomenting, too, with tensions rising between Portuguese- and Brazilian-born residents. After Dom Joao VI, king of Portugal, set up the new court in Rio, he began steadily draining Recife's riches to fund the court's activities and his efforts to expand the Empire. Any glimmers of local pride in the arrival of the Portuguese monarch on Brazilian shores had long ago dissipated; freemasons' lodges—and Recife had a long history of independent-minded secret societies—were incubators for rebellion and republican ideas.[24] Demands from the court, in the form of taxes, showed no signs of abating—indeed, money was disappearing faster than it could be made, either on the sugar or cotton plantations or down the mines. Koster reported that the people of Recife were paying a tax to fund street lighting in Rio, for instance, while Recife itself remained dark. Sugar prices, too, were falling in response to a global spike in sugar production. The devastating drought of 1816 compounded the problem for all sectors of society, unleashing famine; the peoples of Recife were, as Swainson began his adventures, angry *and* hungry, a deadly combination.[25]

Even Swainson finally noticed something of what was unfolding around him—the sentinels placed on the busy streets to keep control of the city, for instance, and the terrible signs of famine. The annual Carnival, scheduled for February, was called off by the governor for fear it might lead to riots from the angry and oppressed (though the Black women who sold fruit, Swainson noted, kept on dancing regardless). Still the naturalist remained most interested in his own affairs, his diaries indicate—in hunting insects, dodging mosquito bites, shooting birds, and also maintaining his health; he was struggling with "an obstinate costive state of the bowels."[26] While sometimes the diaries read like drafts for later published work, at other times they collapse into briefer, more fragmented observations, expressing melancholy, loss, physical weakness, confusion. Differences between these and the published texts indicate the degree to which Swainson worked to reframe his experiences for his anticipated audience. The diaries narrativize his time in Brazil as integrating long periods of uncertainty and stasis, while the published work pursues a confident upward trajectory of achievement, accumulation, advance.[27]

In early March 1817, simmering tensions blazed into open revolt, and Swainson's specimen hunting screeched almost to a halt. His diary recorded few details—only, on March 2, that "The Revolution broke out about twelve o'clock ... Mr. Koster and myself made an ineffectual attempt to ride out."[28] Violence quickly unfolded: the authorities, aware of rising republican sentiment, particularly in the Masonic lodges, issued orders for the arrest of five civilians and five military officers. Brigadier General Barbosa de Castro arrived to take José de Barros Lima, known as the Crowned Lion, one of the ringleaders, into custody; de Barros Lima promptly killed the brigadier general. The next man sent by the governor to capture de Barros Lima and his associates was slaughtered by Captain Pedro da Silva Pedrosa, and full-throated rebellion began. The revolt spread so rapidly that contemporaries suspected it had been fomenting for days. The governor withdrew with his family to a fortress and secured it with loyal men. The rebels, meanwhile, released the city's prisoners, who rampaged through the city streets, shouting "death to the Portuguese!" About sixty people were murdered, including three Frenchmen; it was a dangerous time to be foreign-born in Recife.

The rebels advanced, with eight hundred soldiers, on the fortress; the governor agreed to be sent away, with his family, to Rio. By March 7, the rebels (terming themselves "patriots") had occupied the treasury building and were forming a new independent government, the Republic of Pernambuco. Republican idealism spread rapidly to the north into the interior of Paraíba, too, where several more provisional governments supportive of the rebels were established. The new Republic began working to extend its influence into the south, into Alagoas and Bahia, working to try to secure support from the United States and the recognition of Great Britain.[29]

The man who was able to capture the tiniest variation in a plant's leaf in beautifully precise illustrations struggled to find the words to describe those unsettled, dangerous days. He summarized the period only briefly, in his later "Sketch," as "the memorable revolt of 6th March 1817, of which I was an eye witness" (though he would later write, for himself, more detailed "Notes on the Revolution" with lists of names of the people involved).[30]

The conflict, plus his own ill-health, had practical implications, too, for they prevented Swainson from continuing his journey into the Brazilian interior. He was also unable to depart the country; the port was under a blockade. Forced to shift as best he could, he hunted in his own neighborhood, often limited to "the plants that grew within a few steps of the door." And even here there was danger: he reported "Incessant firing from seven till eight and at intervals all day" on 20th May, which kept him stuck inside. His health deteriorated: "when shall I ever again enjoy the blessing of health?" he mourned, privately.[31]

But the rebellion was, ultimately, short-lived. The rebels were unable to agree on their goals or even core beliefs and were unwilling to mobilize or enfranchise enslaved peoples, fearing the reaction of the plantation owners. Royalist control proved strong to the south, while efforts to engage foreign support failed. The rebels lacked arms, were unable to form a navy, had no unified leadership or vision—while the governor-general of Bahia unleashed an effective propaganda campaign against the revolt. Alagoas and Paraíba fell, then most of Pernambuco, finally Recife and its neighboring towns. The ringleaders were captured and murdered, often in public, and brutally, their humiliating final moments intended to dissuade further revolt.[32]

Swainson evidently found the violent aftermath particularly disturbing; his letters and diaries record acute distress at the plight of the former rebels. The head of one was "carried to Recife . . . for [public] exposure," while others were thrown into prison, "with hardly a shirt or covering of any kind on them, and heavily loaded with chains and irons. They eat out of a large tub, and are covered with vermin, which they cannot free themselves from on account of their handcuffs."[33] His private papers register a broader moral engagement that his "Sketch"—in its sustained focus on himself and his growing catalog of finds—chose not to record.

Swainson gradually began to turn his thoughts back to travel. (His bowels began to improve, too, a welcome change Swainson attributed to a rousing course of Lathams Dinner Pills.) He made preparations for his first collections to sail to England on the *Robert,* together with letters—including some to his good friend and fellow Linnean Society member, a man named William Jackson Hooker, in Suffolk, on the east coast of England. Swainson reported in his diary on June 24 that he'd traveled to the neighborhood of Capunga, where he "conveyed all the collections I have formed (up to this period) contained in four cases to Mr Stewart in a canoe," with a "packet of letters, containing those to Elm Grove, one to Mcleay [secretary of the Linnean Society], Hooker, Latham the surgeon, and others, to go with the cases by the 'Robert.' I returned to Agoa Fria," another Recife neighborhood, he continued, in a moment of unusual optimism, "very well in health and with a light and cheerful heart at having done all I wished."[34]

Swainson left Recife in late June and headed along the Capibaribe River in pursuit of more plants, birds, fish, and insects.[35] Meanwhile his cargo began its long voyage: the *Robert* left Pernambuco on June 28, four days after Swainson left his boxes. Its safe docking in Liverpool was reported in *Lloyd's List* on Friday, August 22.

What happened next, exactly—how, when, and where Swainson's path crossed with a luminous orchid, of a variety so far undescribed and unnamed by a European—has been a matter of debate for two centuries. Only tantalizing hints are left to us in Swainson's diaries and letters and in the "Sketch," published in the guise of a formal letter to Professor Jameson, Regius Professor of Natural History at the University of Edinburgh, in the *Edinburgh Philosophical Journal* (in this way, Swainson identified himself, publicly, as part of a network of leading scholars).

In the "Sketch," Swainson made only passing, general mention of the plants he collected, baffling and infuriating generations of Victorians who came after him. But he provided plenty of information on his route, for this—the catalog of miles passed, places visited—signaled the advance, the onward momentum he aimed to communicate. Thus we learn that he reached "Penedu" (Penedo, at the mouth of the São Francisco River) in the beginning of August 1817. He traveled on to St. Salvador by an eight-day canoe ride and progressed inland, "varying my residence," until March 1818. In April, he left for Rio de Janeiro, where he was thrilled to be treated with respect. The Russian consul Langsdorff, he noted proudly, gave him "every assistance and the most liberal attention." (In his journals he continued to spit furiously about "the total indifference" shown by the British government "towards the advancement of science, but particularly natural history.")[36] Langsdorff sponsored local and foreign scientists alike, and Swainson explored the Organ Mountains in the company of men he met through the accommodating consul. He finally set sail for home in June 1818 on the brig *Betsey*, accompanied by more packing cases stuffed with specimens of plants, animals, fish, insects and more.[37] He arrived in Liverpool the third week of August and set about constructing, for the purposes of blazoning his achievements, the "Sketch."

The vast area Swainson left behind him, the span of his almost fourteen-hundred-mile trip from Recife to Rio, became known by hunters for the lost orchid as "Swainson's hunting ground" and was rifled extensively. With only the "Sketch" for a guide, most Victorians suspected the orchid-rich Organ Mountains were probably the lost orchid's home. But nothing in the piece makes this claim explicitly (that, after all, was not Swainson's aim in publishing it).

Swainson's diaries, printed only toward the end of the twentieth century, transcribed and edited by a family member, make just a handful of references to orchids. On July 27, 1817, not long after leaving Recife, Swainson ascended into the Serra das Russas, mountains approximately one hundred kilometers to the west of the city, traversing along the Rio Ipojuca around modern-day Gravatá and Bezerros. Swainson wrote: "The country continued much the same, being covered with mimosa and low shrubs encumbered with parasitical plants and encrusted with numerous lichens." Orchids were commonly termed "parasites" at the time because of their habit of twining around trees. Swainson also noted that day, without detail, that he "Found an epidendrum."[38] Perhaps this was the lost orchid; it would be easy for Swainson to get the two confused, not least because the taxonomic differences between the genus epidendrum (also epiphytic) and the genus to which the lost orchid belongs would not be pinned down for another five years.

The lost orchid would not have been in bloom at the time, either; it typically ends its flowering in the region by April. If it was there, if he collected it, he would not have experienced its remarkable glamor. But it was not the focus of the gaze recorded in his diaries: that gaze tracked restlessly across the landscape, taking in everything from the huts to churches, goats, sheep, people, insects, soil. Little stayed in his roving spotlight for long. And anyway, he reported finding the flora and fauna of the area underwhelming. "I was much disappointed in observing very little differences in the plants of these hills," he remarked—the best he could say was that the region reminded him of Sicily.[39]

The physical challenges of the route also dominated his attention. These, described in detail in the diaries, are referenced only briefly, judiciously, in the "Sketch" to reinforce his self-presentation as a man of decisive action: "the excessive drought that had prevailed frequently exposed us to great privations," he informed his readers, noting that he and the local men he employed to assist him were forced to drink putrid water, stinking of decomposing plants, from rock pools and crevices.[40] Privately he conceded that his mental health was suffering, too; depression, "melancholy," returned. He also developed sores on his legs and his feet and, at times, asked the men he employed to collect for him, so terrible was the pain. The employees were not immune to bouts of ill health; Swainson recorded that one, a young man named Vincente, had to be sent for medical assistance when he fell ill with dysentery. Swainson noted plaintively that Vincente departed "wrapped in the only blanket I had."[41] Swainson also described, in his private papers, the appalling suffering endured by enslaved people laboring in the plantation labor camps and smallholdings he passed on his long trip. He was deeply shocked by an encounter with a traumatized fifteen-year-old girl from Mozambique, separated from everyone she loved, so lost in "black despair" and "wretched misery" she was unable to function, and ate earth.[42] Swainson explained little of this in his public-facing writings. He did not try to expose the horrors of enslavement in detail or mobilize resistance.

He mentioned orchids on one other occasion in a diary entry, on October 9, 1817, when hundreds of miles to the south of Recife, shortly after leaving Humildes in the Bahia region. Headed toward Feira de Santana, Swainson reported "aloe like plants growing to an immense size" as he made his way up the mountains, with "haylike parasitical plants" and "many beautiful species of epidendrams" that "were in flower"—unlike, perhaps, the orchids he'd collected in July. Again, though, Swainson was just as interested in the color of the soil at his feet and the richly hued macaws overhead as the plants (and this area is not, today at least, a habitat of the lost orchid).[43]

Swainson's wide-angled focus, so evident in the diaries, is indicative of an intellectual approach he blazoned in his "Sketch." Swainson noted with pride that he made "immense collections in every branch of natural history" in Brazil: indeed, by collecting so many, so very different kinds of specimens, he was able, he suggested, "to survey Nature as a whole, [rather] than simply in its minute parts." To us, his long laundry lists of specimens might sound like a chaotic jumble; to Swainson, it was "conducive to the exaltation and expansion of the human mind."[44]

Like many naturalists of his time, he made recourse in his published text to dizzying numbers. These did not, to him, indicate extraction on a terrifying, massive scale, but discovery, new beginnings, achievement. In the service of natural history, he explained, he'd collected 760 specimens of birds (including "two or three new toucans") and more than 20,000 insects, which "may safely be said to exceed greatly a collection of South American insects ever seen in this country." He took drawings and descriptions of "120 species of *fish*, mostly unknown," and sent seeds of many plants to Kew and other botanic gardens. And last but not least he referenced the 1,200 plants in his own herbarium and an "interesting collection of parasitic plants, together with another of cryptogamia," which he'd donated to his good friend, Mr. William Hooker Esquire of Suffolk.[45]

Somewhere in that "interesting collection" of "parasitic plants" was almost certainly the lost orchid. At midcentury, the horticulturalist Joseph Paxton, gardener to the duke of Devonshire, suggested that Swainson used the lost orchid's "stems as a kind of 'dunnage' to set fast certain chip boxes of lichens &c, which he transmitted to" Hooker.[46] Hooker and Swainson shared an interest in lichens, a species of cryptogam; Hooker wrote to Swainson before he left for Brazil, in October 1816, to thank him for some specimens and to ask him for more. Paxton believed—and many after him repeated—that the lost orchid root had been collected only accidentally by Swainson.

I've found no direct evidence for Paxton's claim, though he could conceivably have heard such a story from the great botanist John Lindley, who collaborated with Paxton on the *Flower Garden,* and who would soon meet and name the orchid, as we'll see. But if the orchid was collected from Gravatá and Bezerros in July and was not in bloom at the time of its collection, this might explain why Swainson seemed comparatively uninterested in it—it wasn't collected accidentally, exactly, but it wasn't collected mindfully, either.[47] This space, this narrative gap, would soon allow historians and plant enthusiasts to begin generations of elaborate story making.

The transformation of a plant, collected from somewhere along Swainson's journey between Recife and Rio, into a much-desired commodity ultimately hunted by Ericsson and Bungeroth had begun. Specimens were packed up, most likely in root form, since this was how Swainson sent his other orchids. (He transported three packets of seeds together with "parasitic roots" to the leading botanist Sir Joseph Banks, for example.)[48] The plant probably traveled by brig or two-masted sailing vessel and docked at Liverpool after a journey lasting about two months.

Its next stop was Swainson's friend William Jackson Hooker, whose troubled brewery business was gasping for survival in the economic slump following the end of the Napoleonic wars.

3
TROUBLE BREWING

WILLIAM HOOKER WAS PANICKING about the rising price of grain in the bustling Suffolk market town of Halesworth when he received Swainson's packing case, and inside it, the "parasitic root" from Brazil. Thirty-four years old at the time, Hooker was over six feet tall with thick, dark hair and a crooked nose that had been broken in a school fight.[1] But this was no pugilist; Hooker was passionate about plants.

Hooker's desire to pursue botany was thwarted, however, by financial and familial responsibilities. Married for four years, he and his wife had two toddlers to support; the brewery business, partly owned by his father-in-law, was supposed to be a steady source of income for a young family. Yet Hooker loathed the work, and he had a bad habit of spending the brewery's profits on works of natural history and on publishing his own books on botany. Years later, in a largely hagiographic biography, his son briefly admitted his father's "lavish expenditure on his own unremunerative publications."[2]

Hooker's private letters, meanwhile, record a young man obsessed with the natural world and desperate to expand his collections. Born in Norwich he had, from earliest youth, roamed the coast of Norfolk and the wilds of the Broads in pursuit of rare birds and plants; he discovered a new moss in 1805. He'd also inherited a little property, just enough to convince others he had the necessary gentlemanly status to gain acceptance into the elite world of early-nineteenth-century science.[3] And Hooker seems also to have been, unlike Swainson, good at attracting mentors. He managed to interest several older, well-positioned East Anglians in his career, including the man who would become his father-in-law, Dawson Turner, a fellow of the Royal Society, and James Edward Smith.[4] Elected to the Linnean Society at just twenty-one, Hooker also, importantly, interested Joseph Banks, who among many other honors was an advisor to George III and effectively director at Kew. A towering figure in the early nineteenth century, a man who knew everyone, Banks was the beau ideal of the era's "gentleman scientist."[5] Banks became a supporter of the young brewer.

But Hooker's financial position in 1817–18 was precarious. He was in a very different situation from someone like Banks, a fact the older man seems not always to have recognized. Banks, in his own letters to Hooker, encouraged his protégé to attempt thrilling plant-hunting expeditions in pursuit of rare specimens. He urged Hooker on a visit to Iceland in 1809, for instance, though catastrophe ensued when every specimen Hooker collected was lost on the way home when the ship was set on fire by Danish prisoners on board. (Hooker and his fellow passengers only escaped with their lives when another ship saw their plight and raced to their rescue. They were in freezing waters, twenty leagues off the coast of Reykjavik, at the time.) Banks, characteristically, seems to have felt the whole disaster was a net positive: he cheerfully congratulated Hooker on his escape and assured him that his willingness to face death would "raise him in public esteem" in the future.[6]

But Hooker needed money. The years after the Iceland catastrophe had been, Hooker's son later indicated, delicately, "the most embarrassing of his life." An "unquenchable longing to travel in the tropics" was still urged on by Banks, who assured him that he needed to "submit himself to some sacrifice and adventure in preference to a life of ease." If he, Banks, had stopped at home, he explained bracingly, "I should have been now a quiet country gentleman, ignorant of a number of matters I am now acquainted with." Hooker's friends and relations, meanwhile, recognizing that a "life of ease" was not exactly an option for the brewer, were "unanimous in urging him to remain at home" and study British plants, cheaply and safely, instead.[7]

Struggling to navigate so many competing pressures, desperate to expand his collections of books and plants, by 1817 Hooker had dangerously "crippled his resources." His letters indicate the degree to which money worries were a major motivator, a daily concern. "I am literally in the brewery all day long," a panicked Hooker told his father-in-law in July.[8] A run of disappointing investments hardly helped: Hooker wrote, "Hitherto my income has not [been] sufficient for my annual expenses & now without any prospect of its being greater my family expenses increase." The house needed urgent repairs. Debts were piling up, the price of barley rising dangerously: his bank debt stood at £5,493, he admitted to his father-in-law, a disturbing £508,000 / $652,000 today. In the face of all this, his progress in science was, unsurprisingly, stalling: "1817 is one of the very few years of his life in which he published scarcely anything," his son would later admit. All Hooker wanted was to devote himself to the study of botany—but this was becoming impossible. "I ought & I hope I shall cheerfully sacrifice for the sake of bettering my income," he wrote desperately to Turner.[9]

Then, on a day and date unknown, Swainson's packing case arrived. And while much of what happened next is edged in the thick haze of myth, one thing is certain: Hooker found among the collection of specimens an orchid root—actually, several orchid roots, from Brazil, that prompted his interest and that he decided to cultivate.

Few exotics survived the long journeys from foreign countries to Europe before this period because, as Andrea Wulf has explained, if the wind, salt water, rats, and mice didn't ruin naturalists' specimens, the sailors very well might: seamen were not averse to drinking the spirits in which certain choice samples were stored.[10] The proprietor of Veitch's nurseries would later assert that, at the turn of the nineteenth century, just twenty-five species of epiphyte were recorded at Kew, "most of which had been brought from the West Indies" by Admiral William Bligh, and none of which provoked particular interest.[11] Possibly the lack of attention stemmed from the fact that the plants seemed such poor candidates for cultivation, given that they were considered parasitic and thus unlikely to thrive away from their hosts. Sheet glass for greenhouses was also expensive to produce, given the challenges of its production and the taxes on glass, and the "stoves" required to provide the warmer year-round temperatures that many exotics required were beyond the reach of all but the richest or most dedicated of horticulturalists.[12] Given these obstacles, it simply wasn't worth the expense and difficulty of importing orchids in bulk to Europe.

But Swainson had packed his specimens well. Indeed, by now he considered himself an expert in the preparation of plants and specimens of all kinds (shells, animals, birds, fish, lizards, insects): he noted in his "Sketch" that he had developed "a new process, which will enable a botanist in a tropical climate to dry nearly 400 specimens in three days."[13] He detailed this "new process" at length in a treatise on the subject published in 1822: here he explained, for the edification of other naturalists and enthusiasts, how he'd learned to pack his collections, even in the tropical, damp climes of South America.

It all involved a lot more than Shepherd's tin box. Plants should be sent overseas in seed or bulb form if at all possible, Swainson advised, seeds carefully dried, then scattered between layers of sandy earth. Charcoal should be used if the packing had to be done in rainy seasons, to absorb moisture. Bulbs and roots should be sent in brown paper bags, with moist sugar potentially added as a preservative. On the rare occasion when living plants must be sent, they should be placed in a box or barrel with round holes and located in the middle of the ship, with the top covered carefully against rough seas. Drawings, flowers, and fruits preserved in weak spirits, together with dried plants, should accompany the living specimens, to provide more "materials for a most interesting and valuable work" back home. The way to be sure specimens were preserved from rot was to make a press from a thin deal board and two planks of mahogany, he explained, 1 ¼ inch thick, the screws tightened morning and evening, and fine cartridge paper, to be heated in the sun or by a fire and changed once or twice a day.[14] Swainson's book of preservation methods was a further claim to the regard of fellow natural historians. Practically speaking, it seems to have worked.

The roots of the lost orchid consist, Hooker would later explain, of white "flexuose fibres" that reach—rather weirdly—up toward the air as it grows; the large pseudobulbs are "3-4 inches in length, oblong, compressed, deeply furrowed, dark green, clothed at the base with large withered brown scales."[15] What Swainson sent looked like something from the pages of a gothic novel, a weird object from the boundary line between the plant and animal worlds—or so, at least, many nineteenth-century enthusiasts would suggest.[16]

It arrived just as understanding of how to cultivate such plants had taken another leap forward in Europe. In 1812, an orchid was brought back without earth from Montevideo in a ship's cabin, yet it still flowered luxuriantly much of the way home.[17] The plant's seeming ability to thrive on air electrified enthusiasts, offering tantalizing glimpses of a future in which these remarkable plants could be satisfactorily transplanted to British shores. As the *Botanical Register* explained rather breathlessly in 1817, *Aerides* or "air plants" seemed to possess the astonishing faculty "of growing when suspended, so as to be cut off from all sustenance but that derived immediately from the atmosphere." However, the editors admitted, air alone was not ideal. This was not "the state of existence which suits them best," they conceded; but it was "one [plants] are enabled to endure, as a carp is known to do, [by] being suspended out of water in a damp cellar."[18]

There were yet more signs of a possible way ahead to the successful cultivation of exotic orchids in England. Joseph Banks managed to cultivate epiphytes by putting them in light wicker baskets of threaded twigs with vegetable mold and moss at the bottom. The plant was able to extend its roots out freely in this condition, "yet be kept steady in its station," while the moss gave a little shade and helped keep in the moisture.[19] Even if he had not received his copy of the *Botanical Register,* in the otherwise *annus horribilis* of 1817, Hooker had almost certainly heard of Banks's success.

"Unquenchable" in his interest in exotics, Hooker was surely thrilled to try out the Banksian approach. He had been begging friends and fellow collectors for years in his private letters to send any specimens of exotics they could spare.[20] Hooker had also managed to construct a stove, or heated greenhouse, of his own in the brewery garden—indeed, it was one of the very few things that pleased him about his Halesworth life. "My Hothouse & Greenhouse flourish exceedingly," he wrote proudly to his father-in-law.[21] Greenhouses powered by "fire-heat" tended to become ovenlike, roasting many tropicals to death. But enthusiasts were learning how to provide a moister, gentler warmth by selecting organic substances that burned more slowly, at lower temperatures.[22] Hooker did not leave an account of exactly how he cultivated the plant—and the practice of overheating orchids was reportedly a problem for much of the century—but he was evidently successful. He wrote on a herbarium specimen sheet, now held in Kew, that the orchid "*fl. in hort. nostr 1819*"—"flowered in our garden, 1819."' (See Plate 1.) Indeed, it may have bloomed even earlier, since in his 1827 *Exotic Flora*, likely written closer to the event, Hooker explained that it "blossomed for the first time in Britain in the stove of my garden in Suffolk, during 1818."[23]

Several other orchids arrived from Swainson also, and these too were cultivated by Hooker.[24] But it was the purple and crimson bloom that seemed especially to repay the brewer's efforts. Its beauty almost defied description. "The most splendid, perhaps, of all orchideous plants," Hooker boasted, and for decades afterward, observers waxed lyrical attempting to capture the gorgeous hues of this ethereal flower.[25] What would become one of the world's most famous orchids had arrived in England at an auspicious moment in an auspicious place, for William Hooker was about to emerge from his terrible debts and problems with barley to become one of the most famous botanists of the era, and orchids were about to become a prize many would seek in a century obsessed with the exotic.

It might seem obvious, then, that William Swainson would head back to Brazil to source more of the plants, or at least publish a note explaining where he had found it and the story of its collection. He did not. Hooker wrote a number of times, with pulsing adoration, of the remarkable orchid he'd cultivated. But Swainson said and did nothing. Why he disappeared from the story of the lost orchid became a part of the puzzle, a curiosity in the hunt that fascinated plant historians later in the century.

4

SWAINSON VANISHES

ENTHUSIASTS DEBATED FURIOUSLY the question of Swainson's lack of involvement in the hunt for the lost orchid in the final years of the nineteenth century. "It may well be asked what Mr. Swainson was doing, if alive, while his discovery thus agitated the world," one writer observed feelingly in 1891. "Alive he was, in New Zealand, until the year 1855, but he offered no assistance."[1]

A contributor to the *Standard* newspaper went even further. It was positively fantastic, he spluttered, that William Swainson did not grab the opportunity dangling in front of him: "That he, a skilled botanist, well aware of the excitement his despatch of that parcel had roused in 1818 — aware also of the hard money to be won by repeating it, should have lived in possession of all his faculties for thirty-seven years, and not breathed a word to silence disputes or to grasp the fortune within his reach, is more like some incredible romance than an incident of real life."[2]

In fact, it seems unlikely that Swainson was "well aware of the excitement" his orchid generated. He made no reference to it in any of his letters or his diaries from his new home in New Zealand, where he lived and worked for decades. He had not, of course, actually vanished. In truth, the orchid was not yet the celebrity it would become. While many column inches would be spent agitating about the identity and whereabouts of the packing-case orchid later in the century, in the years after its arrival in England, young scientists like Swainson did not see a lone plant as their path to fame and fortune. Indeed, Swainson did not see botany as his path ahead at all; rather than thinking of himself as a "skilled botanist," he explicitly told Banks, after his arrival back in England, that "Botany . . . became with me a secondary pursuit to my more favourite one of Zoology ^on arriving at Bahia" [the state to the south of Pernambuco].[3] Swainson thought of himself as a naturalist and embraced his own wide-angled approach.

He turned his attention to animals and shells next, producing two well-received books soon after his return, *Zoological Illustrations* (1820–1823) and *Exotic Conchology* (1821–1823). He acquired a new friend, William John Broderip: Broderip seems to have spent much of the next four years trying to get Swainson to learn how to spell and generally informing him (in a way he presumably imagined helpful) that his mistakes and malapropisms and tangled Latin were not helping him advance. Swainson put up with this for a while, then lost his temper and broke off the friendship.

Communication of all sorts seems to have been at the forefront of Swainson's mind in the months following his return from Brazil. He wrote on January 1, 1819, to John Thelwall, an early speech therapist and friend to the Wordsworths, to ask for help with his stammer, perhaps a resolution for the new year. That letter has not survived, but the reply did: Thelwall offered him a residential course of intensive treatment at the staggering cost of 200–250 guineas a year (£19,400–£24,200 or $25,000–$31,100 in today's money), or a shorter course at the hardly more palatable rate of thirty-five guineas a month (£3,750 / $4,800). Swainson did not take up the offer.[4]

After he returned he was also accused of leaving unpaid debts in Brazil, including for the care of several servants, a sick and enslaved boy named Manoël, and a dog. The secretary of the Linnean Society, Alexander Macleay, wrote to upbraid him for mistreating a mutual acquaintance who was left to mop up on Swainson's behalf: Mr. Dickson "was the only one at Rio who paid you any attention," he wrote scathingly, and yet "you left Rio without thanking him and have never written to him since. You may easily conceive that this Communication hurt me exceedingly."[5]

Swainson was still boiling about this "tissue of falsehood" in 1822. It seems likely money he sent to pay his bills went astray, although it is possible someone in the chain of people involved deliberately diverted his money from its intended destination. "My station in society places me above being seriously injured in the estimation of honourable men," he proclaimed hectically to Macleay in February 1822, then rather spoiled the effect by writing to ask if his letter had arrived.[6] Eventually it all died down, but Swainson was bitterly upset, for this was more than mere scandal. His speech impediment and problems with spelling embarrassed and hindered him, for fluency with words was a key accomplishment for the gentleman scientist of the time. Wealth remained valued as a sign of gentlemanly status, and Swainson did not have it. How to make money was and continued to be challenging for those born outside the elite, leaving them scrambling for one of the few paid jobs in a university, museum, or library. Such jobs were paid employment at a time when performing acts for money remained to many rather vulgar, but they were socially acceptable, at least, and Swainson now had a family to support. Getting one of those positions required the support of a well-placed patron, however, and Swainson was bad at attracting them.

To make matters worse, Swainson's application to become Assistant Keeper of Natural History at the British Museum was rejected in that same year, 1822. The development made him especially angry because the appointment went to a man who (according to him) "knew nothing of natural history."[7] Swainson was convinced he had been passed over because he did not have the right connections; whispers of his supposedly ungentlemanly behavior in Brazil surely did not help. Battling ill health, especially headaches, he declared himself ready to throw over life and career alike. He wrote to John James Audubon: "I am sick of the world and of mankind, and but for my family would end my days in my beloved forests of Brazil."[8]

Hooker, meanwhile, was doing rather better. He leveraged his connection with Banks by writing to him and his librarian, Robert Brown, to ask for help. The Regius Professorship in Botany at Glasgow was just coming vacant, and Banks and Brown had been contacted by the Duke of Montrose, patron of Glasgow University, to ask for possible candidates. Banks invited Hooker to meet with the duke at a dinner in town.[9] Hooker's behavior at the meeting must have pleased Banks, for, with Brown's blessing, the brewer was offered the position.

Hooker's son Joseph would later shape his account of his father's life along the familiar lines of post-Enlightenment autobiography. Joseph Hooker's "Sketch" of his father presents 1817 as a brief low point, quickly followed by upward trajectory. In Joseph's telling, Hooker's unusual capacity for hard work, intelligence, and virtue were soon rewarded, and he presented the Glasgow chair as a sign of transforming prospects: "My father's reputation as one of the foremost botanists in this country was confirmed by his success in the Glasgow Chair, and rapidly rose as his successive publications appeared. Very soon he had but one compeer in Great Britain, Dr. Lindley."[10] In this way, the younger Hooker's biography of William is shaped to reveal what Emily O. Wittman has called "the premise of a stable and triumphant sovereign self" that shows "the promise of personal growth and social advancement culminating in insight and self-assured reflection."[11]

When Hooker left Suffolk for Glasgow, Brazilian orchids in tow, he sent "offsets," or naturally occurring offspring, of the packing-case orchid to a prominent collector and merchant in London named William Cattley. He sent a further offset to the Botanical Garden in Glasgow, according to one of the three volumes of his lavish 1823–1827 study, *Exotic Flora*. Both orchids bloomed luxuriantly. Hooker's career success was the product of the same skills and strategies that would bring the orchid to renown. Hooker was effective at connecting himself to key figures, at inserting himself into the fine web of emerging proto-professional networks. Hooker was able to gain the job he wanted partly because he was in close contact with those well-placed to help, just as he was in contact, too, with those well-placed to publicize the orchid. Sending the orchid to Cattley would prove crucial because, as a merchant, Cattley was emblematic of a new kind of collector, one key to the changing landscape of nineteenth-century society and science. Cattley was not well-born and aristocratic; he was a businessman. While Banks helped Hooker get a university position, it would be Cattley who helped usher the orchid along its journey to botanical celebrity.

In 1820–1, however, these patterns and future successes were still far from evident. The private letters of the provincial, middle-class Hooker suggest he felt on the social, scholarly, and scientific margins. He was intensely anxious about the demands of his university role, for instance, writing to Smith to ask for information on lecturing, "about which I know nothing." Told he had to give his inaugural address in Latin, he was thrown into a panic: "I really cannot do it," the former grammar-school boy wrote to his father-in-law.[12] He was constantly anxious about whether there was a new or better job possible, and he wasn't thrilled by the weather, either: "you cannot have an idea of a Scotch hurricane," he wrote to a friend in Suffolk, "spending its rage upon a house."[13] University positions had their indignities, too—he had to stand by the door collecting fees as the students filed in, and he took in more pupils on the side to make extra money.[14]

Still, the move to Glasgow offered a steadier source of income and freed him from worry about grain. If not perhaps quite as resounding a step toward greatness as his son Joseph would later suggest, the move offered release from the worst of his money worries. As Hooker wrote to a friend in 1821, "I must congratulate myself on having got quit of that vile Brewery which was neither suited to my inclination nor my pocket." He added, wryly: "Botany does pay better than stale Beer."[15]

Private letters between Hooker and Swainson reveal just how much the two men were *not* preoccupied with the purple and red orchid in the years after the latter sent it to the former, for all that later-Victorian historians implied that they were. Professional opportunities were their common topic instead: Hooker wrote to Swainson to sympathize over the latter's failure to win the position at the British Museum, for instance, then admitted he had been asked to take on the job himself. Hooker and his wife were sorely tempted by the offer, he explained: they had friends and family in London. Hooker hinted that it was Swainson's interest in the position that played a part in his own eventual rejection of the post: "I was very much influenced by your letter & by being able in the event of my giving up all idea of the thing, to say a good word which I thought you so well merited in your favor."[16] Hooker went on to suggest that perhaps Swainson might have more of his Brazilian plants—and, if so, might be willing to share them?

Hooker's letters also suggest he believed careers were to be made in the study and cultivation of Brazil's botanical wealth—for all that he was not focused on Swainson's packing-case orchid specifically. He had been one of those who encouraged Swainson to visit Brazil in the first place, and Hooker told Swainson on his return, numerous times, that he wanted to examine Swainson's "Brazilian duplicates." He wrote, with unusual forcefulness, in May 1824: "You were good enough to say that could you find the leisure to look [the Brazilian plants] though with attention, you would oblige me by letting me have some of them. Now, my dear sir, if such should be in your power you could not render me a more acceptable service than by sending them to me soon" [emphasis Hooker's].[17]

But Swainson resisted. He was neither willing to send the plants to Hooker nor eager to investigate them himself. Presumably, he did not want to share his collection; perhaps he even feared Hooker would find something remarkable and take all the credit. Hooker's patience began to wear thin. Other naturalists were now active in Brazil, amassing "vast collections," hurrying out their results "to the world, & in periodical works." The future belonged to those who acted expeditiously: "every day, if I may so say, something new is made known of the Botany of that vast country."[18] But by the time Hooker finally got his hands on the duplicates, in 1826, it was too late. The specimens had been "lamentably attacked by insects" and were useless.[19]

Swainson had plenty of other issues to concern him. His wife died young, leaving him with four small children to raise and support. He grew increasingly irascible, his letters revealing a temper that surely did not help him advance. He became estranged from his in-laws and half-brother, who began to despair of his character and fear, deeply, for his future. "Interference only excites opposition, and such opposition seems but to strengthen his determination," Swainson's half-brother Charles wrote despairingly in 1839; "His words are very strong."[20] Swainson also became a confirmed advocate of the Quinarian system, a method of zoological classification organized around the number five. Quinarians argued that all taxa contained five subgroups, and that collections of these five subgroups could be organized into flowerlike circles—unlike, say, the branching, tree-like system of Darwin. But Swainson's commitment to the theory put him increasingly outside the center of the scientific community in Britain.[21] In dudgeon, in the teeth of his family's objections, refusing to discuss the decision, Swainson swept up his children and emigrated in 1840 with a new wife.[22]

Hooker was not impressed. "There is a man who left this country with the character of a first rate *Naturalist*," he wrote irritably years later, in 1854, "(tho' with many eccentricities) & of a very first-rate Natural

History *Artist:* & he goes to Australia & takes up the subject of Botany of which he is as ignorant as a Goose." Swainson's work on taxa and gum trees inspired Hooker's particular irritation: "In my life I think I never read such a series of trash & nonsense."[23]

Meanwhile, the purple and red orchid was flourishing. Cattley's offset bloomed in Barnet, a rapidly suburbanizing village outside London, in November 1821–2. The second offset brought by Hooker flowered in the Glasgow Botanic Garden in November 1824. Both daughter orchids bloomed as lustrously as their parent; the Barnet specimen exhaling an especially seductive perfume.[24]

5
"OLD ANTIQUITY"

JOHN LINDLEY, BORN IN 1799, was raised in an East Anglian provincial town, like William Hooker. George Lindley, John's father, was a nurseryman in Catton, outside Norwich in Norfolk, and he helped foster his son's passion for plants. What George was not able to offer his son—or indeed any of his children—was financial stability; John's sister Mary drudged out a living as a governess.

The nursery business was expanding rapidly in the early nineteenth century, catering to the public's growing interest in plants. The plethora of nurseries springing up all over the nation fueled, in turn, deeper understanding of and interest in horticulture in the population.[1] George Lindley was highly knowledgeable about fruits and vegetables, and he also grasped the public's appetite for the remarkable: the classified ads George Lindley placed in the local press dangled an array of seeds before the readership, from mangel wurzel to a range of rather remarkable-sounding turnips: "NORFOLK WHITE LOAF—WHITE GLOBE—PURPLE GLOBE—GREEN GLOBE—EARLY STONE, OR STUBBLE—DECANTER—BELL—WHITE PUDDING—PURPLE PUDDING—AND YELLOW BULLOCK."[2]

But as a businessman, George was unable to capitalize on the moment. Nurseries required just the right alchemy of location, price, and product to thrive: as the historian Sarah Easterby-Smith has argued, they needed lots of space for cultivation and display yet had to be close to metropolitan centers if they were to be accessible to consumers. Rents on spaces close to the city were high, while cheaper acreage lacked transportation links. Striking the right balance was a challenge.[3]

The Catton nursery hemorrhaged money. Hovering on the edge of bankruptcy, George was not able to afford the expense of sending his intellectual-leaning son to Oxford or Cambridge, nor could he manage a commission for him in the army. To make matters worse, John was blind in one eye, the result of an accident as a toddler.[4] Still, he was devoted to reading and books, earning the nickname "Old Antiquity" from his schoolmates. John Lindley had been—again, like Hooker—educated at Norwich Grammar School, at a time when its headmaster, Edward Valpy, was noted for his expertise in Classics but also for energizing and modernizing the curriculum. Unlike Hooker, John acquired good Latin skills there and, according to his son, polished French and drawing skills from a refugee who had settled in East Anglia after the Terror.[5]

John left school at fifteen or sixteen, traveling to Belgium to gain practical experience in the seed business. He worked for the Dickensian-sounding merchant Mr. Wrench—most likely Jacob Wrench & Sons, a major seed provider in Europe for much of the century.[6] On John's return, he crossed paths with Hooker, who, while still living in Norwich and Halesworth, visited the Catton nursery to buy plants. Hooker was evidently impressed, and the two were in correspondence by January 1818. In a letter of that month, Hooker thanked the teenaged John Lindley for sending him interesting specimens from the nursery and gave him detailed advice on "the quickest mode of drying plants": after pressing them in half sheets of coarse brown paper for twenty-four hours, he recommended spreading all of the half-sheets out on one's "chamber floor for a few hours according to the heat of the weather, generally for the whole night," followed by a couple days more pressure.[7] Hooker also invited John to visit him in Halesworth, where John seems to have worked as a general amanuensis; Hooker often mentions to friends in letters of the period that he intends to ask John to order a book or find a text on his bookshelves.[8] In this way, John Lindley seems to have been absorbed practically, as well as on the page, into Hooker's expanding networks of the knowledgeable and well placed.

I've found no mention of a purple and red orchid, nor of the packing case sent by Swainson, in any of Lindley's or Hooker's voluminous correspondence from this time. Nothing survives to prove that Lindley even saw the orchid in Hooker's stove: if any references were made to it on paper, those letters have not survived. More likely, again, the orchid was not identified as of particular significance to any of the men—recall the vast numbers of plants Swainson sent from Brazil; though Lindley's pivotal interaction with the orchid would come soon.[9]

Like other young men without means or title, Lindley needed mentorship. Hooker introduced Lindley to Augustin de Candolle, a leading Swiss botanist, and to the indefatigable Banks. The latter once again offered an important foothold, for Banks invited Lindley—still childfree, unlike Hooker, and so able to travel—to move to London and become an assistant at his own library and herbarium at 32 Soho Square. It was a pivotal step: Banks opened the door for Lindley into the beating heart of British natural history. Banks's herbarium would become one of the founding collections of the Natural History Museum; it was also a vital, intensely social space of daily breakfasts and meetings, collaboration, conversation, and research that was effectively the center of the practice of establishing, contesting, and navigating norms of the emerging discipline of botany in Britain.[10] Lindley was one month shy of his twentieth birthday when he arrived in London to start his job for Banks in January 1819. Illustrations of Soho Square reveal rooms that were elegant, looming, august, the library stocked with floor-to-ceiling bookcases and thick carpets. In the herbarium Banks kept his vast collection of specimens in Chippendale mahogany boxes, called "cubes."[11]

Lindley took lodgings in walking distance of the herbarium, at 88 Charlotte Street in Fitzrovia. But he kept in close touch with Hooker, who followed his protégé's course anxiously, worrying about Lindley and warning him to watch his step. For Hooker, alert to the power of personality in the professionalizing world of British natural history, spied problems in Lindley's path.

Robert Brown, lead librarian for Banks since 1810, was a brilliant Scot and well-traveled naturalist who, with his friend the botanical artist Ferdinand Bauer, had circumnavigated Australia between 1801 and 1803. But Brown was not an easy character. As Hooker explained in a private letter to Lindley, he "has very nice & peculiar feelings"; a "something,—I know not what to call it." To offer an example of this "something," Hooker told a story of a dinner he had attended at the Linnean Society, where Brown's health had been drunk heartily by the assembled gentlemen. Something about the event set Brown off. For no very clear reason, "He soon rose & left the meeting & has never dined there since."[12]

Brown's good opinion, Hooker suggested, once lost, was hard to regain, and it wasn't easy to know what caused it to be lost in the first place. "Sir Joseph & Mr. B" would be watching the new arrival closely to "be satisfied of [his] qualifications," Hooker warned Lindley, and of his gentlemanly "circumstances" too; much rode on what they decided. Thrillingly, the two gentlemen had hinted at a possible future trip for Lindley, either to Sumatra "or to the Brazils." But Banks and Brown were "too cautious" to intimate where, when, or even whether such a trip might happen.[13]

Presumably, the two were testing young Lindley. He was expected to prove himself willing to do grunt work, thereby to understand and learn more fully the mechanisms by which imported plants moved through space in a globalizing metropole. Hooker and Lindley's correspondence throughout 1819 gives a detailed sense of the day-to-day experience of a young man at the bottom of the pecking order in the Banksian universe, fighting to impress while battling to manage the bustling, complex infrastructures of Regency London—seemingly lost, at times, in near-endless bureaucracy.

On April 22, 1819, for instance, Lindley wrote to Hooker to explain he had been sent on multiple failed attempts to pick up a cargo of plants from the new Customs House on Lower Thames Street. It was a three-and-a-half mile walk from Soho Square along the banks of the Thames, and when Lindley arrived the man in charge of discharging the cargo was nowhere to be found: "The walk you know was a long one & an abominable pair of new shoes so blistered my feet that for the next few days it was not without pain that I was able to hobble to Soho Square," Lindley complained.

His next attempt was no more successful. Lindley found the man he sought at the East India Warehouses in Cutler Street (a nearly three-mile walk), only to be told the parcel he sought had already been collected by someone else, from Devonshire Square. "Losing no time in pursuing my course," Lindley arrived after five o'clock, to find the premises shut. Back he trudged, again, the following morning, another two-and-a-half miles, to be told he must set off a further mile and a half across the river to a warehouse in bustling Southwark, where he learned the plants were already in the possession of the premises' owner. His next foray, to a flat in the Albany in Piccadilly, had to be made twice before he was at long last able to retrieve the cargo. Perhaps unsurprisingly, Lindley seems to have questioned his life course around this time: in the same letter he told Hooker bitterly, "I begin to get quite tired of staying here and I think of returning into Norfolk via Halesworth before long."[14]

Lindley went to visit his family shortly after, in June. He found it intolerable. "After living in such society as I found in Soho Square you may easily enough imagine that I am a long way from being at my ease at home," he wrote to Robert Brown, a statement that was at least partly an effort to remind Brown of his searing passion, above all, for botanizing: "Pray tell me whether I am nearer Madagascar or the E. Indies than I was when I left London." Brown and Banks were keeping their young acolyte dangling about the possibility of the much-longed-for trip, and Lindley was desperate. The easy prose of Lindley's letters to Hooker stands in stark contrast to the awkward grandiloquence of the young man's epistles to Brown: "So restless a spirit am I that I should be perfectly miserable without the prospect of some early opportunity of worshipping Flora in the tropics . . . You can scarcely conceive how very, very anxious I am about the matter."[15] Stressing his passion for "worshipping Flora in the tropics," Lindley was evidently trying (perhaps a little too hard) to present himself as disinterested, a man capable of rhetorical flourishes, a gentleman. Still the trip did not materialize, and still Lindley remained. For all his frustration, Lindley surely recognized that his work in Soho Square, organizing and systematizing, was too valuable an opportunity to give up.

"New" plants like orchids had been producing epistemological as well as practical challenges since the Age of Discovery at least. As more and more of the globe was plundered by Europeans, as more and more specimens were shipped back, the boundaries of what was believed to be known about the natural world were tested to the breaking point. Naturalists fought to produce naming and taxonomic systems to help impose order on what they considered chaos. For centuries, names had been assigned in a fairly haphazard fashion; there was little consensus about how or why they should be bestowed. Sometimes the same plant or animal acquired multiple names, and many adventurers liked to name plants after themselves, their friends, and their monarchs.[16]

But change was well underway, following the publication of Carl Linnaeus's groundbreaking 1735 *Systema Naturae* and its revised editions, which used a binomial naming system based on the number of sexual organs—first the stamens, or "husbands," then the pistils, or "wives." Linnaeus's system promised to impose order of a kind, bringing control to the vast, unfolding complexity of the natural world. Captured on botanical slips and paper-based information technologies, transmitted to print, the Linnaean system was one even amateurs could learn.[17] It became hugely influential in an era of growing interest in plants, even if the first publication of the theory provoked controversy: all that polyamorous plant-frolicking of husbands and wives was a bit saucy for some.[18]

The terms and concepts were not just easy to learn and apply. Linnaeus's system offered a sort of *story* of botanical reproduction, a narrative of courtship and marriage that the emerging genre of the novel was itself working to emplot. In *Bloom: The Botanical Vernacular in the English Novel*, Amy M. King explains that "Linnean botany... is organized around the story of the courtships and marriages of plants, a taxonomical intertwining of natural and social worlds that conflates the workings of the plant's sexual parts with the term *marriage*." And indeed, the close correlation of Linnaeus's theories with story and fiction was explicitly drawn by contemporaries. Linnaeus was literally accused of writing a novel—and (some considered) a dangerous one at that: "there is such a degree of indelicacy... as cannot be exceeded by the most obscene romance-writer." The very success of Linnaean botany lay partly in its accessibility and portability as a fascinating, if somewhat risqué, story making botanical concepts available to a wide array of readers and enthusiasts. In turn, the growth of a "botanical vernacular," King argues, helped the novel itself take on, examine, and work through complicated ideas about bodies and reproduction by deploying terms like the multilayered "bloom" to "signify sexual maturation and availability."[19] At a moment in which the culture more generally was working to grasp what growing up means, and before terms like "adolescence" were fully theorized, the Linnaean narrative of botanical courtship offered a powerful, legible story of change, transformation, and growth.[20]

Linnaeus's theories were quickly challenged, however. Some botanists, including Linnaeus himself, worried that his system missed something crucial—that the focus on sexual organs for organizational purposes separated plants that naturalists suspected might share a closer relationship. Indeed, Linnaeus's system seemed to connect plants that had little in common beyond their shared number of reproductive parts. To organize plants according to a more "Natural," less "Artificial" system would surely be ideal—indeed, to discover this "Natural" system would be, Linnaeus and others in his generation believed, to see the bigger story, reaching back to its very opening lines—that is, the hand of the Creator at the point of Creation.[21] The Linnaean system was a kind of stopgap, then, at once hugely influential, widely adopted, culturally and even narratively transformative, *and* under attack by botanists who fretted that it missed the bigger picture by reducing the botanist to a mere cataloger and by failing to consider plant physiology in a meaningful, more philosophical way.[22]

Linnaeus's theories, and a deep sense of their possibilities, remained influential in the world in which the young Lindley moved. Banks had, since he arrived home on the *Endeavour*, been keen to classify his collections, and he and his librarians—starting with his friend and traveling companion David Solander, a Swedish disciple of Linnaeus's—applied and adapted Linnaean principles while keeping in close contact with naturalists at Uppsala, Linnaeus's home university. James Smith bought Linnaeus's private collections after his death and brought them to England at Banks's suggestion—famously, Smith happened to be breakfasting with Banks when the latter learned the collection was for sale. Banks also encouraged the organization of the specimens at Kew along Linnaean lines. But as the men in Banks's orbit responded to an age flooded with more and more specimens, new intellectual approaches were integrated alongside them to supplement, revise, and speak back to Linnaeus. Brown was especially interested in the work of French botanist Antoine-Laurent de Jussieu, who suggested that sexual characteristics should not take such a defining role in the business of classification—that they should be *weighed, not counted*. Jussieu and followers like Brown advocated for tracking other morphological features of the plants and determining membership of plants within families based on common traits. When Lindley entered Soho Square, then, it was into a world still passionately engaged by taxonomy and by its possibilities—the role of text and paper in bringing order to the vast, complex world—even as Linnaeus's direct influence was on the wane. Banks, for instance, was passing more and more authority over to Brown, who, for all Hooker's worries, Lindley seems to have found a way of managing. Lindley wrote to Brown deferentially, on October 10, 1819, that "access to Sir Joseph's botanical stores and the society of yourself are inestimable advantages."[23]

In Soho Square, the nurseryman's son must have encountered a handful of the exotic orchids presently on British soil. Jussieu had recognized orchids as a separate family in *General Plantarum* of 1789, the work that helped usher in the study of a new "Natural System" in Europe. Lindley would have seen the wicker baskets Banks constructed for epiphytes, and he must have learned more of rare orchids from Brown, who had collected specimens of Australian orchids on his voyages with Bauer—the latter illustrating them gorgeously, in images that play up their glorious, freakish strangeness. Lindley reported himself entranced: he called orchids "the most interesting from the multitude of vegetable tribes ... it would be difficult to select any order superior to Orchidae ... and few even equal to them."[24]

But circumstances intervened to expel Lindley from this elite, buzzing, collaborative world. The elderly Banks, suffering gout for many years, died in June 1820. This posed a problem for all those living and working in his orbit. Lindley was thrust back into an unforgiving universe. Hooker was agog—"Especially let me hear what Brown will do with himself," he wrote to Lindley. "I suppose it will not be easy to learn whether he is disappointed or not with the position that is made for him" (Brown had been left the responsibility of caring for Banks's collections).[25] Lindley could not afford to lose a steady income with his father hovering on the edge of bankruptcy. Determined to help his father repay his mounting debts, John Lindley stood surety for him—a generous decision he would later regret.[26]

The pages of East Anglian newspapers in the 1820s are full of bankruptcies and estate sales. George Lindley's name is nowhere to be seen, thanks to his son's energy and diligence, and a timely offer from William Cattley, the merchant, collector, and acquaintance of Hooker and Banks. Fortunately for Lindley, Cattley had just decided he needed someone to help him catalog and illustrate his collection of exotic plants.

Cattley was astute at identifying young talent; he did not require elite birth. The merchant fully grasped the power of networks that extended deep into the provinces, too, for he bought plants from Hooker himself while the latter was still working as a brewer in Halesworth.

One of these plants—daughter of a Brazilian orchid, that first flowered in Suffolk—was just months away from its first bloom when Lindley arrived to take up his position in Barnet.

6

THE RUSSIA CONNECTION

BACK IN DECEMBER 1819, Hooker wrote from Halesworth to his "dear friend" Lindley, offering to sell plants from his collection, "orchideous plants" in particular, to Mr. Cattley. He explained:

> Mr. Murray, our head Gardener says he shall be most happy to send any orchideous plants we have (as soon as divisable) & any thing else we have ^for Mr. Cattley. We had a great many parasitic orchideous plants from Trinidad which promise well.[1]

Included in that collection, or in a consignment soon afterward, was an offset of the purple and red orchid from Brazil.

Cattley was heavily involved in Russian trade, particularly in grain. Britain had been Russia's main trading partner in the eighteenth century; over the course of the nineteenth, Germany took up this role, but Britain remained an important export market for grain or "corn." A number of key British merchant families from the eighteenth century remained prominent in the period, establishing stronger links in St. Petersburg and Moscow, becoming involved in local trade, business, and shipping, and even morphing into merchant banks.[2]

The Cattleys weren't quite on that scale. Still, theirs was a sizable family affair: William's father John and Uncle Stephen imported tallow and barilla, with offices at Queenhithe. The *Post Office Directory* of 1814 lists four other family members at that time as "Russia Merchts." Cattley & Co, founded by another Stephen, was identified as having "for many years carried on the business of export and general merchants at St. Petersburg" in a court case from 1894.[3] The family was and would remain well established in trade, and William devoted a significant percentage of his income to plants.

Cattley's home in Barnet was at the first stop on the stagecoach route from London along the Great North Road. Dominated by inns, it was easily accessible to the City.[4] It was an ideal location, in other words, for a man who had to work for a living but who wanted space to spread out and cultivate exotics in his evening leisure hours. Cattley was rich, as collectors of rare orchids had to be at the start of the century. But as a businessman in suburbanizing Barnet, he was also living a version of the commuter lifestyle that would become widespread in and characteristic of the century.[5]

Lindley, a talented illustrator, was well positioned to help Cattley produce a work that at once cataloged and celebrated the merchant's collection, that brought his vast selection of plants together into a serially produced, richly illustrated narrative. Lindley had just published his first article and had been elected a fellow of the Linnean Society at the time of Banks's death. Still, he was probably a little shell-shocked at his abrupt exit from the elite enclave of Banks's herbaria, and he seems to have been a bit sore at losing out on the much-desired trip to Sumatra or Brazil, too. Hooker was bracing in his letters: "As things are, it may be as well you did not go abroad, & you are surely a lucky fellow to have met with such a friend as Mr. Catley," he wrote cheerfully.[6]

Cattley seemed, in 1819–20, willing to bankroll young botanists, and Hooker was himself a beneficiary. Hooker reported in a letter to his father-in-law, Dawson Turner, that Cattley was generously offering Lindley a salary to publish a book; he also mentioned that Cattley had purchased the rights to two of his own works. "Lindley is not going abroad," Hooker wrote, in March 1820: "A Mr. Catley (who has lately purchased my *Jung* [*Jungermannia*] & *Mosses* [?]) gives him £400 per annum to publish an "Illustrationes Generum!"[7] Since Banks's death was still some three months off at the time of writing, the letter indicates both that Hooker was happily benefiting from a relationship with Cattley and that Lindley's employment with Cattley was a continuation or extension of a plan on the table when Banks died.

On his arrival in Barnet, Lindley began preparing a spectacular work to be published in parts with Cattley. This was ultimately called *Collectanea Botanica*, enticingly subtitled *Figures and Botanical Illustrations of Rare and Curious Exotic Plants*. Filled with luscious plates, it was, from the first, designed as a visual feast of what was, to European eyes, the strange, the unusual, the different.

There were many gorgeous plants for Lindley to choose from: Cattley's collection drew from all over the world. It included scions of several of the other orchids Swainson had sent to Hooker from Brazil, among them a dwarf yellow *Oncidium* and a waxy-flowered *Catasetum*. When, in November, the purple and red orchid thrust out its luscious blooms, Lindley (like Hooker before him) was entranced.

"There is certainly no plant of which I have any knowledge that can be said to stand forth with an equal radiance of splendour or beauty," he would declare, in the first description of the orchid on paper, in English. Lindley used vivid prose in his attempts to capture not just the flowers but the *presence* of the orchid, the way it occupied physical space:

> For it is not merely the large size of the flowers, and the deep rich crimson of one petal contrasted with the delicate lilac of the others that constitute the loveliness of this plant, it owes its beauty in an almost equal degree to the transparency of its texture, and the exquisite clearness of its colours, and the graceful manner in which its broad flag-like petals wave and intermingle when they are stirred by the air, or hang half drooping half erect when at rest and motionless.[8]

Lindley would include the yellow *Oncidium* and the *Catasetum* from Swainson's import in the *Collectanea*. But it was the purple and red orchid that apparently inspired him to superlatives, "the handsomest species of the order we have ever seen alive."[9]

Just as Lindley was preparing to bring knowledge of this orchid to the world in the *Collectanea*, problems hit Cattley's business that almost derailed the whole project. In a letter dashed off on June 30, 1821, the merchant wrote to his employee:

My dear Sir,

It is with the most painful feelings I am compelled to communicate to you that circumstances have occurred which from their present appearance will render it wholly impossible for me to continue your salary beyond the close of the present year, indeed they are of so pressing a nature that I ought in prudence to discontinue it sooner, but I feel that it would be too sudden to enable you to provide for yourself otherways which I have no doubt you will very soon be enabled to do—

Lindley was not the only person impacted by Cattley's money woes: the merchant informed Lindley he must tell Mr. Wrench (presumably, that same Jacob who helped train Lindley as a teen) that "he must not look to me for security to any amount beyond which will become due at Christmas next." Later in the letter he asked Lindley to come to his house that evening, between 7 and 8 P.M., so the two could discuss how Mr. Wrench should be given this disturbing news. Cattley, evidently harassed, encouraged Lindley to look for another job—though he made it clear he wanted Lindley to complete "what you have in hand" first. In fact, in rather an about-face, Cattley closed the letter by asking Lindley to continue with his current work "as if no change were likely to take place in our business agreements."[10]

The Cattley of the letter seems at once demanding and confused, torn up with angst ("it is with the most *painful* feelings"; "I am *compelled* to communicate") and determined to get more illustrations from Lindley if at all possible. Perhaps, he suggested, sale of the next number of the *Collectanea* might "justify the proceeding further in it"; that is, if the early parts were successful, more might follow, so the project could continue. Cattley the businessman was keen to wring as much out of Lindley as he could.

Hooker had learned of Cattley's financial issues by August of that year. He wrote in agitation, fearing his young friend would be unable to continue devoting himself to botany and that publication of the *Collectanea* would end. He may also have worried that the spigot was about to be turned off. "It is a splendid work which will do you the highest credit in the estimation of men of science," Hooker wrote—but he acknowledged that the state of the field was such, the expense of book publication so high, "scarcely any individual, however wealthy, could afford to continue long" with publication.[11]

On October 31, 1821, after publishing just six parts of the work, Lindley was forced to suspend it. The purple and red orchid was due to be published in the next part. Lindley did not explain to his readers why, exactly, he was forced to call off his work, though since he explained how the project was financed—that it operated not in expectation of a profit, but rather in the hope of recouping costs—it wasn't too hard to pinpoint lack of money as the cause. He would be able to publish just four more numbers in total, he explained, and these would no longer appear monthly but "at uncertain intervals." Thanking all those who had subscribed, Lindley mourned that "causes, over which he has no controul, have rendered it necessary that an earlier termination should be put to the work than was originally contemplated." He closed the part at Tab 31, a yellow orchid whose genus was "communicated to us by Mr. Brown."[12]

Somehow Cattley and Lindley seem to have worked out a plan, or Cattley's finances recovered a little. Another two parts, with ten plates total, would eventually be published. The publication of part 7 is hard to date, though it was somewhere between late 1822 and 1824; part 8 came out—after more delay—in 1826.[13] Part 7 included the purple orchid from Brazil, and it was a triumphant return to form. Lindley had made good use of the extra time he had been given, if reluctantly, to reflect.

Lindley's months in Banks's herbarium with Robert Brown, on top of his years with Hooker and his introduction to the works of de Candolle and de Jussieu, had primed him to pay attention to the plant's morphology, to weigh the reproductive parts together with other features of the form. (His training also pushed him to interpret the plant in relation to other specimens in London, of course, rather than, say, those of its original habitat). Lindley would publish in 1830 *An Introduction to the Natural System of Botany*, a work that aspired to overset the "Artificial" system of Linnaeus in Britain. In the "Natural" system, Lindley argued, "the affinities of plants may be determined by a consideration of all the points of resemblance between their various parts, properties, and qualities . . . hence an arrangement may be deduced in which those species will be placed next to each other which have the greatest degree of relationship."[14] Lindley felt he spied a relationship connecting the purple orchid with another in the collection in Barnet, a South American exotic currently named *Epidendrum violaceum* that had been imported by the nursery Conrad Loddiges & Sons in Hackney (cheered in the *Botanical Cabinet* as an augur of supposedly "unknown treasures" in South America).[15] There was a difference: the latter was *bifoliate,* or two-leaved, while Swainson's plant was *unifoliate,* meaning it produced one leaf per pseudobulb. (Pseudobulbs in orchids look like swollen stems and are used to store nourishment.) But the plants shared enough resemblances, Lindley considered, to be "placed next to each other": both had large, leathery leaves, showy flowers with a particularly remarkable labellum, or lip, and four pollen masses in two pairs. The plants thus "obviously offered such striking generic resemblances . . . and at the same time so many beautiful specific differences," Lindley wrote, "that we no longer hesitated to establish upon the two a new genus."[16]

And so Lindley named the purple orchid. As George Eliot would put it, "The mere fact of naming an object tends to give definiteness to our conception of it—we have then a sign that at once calls up in our minds the distinctive qualities which mark out for us that particular object from all others."[17] The name Lindley chose for the genus—*Cattleya*—recognized his merchant patron; to designate Swainson's plant specifically, Lindley used the term *labiata,* referencing the orchid's pronounced crimson lip—and hinting perhaps that, from the very moment of "[marking] that particular object from all others," Lindley correlated the orchid, soon to be the focus of such intense pursuit, with women, sexuality, desire.[18]

The full name for the orchid in the packing case thus became *Cattleya labiata.* Lindley changed the name of the other orchid, *Epidendrum violaceum,* to *Cattleya loddigessii* while he was about it, to honor the nursery that imported it and capture their relationship "next to each other."[19] He went on to add two more species to the new genus, fellow Brazilian orchids *Cattleya forbesii* and *Cattleya citrina,* and further distinguished *Cattleya* from other "air" orchids, specifically *Coelogyne.* In this way, Lindley mapped his new genus into the Orchid family and its wider tribe Epidendrae.

The orchid Swainson plucked from the Brazilian canopy now had a name in Europe. That name recognized neither the man who took it nor the man who brought it to bloom in Britain, still less the community and landscape from whence it was taken. Its name pointed out the story Lindley wanted to tell about it, the "affinities" he considered important, the approach to the natural world he wished to advance. *Cattleya labiata* positioned the orchid in a complex web of family relationships, rather like a character in one of the many, many nineteenth-century novels whose name is forged by the author to point out a hidden backstory, concealed identity, and meaning.[20]

Why was Cattley given, as it were, the patronymic? Even in the nineteenth century, writers pondered the paucity of information concerning "the man after whom Lindley named one of the most magnificent of the genera of Orchids."[21] Lindley stressed that he deeply admired his "patrono nulli secondo" (advocate second to none) in his first book-length publication, on foxgloves, *Digitalium Monographia* (1821).[22] He went on to explain in the *Collectanea* that he named the genus after Cattley because "it has given us an opportunity of paying a compliment to a gentleman, whose ardour in the collection, and whose unrivalled success in the cultivation, of the difficult tribe of plants to which it belongs, have long since given him the strongest claims to such a distinction."[23] Perhaps Lindley's gratitude was aroused because Cattley helped foot the bill for the publication of both the *Digitalium* and the *Collectanea*, two volumes that would otherwise have been available by expensive subscription alone, and rescued him from a precipitous end to his career after Banks's death. Perhaps he was hoping to keep in Cattley's good books, and so gain financing for more parts of the *Collectanea*. Still, it is a little surprising he did not name it after Swainson.

It may seem odd, too, that Hooker did not attempt to name the orchid he would describe, in 1827, as "The most splendid, perhaps, of all orchideous plants." Hooker seems to have chosen early to defer to Lindley on matters of taxonomy: in a letter of January 29, 1818—a year before Lindley went to London—Hooker wrote from Halesworth to the Catton Nursery to ask his young friend to help him work out the (given) names of plants that had just arrived from the Cape of Good Hope, and sent Lindley several catalogs to help with the process. Hooker seems to have fully accepted Lindley's subsequent naming of the purple orchid, listing "CATTLEYA Labiata (*Lindl*)" in his 1825 *Catalogue of Plants Contained in the Royal Botanic Garden of Glasgow*. The earliest reference I've been able to locate to the *Cattleya* genus in Hooker's papers is in the 1822 "Herbarium Hookerianum," held at Kew. Here, in a leather-bound, handwritten ledger of all 6,372 plants in Hooker's collection, there is mention—though it looks like an addition, made in a second hand—of *Catleyia Loddigesii Lindl.* This indicates awareness, at least at the time of writing, of Lindley's renaming of *Epidendrum violaceum* and adherence to Lindley's designation of the *Cattleya* genus.[24] If Hooker ever gave the orchid another name, when first he flowered it in Halesworth, when he sent it on to Cattley, that name has not survived.

Why do these spaces, these little gaps and uncertainties, matter? Because, like Swainson's silence, they provoked questions; because they left spaces for deduction, like the lacunae narratologists have long identified as integral to the business of reading and interpreting narrative. Mid- and later-Victorian historians would presently "mobilize [their] knowledge and experience to supplement what is left unsaid" in the tales told of *Cattleya labiata*.[25] Joseph Paxton would claim the orchid was mixed in accidentally with lichens, journalist Frederick Boyle would assert that the lost orchid was sent to Lindley, not Hooker, as if to explain why the latter took no role in the naming—Boyle even called this a lone "certain" fact in the midst of "legends and myths."[26] To this day, scholars assert that Cattley received his plant directly from Swainson, perhaps to explain Cattley's close identification with the plant. This is directly contradicted by Hooker's 1827 *Exotic Flora*, yet Hooker's words in that text—the original "blossomed for the first time in the stove of my garden in Suffolk"—were attributed to Cattley in the *Orchid Review* in 1893, a mistake repeated often since, even though Cattley lived in Barnet, nowhere near Suffolk.[27]

By 1822–1824, then, the purple and crimson orchid was not only physically present in Britain but positioned by botanists in larger European scientific and intellectual narratives. Three specimens of a plant whose ancestors twined around trees in Brazil, drawing in humid air at high altitudes, were now known as *Cattleya labiata,* thereby identified with a British collector, named by a British botanist, and situated in British and European knowledge systems.

Now named and identified, references to it began to appear with increasing frequency in other European tomes and periodicals also. Its form, coloring, and growth habits were described in botanical Latin and English; it was visually represented in drawings and etchings in magazines aimed at plant enthusiasts. It was potted up and shown off in collections as a desirable object, taken to competitions and displayed as a botanical specimen. Cattley showed his at the Horticultural Society in

1829, for instance, where it appeared alongside two Spanish onions and a pineapple.[28] In the botanic gardens in Glasgow, its sibling offered a wider public a tantalizing glimpse of a plant from the other side of the world.

With each fresh appearance in print and on botanists' tables, in magazines, botanic gardens, horticultural competitions, nurseries, and auction rooms, the new orchid's meaning was renegotiated, its identity reshaped. Almost from the moment of its arrival, *Cattleya labiata* was a tangible signal of, and a means of normalizing and celebrating, Britain's global reach and growing technical and economic power—including in young nations like Brazil, seeking to overthrow colonial authority. Brazil finally declared independence from Portugal on September 7, 1822, five years after the independence bid of 1817, following further rebellions, rioting, and rising racial and community tensions.[29] Britain, supposedly affirming Brazil's independent status, thereafter viewed it as open to more direct influence, to more resource extraction and economic exploitation.

The packing case orchid was on the road to becoming, in Marx's classic formulation, a commodity. Bought, sold, and exchanged, its value would be connected not just to the thing itself but to the vast networks of labor and law required to produce it—the sourcing, packing, shipping, transportation, cultivation, marketing, and sale, all of which helped foster the flourishing of a commodity trade in orchids and other rare plants. As a commodity it would acquire, through its consumption and use, "use value" (to borrow again from Marx, born the year the orchid reached British shores). It would come to satisfy human fancies, urges, and needs—to signal wealth and power, or connoisseurship, or modernity, or attachment to the past, or scientific acumen.[30]

These meanings and more were still emerging in collections and on paper in the 1820s. At different moments, to different groups, in different places and kinds of publications, what the orchid represented would shift. Soon, *Cattleya labiata* would become known as the "lost orchid" and would be sighed over as a vestige of an ancient, almost mythic past. Very soon—by the 1830s—lives would be risked in its pursuit. The hunt for the lost orchid would be pursued with such determination later in the century because it was painted as the apotheosis of an alluring "exotic" almost the moment it arrived.

PART II

EXOTIC

She said that the orchid-house was so hot it gave her the headache. She had seen the plant once again, and the aerial rootlets, which were now some of them more than a foot long, had unfortunately reminded her of tentacles reaching out after something; and they got into her dreams.

—H. G. WELLS, "THE FLOWERING OF THE STRANGE ORCHID" (1894)

7
FIRST SIGHT

REPRODUCTIONS OF THE ORCHID, shrunk to today's modest book-size proportions, do not do it justice. On the massively tall, thick pages of Lindley's lavish *Collectanea Botanica*, this first European image is breathtaking. Light seems to shine through its deeply green leaves: it is as if the orchid stands between us and the sun. Lilac-pink petals are gorgeously, delicately translucent, while the veins on the lip, or labellum, and the pollen masses and dissected flower parts that occupy much of the plate's bottom third are painted in intense detail. (See Plate 2 and, for a close-up of the flower, Plate 3.)

The illustration, made from Cattley's plant in Barnet, was drawn for Lindley by John Curtis, son of a flower-grower. It was engraved by a steward's son, Charles Fox. Both collaborators were, like Hooker and Lindley, former Norfolk residents who moved to London to gain access to professional opportunities. Curtis would soon become a Linnean Society member; Fox later engraved a painting of Robert Brown. Lindley probably met them at one of Banks's lively soirées in Soho Square.[1]

Cattleya orchids remained "exceedingly rare," the *Botanical Register* noted in 1825, pointing to just two of species *C. labiata*, both owned by Cattley.[2] A full decade later, the *Floricultural Cabinet* placed it on a select list of plants that "cannot be procured under from five to ten guineas each [about £650–1,100 / $550–1,200 today]; indeed many of them cannot be purchased at any price." *Cattleya crispa* could be bought at two or three guineas from the major nurseries in London, while enthusiasts could pick up *C. forbesii* "easily from any private Collection; or for about half-a-guinea each on average, from the London Nurseries."[3] But *C. labiata* remained out of reach.

Images like the one in the *Collectanea*, in other words, were the only way most early enthusiasts could catch a glimpse of the plant. No one sought this orchid in Brazil in 1822, nor would they for a decade. For the orchid to become a prize worth dying for, an orchid that would send Ericsson and Bungeroth after one another in deadly pursuit, it first had to become a widely known object of desire among the buying public.

The first stage in the hunt for the lost orchid arguably begins, then, in the pages of magazines, periodicals, and book-length botanical works of the second quarter of the nineteenth century. It was here that knowledge of *C. labiata*, as a botanical specimen, was first disseminated in Europe, and it was here that the plant was presented to an interested, increasingly literate public as an especially luscious and desirable exotic. Lindley's first lustrous, haunting image offers signposts toward the eventual celebrity it would become.

AS WE SAW EARLIER, the full subtitle of Lindley's *Collectanea* celebrated the "figures and botanical illustrations of rare and curious exotic plants, chiefly cultivated in the gardens of Great Britain." The words "EXOTIC PLANTS" were set in all caps in the center of the title page.[4]

"Exotic" is a term typically used in horticulture and botany to mean nonnative plants. Used in this way, it may seem strictly descriptive, yet the term carries implications of the unfamiliar, the foreign, the alluring. Edward Said's pathbreaking *Orientalism* explores Western writers' construction of the "Orient" as an "exotic locale," a place of "strangeness ... difference" and "exotic sensuousness." Appearing to signify real people and places, particularly in the Middle and Far East, the discourse of the exotic Orient was built to glorify the ordering, sensible, intellectual "Self" by romanticizing, and setting the narrating self apart from, the strange, uncontained, sensual "Other." "The Orient was almost a European invention," Said argues, "and had been since antiquity a place of romance, exotic beings, haunting memories and landscapes, remarkable experiences."[5]

First Sight

But the idea of the exotic spilled out of the Orient, other scholars have since noted, into (for instance) the south and east of Europe as well as the tropics. It stalks Gothic novels, haunting hidden rooms and ruins as late eighteenth- and nineteenth-century writers associated other landscapes, other people, and especially other women with not just strangeness and difference, but also dangerous sexuality.[6] There are many forms and places of the exotic. Many things and people can be represented as exotic, but use of the term is always rooted in a response to someone or something unfamiliar to the narrating self and the anticipated reader. Peter Mason argues that this is the definition of the exotic: "it is the very act of discovery which *produces* the exotic as such."[7] To respond to the exotic is to respond to something we consider alien; drawing our fascinated gaze, its strangeness provokes us to translate and contain it. To that end, for instance, Gothic novels typically end with the destruction of the exotic Other and so the restoration of normalcy. At the end of Sheridan Le Fanu's 1872 Gothic novella *Carmilla*, about a gorgeous female vampire from Styria, a stake is driven through Carmilla's heart, her head is cut off, the whole body is burned to ashes, and the ashes are cast into the river—just to be sure she is dead.[8]

Fantasies of alluring strangeness spill into nineteenth-century representations of exotic plants from the first. The associative connection of "exotic" with "rare and curious" is made explicit on Lindley's title page and points to the ways in which the fascination of the plants lies, imaginatively, in their strangeness and difference. Associations of orchids with sex and alluring sensuality were already well-established, not only because *orkhis*, the Ancient Greek word from which "orchid" descends, signifies "testicle." (Some orchid species possess twin tubers). Vanilla was widely believed to be an aphrodisiac, and Louis Liger's 1704 *Le jardinier fleuriste et historiographe* claimed to tell an ancient myth in which Orchis, son of a satyr and a nymph, was turned into a flower as a consequence of his disorderliness and sexual misdeeds. The story was almost certainly made up by Liger, historian Jim Endersby points out, noting that Orchis does not appear in classical texts. Still, it was much repeated in the nineteenth century and beyond, surely because it "encapsulates the qualities that Western cultures have come to associate with orchids"—dangerousness, difference, sexuality. "It is no coincidence that tropical orchids entered the European imagination at the same time as Liger's myth," Endersby concludes, reminding us that Linnaean terminologies reinforced the sexuality of the plant; "The eighteenth century gave the orchids their rational, scientific name, but did so in the Linnaean language of the marriages of plants. Paradoxically, the supposed age of Enlightenment also saturated orchids in images of sex and (thanks to the fate of Orchis) of death."[9]

Lindley retained and reinforced a language of strangeness, allure, and difference in his work. His entries on orchids suggest that "many inexplicable phaenomena" are associated with exotic plants in ways that presage their later-century presentation in eerie fiction by H. G. Wells.[10] In the 1894 "The Flowering of the Strange Orchid," a new purchase drains the blood of its owner through tentacular roots; orchids are not quite *that* dangerous in Lindley's writings, but their oddness, their difference is nonetheless presented as a significant part of their attraction.

Lindley attempts to manage the paradox Endersby notes, that Linnaean scholarly discourse both orders the plants and "saturates" them with sexual meaning, by presenting his own text and images to identify and, to some extent, explain the oddness of these newcomers. Transforming them into managed, curated specimens, he converts the "curious" and "inexplicable" into the legible, the ordered, the familiar.[11] This process unfolds through judicious use of scientific conventions in the texts and in the carefully arranged images accompanying them.

Botanical illustrations are not just literal visual transcriptions of a plant, though they may present themselves as such. The botanical image is not a neutral depiction of reality.[12] Just as the camera is positioned to shape our reading of the action in film and in photographs, so illustrators choose, position, and curate their subject to convey and prompt the production of a certain set of meanings in and by the observer. Value-laden choices inevitably stand behind what parts are made the focus of an image. Roots were of great importance to sixteenth-century apothecaries and herbalists, for instance, and so were often included in images from that period. The flower moved center stage later, in the eighteenth century, as Linnaean principles began to shape illustration conventions. To Linnaeus, the number of sexual organs hooded within the flower were critical to understanding and classifying a plant, while the roots were less revealing, so were often left out or rendered in only partial or incomplete form.[13]

In Lindley's *Collectanea*, *C. labiata*, without roots, is set against blank white space. Gill Saunders notes in her study of botanical illustration, *Picturing Plants*, that the use of the blank background was itself an illustrative convention by the late eighteenth century, again established to support the Linnaean focus on the flower.[14] De-contextualized, isolated in a blank field, the plant is visually divorced from the plants that surrounded it as it grew. The specimen's new habitat is the printed book—and the European publishing and intellectual traditions that produced it.

Lindley references those traditions in text also, mentioning European scholars by name, using European scientific conventions to designate the orchid's place in the "Natural Order" and in the Linnaean system.[15] He cites Robert Brown, and the illustration visualizes his own Brownian, post-Linnaean intellectual approach, in that flower and leaves are in dialogue with dissections of the column and pollen masses.[16] Lindley's extensive use of botanical Latin further positions the plant in familiar scholarly tradition.[17] Latin had been used since Pliny the Elder at least for the business of describing and identifying plant life: through the Middle Ages, it continued to be used by those who wrote about plants because Latin was the language of writing and scholarship. It gained new meaning and purpose in the eighteenth century as Linnaeus's theories centered parts of the plant, its sexual structure especially, for which no terms existed. Many of the plants and parts of plants Linnaeus and his followers sought to describe in the eighteenth and nineteenth centuries were unknown to the Greeks or the Romans in classical times—so, to stay relevant, botanical Latin was forced to evolve.[18]

Rather than dropping Latin entirely, that is, botanists in the late eighteenth century—including Linnaeus—chose to adapt it because it enabled international communication between educated men on the richness of an environment of which they were becoming collectively aware. As a *lingua franca*, botanical Latin allowed rapidly changing information to be easily shared across nations. Yet it excluded even as it included: women, the poor, and people of color were unlikely to have the training to enter the conversation. And understanding botanical Latin required ongoing immersion in its intricacies, so that Lindley's use of it in the *Collectanea Botanica* communicates to readers both his desire to connect with scholars across nations but also his understanding of the boundaries defining the world of an elite.

Still, there are whispers of change, an urge to connect with a broadening readership. If the botanical vernacular spread into and through literature, as Amy M. King argues, the democratizing forces of popular literatures also spread into and through botany: the field and practice were becoming more accessible to those without elite birth, to those with less spending money, to artisans and commuters, who preferred cheap, portable texts—like magazines—as well as to those more familiar with new forms of entertainment, including the narrative and visual pleasures of mystery, sensation, melodrama, spectacle.[19]

Hints of this are evident in the ways Lindley makes his information available. His work offers new enticements at a time when more and more kinds of people were meeting in what Sarah Easterby-Smith calls the "broad landscape" of contemporary scientific engagement.[20] For instance, after describing *C. labiata* in botanical Latin in the *Collectanea*, Lindley transitions to English and redescribes it. He explains how the plant arrived in Europe and focuses on the roles of Swainson, Hooker, and Cattley in its discovery and cultivation, presenting it, from the first, as the product of British achievement. Lindley—obviously conscious of the need to keep the Cattley tap open and flowing—works hard to make a widening spectrum of readers feel welcome and engaged. The addition of luminous, richly hued plates was a further way to attract readers.

Color plates had long been expensive to produce and were hard to keep consistent, thus most botanical plates were, until late in the eighteenth century, black and white. Many botanists felt that poor coloring was worse than no color at all, and that anyway the morphology of the plant was or should be the focus. But Lindley lived in an era fascinated by looking and spectacle: this would be a century driven to produce daguerrotypes, photographs, microscopes, telescopes, stereoscopes, kaleidoscopes, magic lanterns, and more. Many of these tools still lay in the future. But surely Lindley was aware that buyers wanted color; botanical magazines like John Claudius Loudon's *Gardener's Magazine*, which did not include color plates, found themselves unable to compete with showier rivals and were forced to close.[21]

Color was typically added to botanical plates by hand, in a wash, on top of the outline. Some presses deployed a sort of early assembly-line process, where different colorists added each fresh layer of color. Samuel Curtis, responsible for printing *Curtis's Botanical Magazine* from 1811 (and son-in-law of the founder), used his own family members for this, so that he and the illustrators could oversee the process carefully. Four of his daughters worked on the images, though as many as thirty colorists were employed by the periodical.[22] The production of such images was, in other words, intensely collaborative, involving not only the illustrator and engraver, but colorists, printers, and editor. This was partly why colored illustrations tended, for centuries, to be regarded as less highbrow, less "pure" than black and white. Some audiences preferred the notion of art created by a solitary genius, and others were suspicious of works produced in a workshop.[23] Color was also long viewed as a crutch in printmaking, an effort to distract the eye and hide mistakes that true masters did not make. Erasmus celebrated Dürer for the ways in which he conveyed even "that which cannot be depicted" in just black and white, for instance, without recourse to the "blandishments of color."[24]

But technological improvements in printmaking in the nineteenth century, combined with a desire "to test the limits of new media" and "the drive to reach new audiences either for ideology or for profit" produced fresh attitudes to color.[25] The "blandishments" of color were becoming a positive, associated not with lack of skill but with novelty. *C. labiata* reached Europe just as attitudes to color were changing, and the particularly rich hues of the plant made it deeply appealing to publishers (and colorists) looking to attract wider audiences and show off new capabilities.

Full color plates (five per part) accompanied each *Collectanea* entry for those who could afford twelve shillings—approximately £62 today, or $79. The slightly less affluent purchaser, paying eight shillings or about £41, or $52, got black and white. At one point Lindley hoped to make even more expensive, lavish images available to his buyers: "It was originally intended that a few copies should have been prepared with plates more highly finished." That plan fell through—another sign, no doubt, that Cattley's purse was not bottomless. Still, the presence of color remains significant and suggests Lindley's recognition that exotics in rich hues were attractive to the era's consumers.

That said, a hand-produced process did not lend itself to precision or consistency. And color is, as Goethe pointed out, at once "real," a matter of pigments and light waves, and experiential, rooted in perception. Not every reader of botanical magazines had access to the same images, the same hues—and Lindley seems to have recognized this, too, for his textual descriptions in the *Collectanea* employ a language of spectacle and color that both build on and recognize the era's growing taste for the visual, and also supplement, complement, even stand in for elements of the illustration. Of a species of *Bromelia*, for instance, he begins:

> Let the reader imagine to himself a plant about three feet high, formed of a mass of leaves shaped like those of the common pineapple, but from three to five feet long; let the lowest be of a dull dark-green, and the uppermost of an intensely brilliant scarlet; the whole surmounted by a pyramidal tuft of flowers of the most exquisitely delicate crimson; and a faint idea may then be formed of the superb appearance of *Bromelia fastuosa* in perfection.[26]

"Let the reader imagine": Lindley appeals to the inner eye in tandem with observation and intellect. Shape is evoked through comparisons (to a pineapple, the pyramid), color delineated in adverb-heavy descriptions ("dull dark-green," "intensely brilliant scarlet," "exquisitely delicate crimson"). The reader unable to purchase the color-washed print version is encouraged to imagine the rich hues of the plant, and the reader possessing the color plate is given access to information about scale and size in addition to color. Words like "tuft" further supplement the picture by intimating texture, so that the botanical illustration is only one means among many of appealing to our imaginations and igniting our senses. Throughout the book, orchids are presented as the apotheosis of the exotic—"the most interesting" of the whole "multitude of vegetable tribes," Lindley contends, because of their sensory thrills, their "general elegance . . . durability of blossoms, splendid colours, delicious perfume" and "extraordinary structure."[27] Of forty-one entries on exotics, seventeen center orchids. *Collectanea Botanica* was designed to generate wonder, to invite fascination at the sensory overload on offer from exotics of all kinds, but especially orchids. Lindley framed color not as a "blandishment," a lure away from intellectual content, but as a portal *into* it. Richly hued exotics were especially likely, he suggested, to lead readers to new, deeper insights, to opening of the inner as well as the outer eye; to awakening.

This supposed process of transformation through fascinated looking shaped the story, the narrative of exotics unfolding in print texts in the

1830s and 1840s. As more and more middle-class enthusiasts entered the market, color and spectacle on the page and in the plant-pot was often mentioned as a crucial mechanism for aiding understanding and thereby intellectual and spiritual growth. The *Floral Cabinet, and Magazine of Exotic Botany*, for example, foregrounded illustrations from the first issue in a preface that celebrated "delightful science."[28] Delightful science was pleasurable science: "no inconsiderable degree of light will be shed" by information both "useful and entertaining." The pleasures of looking, the editors explained, at rare plants and "botanical novelties," combined with information in easy-to-access English, would make discovery and the gaining of information neither dry nor difficult to understand, but fun. This was not intended as an introduction to botany nor a source of specialized scientific knowledge. The *Floral Cabinet* aimed to delight the senses of readers through gorgeous plates while offering information "in popular as well as botanical language": the spoonful of sugar would make the medicine go down with ease.[29]

The *Floral Cabinet* presented its approach as innovative, but other magazines and periodicals were doing much the same thing.[30] Newspapers and novels were rapidly adding illustrations: by the end of the century, the *Strand* magazine would insist on an image per page. As printing press technologies advanced, and with the repeal of the Stamp Duty in 1855, the costs of producing and disseminating images and newsprint dropped. Meanwhile, as more of the population gained literacy skills, more workers were looking to pursue middle-class interests. Illustrations enticed, thrilled, and educated these new reader-enthusiasts, as novels, periodicals, and self-help manuals worked to "[translate] the visual into the verbal for the non-specialist spectator," as Kate Flint has put it.[31]

The richness and color of *C. labiata* made it especially likely to launch the kind of intellectual journey the *Floral Cabinet* hymned for its readers, texts suggested, through a kind of novel-like *Bildung*—from pleasurable looking, to learning, to new and deeper understanding. One of the first plates in 1837 lauds it as an orchid of truly "splendid and matchless color," "unquestionably the queen of orchidaceous plants." Almost four pages of closely written text (on the plant's naming, its tribe, the locations of the genera, its biology, and growth habits) are introduced with a richly colored plate in which the glorious flower is prominently centered on the page. Text reinforces the loveliness by taking the reader's eye around the plant, encouraging slow, attentive, appreciative observation:

> The lovely colour and transparent texture of the sepals and petals; the rich and elegant markings at the base of the lip, with the splendid and matchless colour of its disc; and finally, the graceful arrangement of its large and spreading flowers, must strike with admiration every beholder who is not actually insensible of the charms of nature.

"But who," the text continues, amplifying its last point, "can be insensible of such transcendent beauty?" The passage moves from botanical detail to aesthetic appreciation, from describing the "large and spreading flowers" to invoking an emotional connection with the image. *C. labiata*'s rich coloring will or should take the viewer to the edge of transport: "Who," the text demands, "can behold with indifference this 'herb of glorious hue'?" The modern viewer is imagined as world-weary, on the edge of "insensibility" and almost closed off to feeling, but the illustration of *C. labiata* is meant to awaken us, provoking us out of our "indifference" into a state of questioning and awe.[32]

John Keats's sonnet "On Seeing the Elgin Marbles," written in 1817, as Swainson journeyed in Brazil, conveys a strikingly similar experience of looking. The speaker begins on the edge of insensibility: "mortality / Weighs heavily on me like unwilling sleep." But an encounter with "grandeur," in this case the "wonder" of the marbles, provokes near-overwhelming feeling, a sense of something "dim-conceived," the "shadow of a magnitude." Looking at a rare and curious object can, both Keats and this representation of *C. labiata* affirm, lead toward greater insight, both factual knowledge *and* an intimation of that which passes understanding. The editors conclude the Preface to the first volume of the *Floral Cabinet* by suggesting that looking at exotic plants may help viewers gain insight into the truly "remarkable": "delightful science" will "lead the mind to the contemplation of the wonders of the vegetable world." By noticing these wonders, viewers may even glimpse the hand of God in the Creation, the "glory of that Almighty Being who created 'All the fair variety of things.'"[33] Looking at *Cattleya*, then, is a means of gaining important information, but it is also a means of helping us reach beyond our ordinary, dull lives, toward the divine.

C. labiata was often represented as a plant anyone could admire, learn from, be awakened and transported by over the next decades. As Joseph Paxton put it in the *Magazine of Botany* in 1841, it could not be "witnessed without the most enlivening emotions."[34] Its rich coloring was emphasized as Latin took ever more of a backseat. In the *Botanical Cabinet* in 1833, for instance, a magazine that was half advertisers' catalog for the Loddiges nursery, only a few lines of Latin are even included. The orchid's Linnaean classification is given at the top in ways that merely gesture, in passing, at scientific credibility.[35] Even in texts written by authors with rigorous botanical training, Latin began to recede. Hooker, in his *Exotic Flora*, includes the codes of scientific seriousness, such as Latin, and dissected parts of the orchid at the base of the illustration, with the main illustration of the plant positioned against a blank background. Yet far more English is used here than in the *Collectanea*, so that readers needed less and less training to understand the "delightful science" on offer.[36] Hooker included two separate indexes, one listing the botanical names of the plants, the second their "English names"; he called Swainson's plant, in the latter, the "Splendid-flowered Catleya." (See Plate 4.)

The visual field, meanwhile, in *Exotic Flora* is dominated by massive, pouting, sensual blooms; the leaves of the lost orchid twist and curl and thrust at the reader. If John Curtis's illustration for Lindley presented a plant sunning itself gloriously, Hooker's engraving (by Joseph Swan of Glasgow) pushes its loud showy labellum in our faces. Hooker's text description also emphasizes the richness of the color—the petals of the "very large and spending flowers" are "of a delicate and yet bright hue, between lilac and rose coloured," while the lip is "beautifully notched and waved: its general colour is pale pink, the inside deep purple, forming a large uniform and very bright coloured blotch at the extremity, while the lower portion is picked out with whitish branching veins."[37] In the image by George Loddiges, too, engraved by George Cooke, the gorgeous bloom (now entirely unencumbered by dissections) plays a richly commanding role. (See Plate 5.)

The pattern traced here in print representations of *C. labiata*—of more emphasis on color and spectacle, more English description, less botanical Latin and dissections, and an enfolding claim that seeing exotics enlivens and awakens us, eyes, bodies, and souls—is evident in depictions of orchids generally in the period. But *C. labiata* was especially popular, a sort of botanical pinup. By the time of Lindley's second representation of the plant, in 1836, an image produced by his friend and children's governess, the talented Sarah Drake, the botanist explained that it had already "been represented in most of our Botanical periodicals." Yet "all the plates . . . are deficient in the richness of colour that is so peculiarly characteristic of the species, and that constitutes its chiefest ornament," he contends (see Plate 6). Lindley claims the plant tests the very limits of printing technology, which struggle to reproduce its spectacular beauty. Thus Lindley uses text once more to supplement the illustration, bringing to the reader's notice features of the plant, like size and texture, that can't easily be communicated in a two-dimensional drawing. He explores subtleties of the plant's occupation of space—its "graceful manner," the way it behaves when "stirred by the air"—and repeats that knowledge of the existence of so remarkable a species "cannot be too widely diffused" in the population, so that all of us, as a community, may be stirred.[38]

C. labiata became an increasingly common sight in print in the late 1830s and 1840s. In *Curtis's Botanical Magazine* in 1842 Hooker termed it "often described" and "well known."[39] Still, few had access to the living orchid yet. For *C. labiata* multiplied slowly; propagation by division and separation of offsets from the parent is only possible, Loddiges explained, "now and then." The *Botanical Cabinet* provided notes on cultivation—it "requires the stove," is finicky and vulnerable to woodlice—but this was hardly useful information to the majority of those who viewed the plate in 1833, even those with their own stoves. Lindley also described it, in the *Collectanea*, as one of a tribe of "difficult" plants.[40] A mere handful of collectors possessed a specimen of the orchid, even by the mid 1830s—George Barker of Springfield, near Birmingham (according to the *Floral Cabinet*), Earl Fitzwilliam at Wentworth House in Yorkshire (according to Lindley in *Edwards's Botanical Register*), the Duke of Devonshire at Chatsworth (described in *Paxton's Magazine of Botany*, 1838), and the young but short-lived Lord Grey of Groby. The Veitch nurseries' manual on orchid-cultivation noted, as late as 1889, that "propagation by division" remained the only "available method" for producing new orchids and that this was "no more than sufficient to secure a limited number of plants."[41]

Rather like another exotic figure in the nineteenth-century imaginary, then—the exotic woman, locked in a Zenana, hidden behind the veil—

C. labiata was represented as a desirable figure of sensual allure, made available by modern print technologies even as "true" accessibility was tantalizingly out of reach. Paper illustrations of *C. labiata* served to whet an appetite that could not easily be slaked. It was an alluring spectacle *and* an object of science, a delight for a broadening swathe of the public *and* a meeting ground of the serious, a pleasure of modernity *and* a rare *objet* for the initiate, a market commodity and a glimpse of the divine hand of God. Appetite for this exotic was only growing; it was high time, in other words, that someone ventured in search of some more.

8
FALSE TRAIL?

TWENTY-SIX-YEAR-OLD George Gardner set out for Brazil with high hopes in 1836. The working-class son of a gardener, George was educated in a small parish first, moving with his family to the city of Glasgow in 1822. Here he was admitted to the local grammar school, where—like Hooker and Lindley before him—he gained a classical education, an important signal of gentlemanly accomplishment still necessary for aspiring young men of science.[1]

Passionate about botany, Gardner started out pursuing the stable, safe profession of medicine. This was, at the time, "the least prestigious of the learned professions" because it was dangerously associated with trade and commerce—doctors, after all, sold drugs.[2] Still, it was a steady choice for a talented young man of low birth. And Gardner worked hard, at least according to his biographers: an 1855 author recalled that he "distinguished himself in the prize list." In 1830 he joined the Glasgow Medical Society, where he was remembered as "unremitting" in his attendance at the Royal Infirmary.

But Gardner's fascination with the natural world lingered. On a ramble around the still waters of Mugdock Loch, to the northwest of Glasgow, Gardner spied a small, rare lily that resembled a buttercup—*Nuphar pumila*. Gardner screwed up the courage to approach William Hooker himself.[3]

The former brewer was by now a key figure for a new generation of hopeful botanists. Hooker had started running field botany classes at Glasgow in 1821, one of three professors to come up with the idea almost simultaneously (the others were at Edinburgh and Cambridge). These classes were, argues David Elliston Allen, "possibly the most important development ever to have taken place in the history of organized field botany," for the classes, mostly attended by medical students, sparked excitement, even career-changing enthusiasm. Many of the young men who attended went on to become leading botanists and plant hunters, while their "*esprit de corps*" "bubbled over into the founding" of scientific societies.[4] The alumni of the field botany classes subsequently developed systems for plant exchanges, helping promote active, supportive research communities that grew in influence as scientific subfields evolved into professional communities.

Gardner was one of the medical students swept up in a new craze, finding a new outlet for his interests and the support systems he needed to transition from medicine into natural history. Hooker took Gardner on botanical expeditions around the Scottish Highlands; perhaps it was on one of these that the two discovered a shared interest in cryptogams, the peculiar plants that had decades earlier provided Hooker common ground with William Swainson.

Gardner went on to earn his diploma as a surgeon "with high marks of distinction." But under Hooker's encouragement, he brought out his first book on plants, too, *Musci Britannici, or Pocket Herbarium of British Mosses*, in 1836. The work was well-received and brought Gardner to the attention of Lord John Russell, a Whig politician and sixth Duke of Bedford. The duke and Hooker recommended to Gardner an expedition to Brazil.

The British imperial gaze was turning ever more intently to Latin America in the wake of the independence movements of the 1820s. Brazil, as we've seen, declared independence from Portugal in 1822, leaving the young nation, at least as far as the British were concerned, open to more influence. The expenses of a military campaign for full takeover were judged too high, and might provoke dangerous anticolonial spirit and activism, thus the British largely chose to pursue commercial expansion instead, crafting a mythos of Brazil, and other Latin American countries, as both "free, and . . . English" (as Foreign Secretary George Canning famously put it). Essentially, it presented Brazil to the British public, and to the world, as a model of a free nation, full of opportunities, under beneficent British protection.[5]

The duke offered Gardner passage out to South America on one of his own son's ships and, when Gardner modestly declined, sent him fifty pounds anyway to help fund the trip. With money in his pockets Gardner took the plunge, departing Liverpool for South America on May 20, 1836.

The young man caught his first glimpse of land seventy miles to the east of Rio de Janeiro on July 22. As the ship sailed along the coast, Gardner's eye was glued to his telescope. "It is quite impossible to express the feelings which arise in the mind while the eye surveys the beautifully varied scenery," he later wrote; nothing else ever "left a like impression upon my mind." Like many another English botanist, too, Gardner described thrilling emptiness ahead of him: "whole provinces, particularly in the north, still lay open as virgin fields."[6]

When European botanists framed their journeyings as voyages into emptiness, they of course ignored and as a consequence erased from the narrative the Indigenous people, enslaved residents, and European- and African-descended residents already there.[7] In his published narrative, Gardner also failed to mention the explorations of Swainson, Koster, and another friend of Swainson's, William Burchell, just as Swainson had managed to forget Koster.[8] However, a little surprisingly, Gardner *did* mention the earlier South American journeyings of one particular British explorer named Charles Waterton. In 1825, Waterton—an eccentric who claimed as ancestors multiple saints, most recently Sir Thomas More, and members of several royal families—published *Wanderings in South America*. Waterton visited Pernambuco in the year 1816, mere months before Swainson, and traveled in Guyana ("British Guiana" was the one part of mainland South America under British control). Here Waterton managed his family's sugar and coffee plantations.

Wanderings foregrounded, to a quite extraordinary degree, the author's own vigor and physical prowess. Gardner seems to mention the work largely to stake out how very differently *he* interacted with nature from his precursor. The gardener's son makes plain to us, on nearly every page, that he (unlike Waterton) is a serious, sober-minded scholar, a man of the mind, not the body. One of Waterton's most memorable scenes involved a hand-to-hand encounter with a caiman, cousin to the alligator.

False Trail?

In full action mode, Waterton presents himself climbing the mast of his boat to scrutinize the beast:

> By the time the Cayman was within two yards of me, I saw he was in a state of fear and perturbation; I instantly dropped the mast, sprung up, and jumped on his back, turning half round as I vaulted, so that I gained my seat with my face in a right position. I immediately seized his fore legs, and, by main force, twisted them on his back; thus they served me as his bridle.

Waterton was shaken by the creature, but not stirred, or so his tone of cool detachment suggested: the caiman, "probably fancying himself in hostile company... began to plunge furiously, and lashed the sand with his long and powerful tail. I was out of reach of the strokes of it, by being near his head. He continued to plunge and strike, and made my seat very uncomfortable. It must have been a fine sight for an unoccupied spectator."[9]

In his own work Gardner explicitly references this scene. But he gently pokes fun at it, presenting himself as an attentive observer and a methodical collector of data, not one given to wrestling caimans. Alligators, when they appear in his published work, inhabit a lake, quietly, cohabiting with the blossoms "and broad floating leaves of a water-lily (*Nymphoea ampla, DC*) and intermingled with them, the yellow flowers of *Limnocharis Commersonii*, and a large *Utricularai*." Gardner's use of botanical Latin also positions him in the passage, in his encounter with alligators, with the scholar.[10] Working to embrace and inhabit the emerging identity of professional scientist, Gardner—born into the working class—signals to readers that he is cerebral and learned.

Yet the two works of South American journeying are not as different as Gardner tries to suggest. Like Waterton, Gardner presents himself as a new arrival to a nation supposedly vacant, "undiscovered," omitting not only vast and diverse communities of local and Indigenous peoples but also their skills, expertise, and knowledge. Gardner makes only passing reference to the actions and decision-making of his guides, for instance.[11] The collaborative "we" appears rarely, and typically only when he begins to theorize, from data, about larger principles ("thus we find..."). Gardner reports with seeming satisfaction that he fired an assistant, Pedro, even though Pedro helped him through illness, and they had become "more like companions than master and man." Pedro, it appeared, wanted to set off on the next stage of a journey two days later than Gardner: "this being more than I could reasonably bear, I instantly discharged him."[12] Gardner hired a traveling Englishman instead.

Indeed, the account makes plain the degree to which Gardner deployed and reinforced British and European networks as he embarked on his botanical collecting. Gardner does not explicitly present himself as an imperial agent, but his journey, like Waterton's, is clearly enabled by what Cannon Schmitt calls a "national-imperial infrastructure"—of ships, information, financial systems, letters of introduction to English-speaking patrons, bankers, and merchants. Botanists like Gardner and naturalists who looked for exploitable commodities firmed up those networks as they traveled, much as footpaths become more well-trodden and accessible to the next walker with each use. The trips became easier and more profitable, which encouraged others to follow suit.

Tellingly, the full title of Gardner's work is *Travels in the Interior of Brazil: Principally Through the Northern Provinces and the Gold and Diamond Districts*. Gardner points out the resources that could be extracted in the future, noting (for instance) how close iron ore was to the surface of the ground, the quality of the timber and how easy it was to access, as well as the locations of precious ore and stones. The *Travels* works to advance Gardner's claims to further support, then, partly by pointing out the ways in which the interests of nineteenth-century science and empire intersected.[13]

For all that Gardner presented Brazil as an undiscovered nation in print, open and ready for the extractive capabilities of Europeans, glimmers of active resistance surface in his private letters. On August 25, 1837, Gardner reported in a letter to Hooker from Rio that he had sent "2 boxes of orchidae & cacti" to the Duke of Bedford and another two to Mr. Murray at the Glasgow Botanic Gardens. But the pursuit of the cacti, he told Hooker, nearly led to disaster. Gardner hired a boat in order to get a better look at a species of cacti on a rockface, and he rowed into the bay with John Miers, a British naturalist from Chile. The two were in the midst of stripping specimens from the rock when they were approached by a local man in a canoe who told them to leave. "This we immediately prepared to do," Gardner later explained, but their boat had drifted out of reach.

As the naturalists battled to reach the boat, the local man lost patience. Disappearing into his cottage, he emerged, running, "armed with a large bludgeon, with which he immediately aimed at my head. Had it struck my head," Gardner continued, "it must have killed me; I fortunately, however, raised my left arm, which received the blow." The two men were only able to escape the "brutal fury by leaping down, at great risk, a precipice." Gardner's left arm was damaged in what he termed that "most brutal assault."[14]

Gardner largely focused on the dangers of the *landscape* in his published book—the storms, for example, that produced muds, or the bolting horse that thwacked his head against a tree, leading to severe concussion.[15] Resistance by living people he tended to erase or treat only briefly. Gardner termed a local guide, for instance, roused to anger against him and Pedro, "a lazy talkative fellow."[16] Convinced of his right to the plants, in private Gardner presented those who attacked him as illegitimate actors, blocks in the way of enlightenment and advancement. In public-facing writings, he wrote out their resistance, focusing on the ways in which local people leaned on and supported him.[17] He prescribed medications to and operated on local people, he explained, accepting food in exchange for his services; in private letters, he admitted accepting their cash. The need to present himself as *un*motivated by financial need, as "gentlemanly," as well as being a successful European agent, continuously shaped his public self-presentation.[18]

By day, Gardner journeyed into the forests, mountains, and along the coastline around Rio, hunting the botanically remarkable. By night he wrote letters to his supporters, including Hooker, and drafted his book. He noted that others were said to be botanizing in the area: a Belgian gave him particular cause for concern.

Then, in November 1836, Gardner spied something familiar in the Tijuca mountains.

The Pedra da Gávea is an extraordinary granite and gneiss mountain, known as Topsail Mountain or—to English sailors at the time—"Lord Hood's nose" because it apparently looked from the side like the sloping features and profile of Samuel Hood, commander-in-chief of the Mediterranean Fleet. It had, as Gardner put it, a "nearly perpendicular precipitous face."

On the face of that mountain Gardner glimpsed a species of orchid he thought he recognized growing in profusion: "Its large rose-coloured flowers were very conspicuous, but we could not reach them." Gardner believed it was *Cattleya labiata*, but he couldn't be certain; given the danger, he was forced to retreat. A few days later, as he was exploring the facing Pedra Bonita mountain, he saw something which, the closer he got, he was certain he knew:

> On the edge of a precipice on the eastern side, we found, covered with its large rose-coloured flowers, the splendid *Cattleya Labiata*, which a few days before we had seen on the Gavea.[19]

This time, Gardner was desperate to collect the "splendid" mass of orchids. But recovering the plants would again be difficult; he faced, he explained in an article in the *Journal of the Horticultural Society*, "no small risk of falling over."[20]

In an eighteen-page private letter to Hooker, dated December 18, 1836, Gardner detailed the danger at far greater length. It was pouring with rain, he explained, the granite mountainside slick. Heavy November cloud reduced visibility. Time and again Gardner fought to reach the plant but was forced back because of "the wetness of the rocks, the strong breeze that was blowing, and the haziness of the atmosphere, for the top of the mountain was enveloped in clouds."

Gardner emphasized to Hooker his utter determination in the teeth of these difficulties. Easing himself along the heart-stopping precipice, Gardner used "a long stick, with a hook at the end of it" to gather the "whole mass" of orchids—"as much as we were able to carry." The downward journey, laden with plants and "drenched to the skin," was almost as dangerous as the ascent.[21] This account of real peril is, again, not to be found in the book-length work, part of its project (perhaps) of showing readers that he was never seriously threatened, that he was dependable and a financially safe bet for any patrons seeking to fund future investigations.

The plants, stripped from the mountainside, went straight to his backers. Gardner proudly sent a case of *Cattleya* orchids to the Duke of Bedford, "one of the most liberal supporters of my mission to Brazil," he told readers.[22] He sent another to Stewart Murray at the Glasgow Botanic Gardens. A third went to a Mr. Skirving in Liverpool, while a fourth was sent, by stages, to James Godfrey Booth, a Scottish seed merchant in

Hamburg. Gardner collected more possible *C. labiata* later on his trip also, near a hamlet called Sapucaya along the Rio Parahyba (Paraíba), between the provinces of Rio de Janeiro and Minas Gerais.[23]

Comparison of Gardner's published observations with his private letters again reinforces how carefully he worked to calibrate his public self-expression, as well as how he worked to navigate the different demands of the letter form and nonfiction—in this case, in the genre of a natural historian's published travel journal. We gain particular insight into how his botanical journeying and plant collection was narrativized for audiences because the eighteen-page letter about the finding of *C. labiata* was subsequently published. Hooker was, from 1826, editor of *Curtis's Botanical Magazine* (and, from 1835, the *Companion to the Botanical Magazine*): he printed Gardner's private account of the apparent rediscovery of Swainson's orchid. But Hooker made notable changes in the editing process, *re*-presenting his friend's orchid-hunting for readers.

When Gardner wrote the letter to his mentor, he narrated his experiences of plant hunting and travel, but he also shaped the narrative of those experiences in accordance with what he took to be the conventions of the letter. "To write letters well and elegantly is a part of the accomplishment of a gentleman which it is very important and very difficult to acquire," explained one author of a contemporary advice text, and etiquette guides invariably included whole chapters on how letters should be constructed and presented, from the forms of address, to the weight and color of the paper, to the color and kind of wax seal, to the more complicated business of developing a correct "air."[24] Letters should be written with great care, even while traveling, yet must have the air of being "rapid and unstudied," explained the 1856 *Etiquette for Gentlemen,* while the author of the 1839 *Advice To a Young Gentlemen* distinguished business letters from "letters of courtesy" by stressing that "the last perfection of the latter is to be gracefully fresh."[25] The final product must, to pass the test of gentlemanliness, be "polished, graceful, nice."[26]

The gardener's son was perhaps particularly careful using epistolary conventions in his letters because Hooker was by now—the mid-1830s—a leading author and editor (and would become director at Kew in 1841). Gardner needed Hooker to believe that he had the social as well as intellectual and practical skills necessary to thrive in the world of Victorian science. Even as late as the middle of the century, "claims to expertise and credibility" could not be verified, Endersby reminds us, "so good manners, courtesy, and an aura of respectability had to do the work that would eventually be done by formal qualifications and institutions or professional affiliation."[27] When Gardner opened his letters "my dear sir" and closed them "your very obedient servant," when he sent best wishes to Hooker's wife and family, and offered "compliments of the approaching season," Gardner demonstrated that he had learned the conventions of the polite letter, thereby telegraphing that he had the necessary accomplishments to move adroitly in the print, social, and intellectual environs of the gentleman-scientist. These are very different from the letters Digance and Ericsson sent Sander from Brazil, decades later.

In describing *C. labiata*, Gardner's account of his collecting practices conformed to many of the genre conventions of the book or journal article, so that the letter seems a sort of early draft of the book. Gardner also, like Swainson, made reference repeatedly to numbers: he detailed the number of cryptogamia he collected, for instance—four hundred and eighteen—and the number of specimens—over six thousand. He worked to foreground his achievements and stressed that he was justifying the faith that had been put in him: Gardner repeatedly mentioned that he was fulfilling his subscribers' demands. He took care to mention the other men of science with whom he was in dialogue; the names of the rich and well placed in Europe were used in the letter, as they are in the book, to signal his connection to them.

But, for all Gardner's carefulness, comparison of the handwritten document and Hooker's edited version in print suggests the older man thought the original could be improved. Hooker erased repetitions and changed phrases he seems to have found wanting, suggesting that he grasped his protégé's need to appear gentlemanly and felt the gardener's son had not quite pulled it off. He added detail to convey what he evidently considered greater scientific exactitude, too. Gardner remarked in his letter, for instance, that on first seeing a "large patch" of *C. labiata*, he managed "with some difficulty and no little risk . . . to reach it, and obtained a few flowers, and several roots of it." Hooker turned the jagged, dactylic "no little risk" into the smoother, iambic "no small risk" (which Gardner would later reproduce in his article), then rounded out the sentence with botanical terms: "I . . . obtained both flowering specimens and living bulbs of this beautiful Epiphyte." Hooker made Gardner more polysyllabic—perhaps, to the ear of the time, more "polished, graceful, and nice"; he also made his collection practice more thorough. George Bentham, future president of the Linnean Society, stressed how carefully a specimen must, by this period, be preserved—"A botanical **Specimen**, to be perfect, should have *roots, stem, leaves, flowers* (both open and in bud) and *fruit* (both young and mature)." Hooker perhaps aimed to make Gardner's practice more thoroughly adhere to the very best standards of the day.[28]

And yet, to my ear at least, there is something strange about Hooker's final words. Why *should* Gardner have thrown in that *C. labiata* was "a beautiful Epiphyte"? The phrase seems to weigh down an already long sentence with information that Hooker, an expert, hardly needed to be told. Again, this addition was probably made with knowledge of the journal's audience in mind. Hooker surely suspected that some readers were less knowledgeable than others. By casually describing the plant as "this beautiful Epiphyte," Hooker quietly informs a reader who did not know yet how to categorize *C. labiata* or grasp its family and growth habit.

If correct, this suggests Hooker's grasp of the understanding of his readership, the still-inchoate, emerging botanical community in the 1830s. That audience encompassed a spectrum, from gatekeepers who insisted on standards to self-taught enthusiasts learning as they went by reading the new magazines. Hooker—a little older than Gardner and considerably more established—reprises the scene of discovering *C. labiata* carefully and flexibly to make it accessible and relevant to this widening audience. The change is revealing, for the orchid was about to become an international superstar, its megawatt reputation spurred on by the middle-class consumers' hunger for exotics—just emerging, but vital for an editor like Hooker to accommodate.

Gardner returned home on July 10, 1841, proud of his collection and relieved so many of his plants arrived intact: "the numerous collections shipped to England, from time to time, all arrived safely."[29] Gardner had amassed thousands of specimens, including at least four cases of what he was confident were *C. labiata*. The orchid was surely, now, comfortably established in Europe.

True, Gardner saw worrying signs of threat to the orchid population's survival on his trip: when he went back to the site of his first collection of *C. labiata* orchids a year later, he expressed himself aghast to discover that much of the Tijuca forests had been stripped. On the Pedra Bonita, he collected orchids from tree trunks that were in the process of being felled, ready to be turned into charcoal. By the time he returned, the area seemed a blasted, dystopian wasteland:

> The forest, which formerly covered a considerable portion of the summit, was now cut down and converted into charcoal; and the small shrubs and *Vellozias* which grew in the exposed portion, had been destroyed by fire. The progress of cultivation is proceeding so rapidly for twenty miles around Rio, that many of the species which still exist, will in the course of a few years, be completely annihilated, and the botanists of future times who visit the country, will look in vain for the plants collected by their predecessors.[30]

Gardner sounded the alarm, too, in his article for the Horticultural Society's *Journal*, on a strange lack of rain following a period of rapid deforestation. But he did little to connect such developments to his own extractive practices, and his *Cattleya* orchids were dispersed to collectors.

Hooker, once promoted to director-general at Kew, continued to champion his protégé's career. He recommended Gardner to the position of superintendent of the Botanic Garden in Ceylon (now Sri Lanka—and, at the time, reeling from the brutal British treatment of those who'd attempted to protest heavy taxation in 1848). Gardner, dedicating the *Travels* to his mentor, took up the post.[31]

According to an obituary in the *Ceylon Times*, Gardner visited a friend, colonial judge John Reddie, on March 9, 1849. The next morning, they rode on to take tiffin with Lord Torrington, governor general of the state, and his wife. Gardner must have congratulated himself on the choice he had made, along the drive, to pursue his passion for botany. Following his self-proclaimed credo—"Difficulties only appear insurmountable, when they are not looked boldly in the face"—had apparently paid off.[32]

Lady Torrington remarked that the botanist seemed in especially good spirits. At about three in the afternoon, the thirty-nine-year-old clutched his arm and fell forward, murmuring that his life was ended. The Torringtons called for two doctors, but Gardner could not be revived.[33]

9
ORCHID VARIETIES

GEORGE GARDNER DOES NOT seem to have doubted—either in his letters, in his *Travels in the Interior of Brazil,* or in the articles he wrote shortly before his death for the *Journal of the Horticultural Society*—that the plants he collected on the mountainside outside Rio were *Cattleya labiata*. But the specimens he sent back to Europe were more rose-colored than lilac or purple, and many of his colleagues were less sure.

Color alone was not necessarily a reason to doubt them, however. Plant hues may be affected by climate, soil, or maturity. The difference between rose and lilac can also be a matter of perception. In an era well before color photography, too, when plants were located in widely dispersed collections, few specialists could set plants from different imports side by side for comparison. How could one possibly compare a plant seen in the Glasgow Botanic Garden, say, with another examined three or ten years later in a private collection?

New *Cattleya* orchids were arriving almost weekly from all over South America, and many of these also looked very much like the packing-case orchid and Gardner's new plants. Were they the same? More importantly, what did "the same" mean in the natural world, when specimens—even those growing in the same population—showed palpable differences?

What did it mean if one orchid looked very much like another—in its size, coloring, or leaf structure, in the number of petals and pollen masses—yet the plants grew miles apart? What if they looked similar but flowered at different times of the year? (Swainson's plant was, unusually, November-flowering.) If one set of blooms were larger or smaller, if a plant produced fewer or more flowers, were these details meaningful? How far could differences in individual specimens be attributed to the skill of the cultivator, the challenges of cultivation, and the plant's experiences of living in captivity?[1] What impact did transportation and transplantation have?

Botanical descriptions of *C. labiata* from the period attest to ongoing, powerful disagreements, even among specialists inclined largely to a so-called "Natural System."[2] William Hooker himself struggled to determine the relationship between the new orchids pouring into the Customs House and the purple cattleya that first grew in his stove. A large, lovely pink orchid from Venezuela, for instance, flowering in the home of a Mrs. Hannah Moss outside Liverpool, was called by Hooker in 1839 *C. mossiae*. (See Plate 7.) The orchid, Hooker mused, bore "a general resemblance" to the "Purple-flowered CATTLEYAE," *C. labiata* in particular, but it was bigger, with differences in color, petal shape, and leaf.[3] He explicitly distinguished *C. mossiae* from *C. labiata* in *Curtis's Botanical Magazine*.[4]

But Lindley—Chair of Botany at University College London from 1829—disagreed. In the *Botanical Register* he suggested that "Mrs. Moss's Cattleya" was "a mere variety" of *C. labiata*, and he called Hooker's whole designation of *C. mossiae* into question. The correct appellation for the pink Venezuelan orchid, he wrote, was *C. labiata* var. *mossiae*—it was, in other words, a variety of *labiata*. The Venezuelan orchid was not a new species in the genus *Cattleya*, but a variety of a plant already known.

Lindley, Hooker conceded uneasily a few years later, "may be right, and if tried by the same standard, a considerable number of Orchideous plants will have to be abolished; for there is no question that, in a state of cultivation at least, plants of this family are liable to very great variation."[5] But if there is such "very great variation" in this and in other species, what did "species" even mean? Where should lines be drawn, and what organizational systems were most persuasive? Hooker had encountered another purple Brazilian orchid in the late 1830s at the Glasgow Botanic

Garden that arrived under the name *C. crispa*. That designation was surely incorrect, he determined: "that plant, as represented by Dr. Lindley, is so very different [from this], that we cannot consider the two to be the same." But what was it? Could it be another variety of *labiata*? He finally decided that it was a variety of a different *Cattleya*, the *intermedia*, first identified in 1828. But larger questions lingered and would, eventually, have to be answered:

> I am indeed by no means clear about the limits of the species of CATTLEYA: I mean particularly, the large purple kinds resembling the original *C. labiata*. To me the Genus appears . . . subject to great variation in the size and form of the flower, and the relative length and breadth of the leaves.[6]

Confusions continued through the 1840s: another deep purple orchid with a large lip was collected by a hunter named Charles McKenzie and introduced by Low's nursery in Clapton. The new orchid strongly recalled the original flower. From Venezuela, it was named "var. Atropurpurea" in *Paxton's Magazine of Botany* in 1841, and called a "very near ally" of *C. mossiae,* a "variety" of *C. labiata* ("probably"), with the former now judged merely a "deficient" variety, its fainter coloring evidence of deficiency.[7] But others disagreed, and when Hooker next described *C. labiata,* in 1843, an illustration by Walter Hood Fitch shows a pink orchid with a strong look of *mossiae* blowing its vast trumpet toward us. The purple-and-crimson coloring of the orchid that bloomed in Hooker's stove was presented as a less crucial part of its identity; location and habitat, too, carried little weight, since the orchid Fitch painted hailed from Trinidad.

Even Lindley, in 1844, sounded less sure than he had twenty years previously about the *mossiae/labiata* relationship. "The varieties of what has been called C. Mossiae are numerous," he conceded, "and seem to prove that no reliance can be placed on the supposed distinctions between it and C. labiata. It must however be confessed that the question is open to further consideration."[8] Confronted by a lobed cattleya, imported by Loddiges from Brazil, Lindley's entry in the *Gardeners' Chronicle* in 1848 explained that this purple orchid, with "rich crimson veins on the lip," was surely "nearly allied" to the plant they knew well and was perhaps another variety. But the lobing of the petals and lip still suggested meaningful difference: "At all events it is as well marked a form of the genus as

C. Mossiae," Lindley concluded; "for the purposes of cultivators, [it] may be looked upon as a distinct species."[9]

At midcentury, in *Paxton's Flower Garden,* Lindley and Paxton evoked the *many* "Varieties of the Ruby-lipped Cattleya" with an illustration of a white and a pink cattleya literally growing together on top of a bark chunk, so that the plants seem to spring from the same root. (See Plate 8.) *C. mossiae* was an "alias" of *C. labiata,* they now determined, in spite of the two orchids' different places of origin: one orchid, two personae. As the illustration visually reinforced the idea of shared roots, the text used three-part names to maintain the linguistic root "*labiata*": "We . . . present those now figured with the names of the White Ruby-lipped Cattleya (*C. labiata candida*) and the Blotched (*C.L. pieta*)." Gardner—that "lamented botanist"—is given full credit for rediscovering the orchid outside Rio de Janeiro, and the many *mossiae*-like variations are attributed to the fact that these orchids tend to grow thousands of feet above sea level, a height that impacts their coloring. The authors indicate that any further effort even to attempt to list the varieties of *C. labiata* would be a fool's game: sounding almost impatient, they conclude, "It would be useless to attempt an enumeration of the varieties that exist of this plant, unless for the purposes of a Florist."[10]

Jim Endersby wryly remarks that academic botanists with herbaria—meaning collections of dried, preserved plants mounted on paper, stored in cabinets—tended to prefer to "lump" plants into fewer, bigger species, rather than "splitting" them into many new species, perhaps partly for practical reasons—they couldn't cope with any more boxes and drawers.[11] Still, the goal of botanical classification systems, as shaped by those in the metropole (like Lindley, Hooker, and Paxton) was to "give clear, unambiguous names that improved communication within the empire."[12] Larger, looser categories posed risks: they might fail to provide clarity and establish the requisite order. Valuable resources and opportunities might be overlooked: a new plant might be misread as a familiar one. A rare plant might pass by unheeded.

Those fears tugged at accounts of *C. labiata,* too, in the unsettled, ceaseless attempts to define. Doubts about Gardner's import—the suspicion that what he had brought back was not, in some still-to-be-determined sense, the same as what Swainson had sent Hooker—left open more lacunae, more spaces in the narrative that mobilized yet more efforts at control, order, and sense-making. The most determined of these would unfold a full half-century later: in 1897, Louis Forget—a Frenchman and Sander collector—published an article in the journal *Le jardin* asserting that Gardner's orchid from the mountains outside Rio was actually "Laelia Lobata," also known as the "Lobed Cattleya," or the "Cattleya lobata" imported by Loddiges in 1847.[13] Robert Allen Rolfe, a curator at Kew, stumbled on Forget's article in 1907. Realizing that Gardner's dried specimens from the mountains outside Rio were preserved at Kew, Rolfe soaked and examined them and decided Forget was correct. The specimen, like *C. lobata*, had eight rather than four pollinia. Forget also asserted that the orchids Gardner collected from Sapucaya, later in his trip, were probably *C. warneri*, another purple orchid, which was known by the late nineteenth century to be common in that region. Although Rolfe was unable to confirm this definitively—there were no specimens from that import preserved at Kew—he agreed again, on the basis of *C. warneri*'s growth habits, that Forget was likely right.[14]

These calculations still lay half a century in the future when Gardner's plants and their descendants began circulating in European nurseries and collectors' greenhouses. At the midpoint of the century, all most collectors knew for certain was that more plants were becoming available under the name *Cattleya labiata*. Even as the orchid's precise identity, its location in taxonomic systems, in "cubes" and on paper, remained in flux, the name itself increasingly signified *the most beautiful orchid*, the first in the land, the most wondrous exotic of them all. *C. labiata* was regularly called, in periodicals, the "queen" of the genus, a sort of magnificent Victoria Regina of the plant world. As Paxton's *Magazine of Botany* put it,

> It is easy to conceive that *Cattleya labiata* must have created an extraordinary sensation among floriculturists when it first developed its magnificent flowers. . . . Like an inimitably beautiful object in an attractive landscape, it stands forth in its princely array, altogether unapproachable by any of the numerous candidates for favour by which it is surrounded. . . . To say of any single species in a richly painted and lovely tribe like Orchidaceae, that it rises far above the rest in stateliness and splendour, is assuredly no mean praise.[15]

Who would not want to own a little of this? Who would not want to possess a plant characterized as "extraordinary," "magnificent," "inimitably beautiful," "altogether unapproachable"? As Gardner's plants entered the market, more enthusiasts could begin to hope to own a so-called queen of the genus. And if some specialists wondered about the new arrivals—if the identity of these orchids were in doubt, like the mysterious heroines of sensation fiction, a popular genre just exploding onto the literary scene—these doubts did not, as yet, dominate discussion. The growing acceptance of the idea that there were *many* varieties of the ruby-lipped *Cattleya* would soon make the exotic seem thrillingly available to middle-class people.

COLOR PLATES

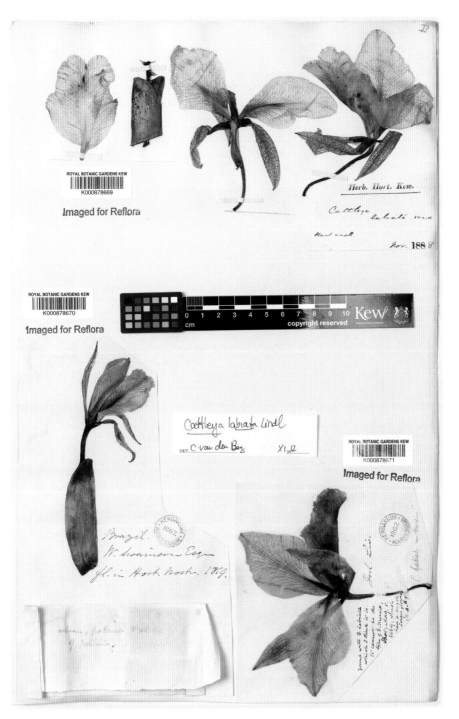

1 On a Kew Herbarium Specimen sheet (K000878670) William Hooker notes: "W. Swainson Esq., fl. in Hort Nostr. 1819," or: *flowered in our garden, 1819*. Board of Trustees of the Royal Botanic Garden, Kew.

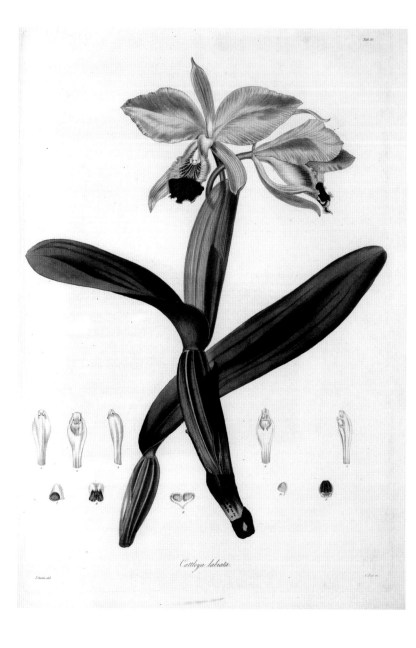

Cattleya labiata.

2   (*left, facing page*) The first illustration of *Cattleya labiata* appeared in John Lindley's *Collectanea Botanica* (London: Richard & Arthur Taylor, 1821–1826), Tab 33: "Without exception, it is the handsomest species of the order we have ever seen alive." RHS Lindley Collections.

3   (*above*) A close-up of the illustration from the *Collectanea Botanica* by John Curtis. RHS Lindley Collections.

4　This illustration, by Joseph Swan of Glasgow, was produced for William Hooker, who first flowered the orchid in Europe in his Suffolk garden. *Exotic Flora*, vol. 3 (London: Blackwood, 1825), Tab 157. Peter H. Raven Library, Missouri Botanical Garden.

5   The first illustration of the lost orchid in a nursery catalog—by the great George Loddiges in Hackney. In *Botanical Cabinet* 20 (1833): no. 1956. Peter H. Raven Library, Missouri Botanical Garden.

6 (*left, facing page*) This, the second illustration produced under John Lindley's watchful eye, was prepared by Sarah Drake from a flower "in the garden of the Horticultural Society in October last." *Edwards's Botanical Register* 9 (1836): Tab 1859. Peter H. Raven Library, Missouri Botanical Garden.

7 (*above*) "Mrs Moss's Superb Cattleya" was found to be surprisingly like the orchid from Brazil, even though it hailed from Venezuela. *Curtis's Botanical Magazine* 12 (1839): Tab 3669. Peter H. Raven Library, Missouri Botanical Garden.

8 (*left, facing page*) In this illustration of "Varieties of the Ruby-Lipped Cattleya" from *Paxton's Flower Garden*, vol. 1 (London: Bradbury & Evans, 1858), 118, two varieties seem to spring from the same root. Peter H. Raven Library, Missouri Botanical Garden.

9 (*above*) George William Job, "The Bayswater Omnibus" (1895). Victorians were bombarded with advertisements on all sides—including, as here, on the bus. Museum of London.

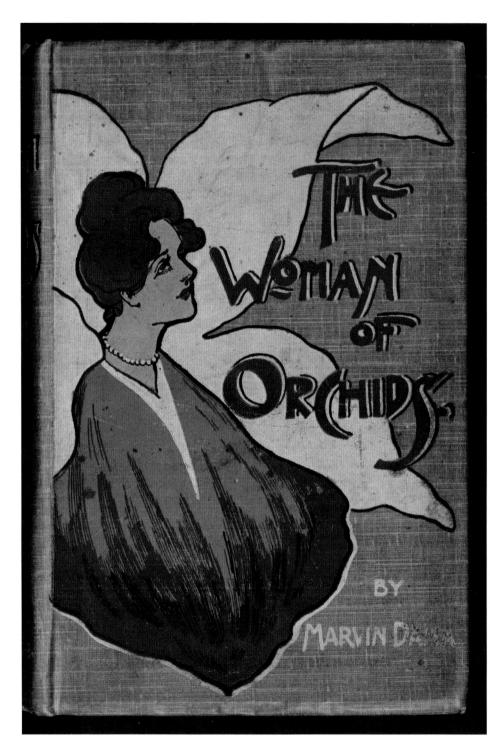

10 (*left, facing page*) On the cover of *The Woman of Orchids*, by Marvin Dana (London: Anthony Treherne, 1901), the association of woman and orchid as objects of desire is made particularly clear. Department of Special Collections, Stanford University Libraries.

11 (*above*) Estêvão Silva's 1890 "Orchideas" (Museu Nacional de Belas Artes, RJ) was—according to Lucien Linden—a crucial clue that led to the rediscovery of the lost orchid. National Museum of Fine Arts, Rio de Janeiro.

12 (*facing page*) On this Kew Herbarium Specimen sheet, K000878662, rest some of the plants Sander's men sent back in 1891. Cássio van den Berg and his team confirm the plant's identity on a typed 2012 sticker as *Cattleya labiata*. Board of Trustees of the Royal Botanic Garden, Kew.

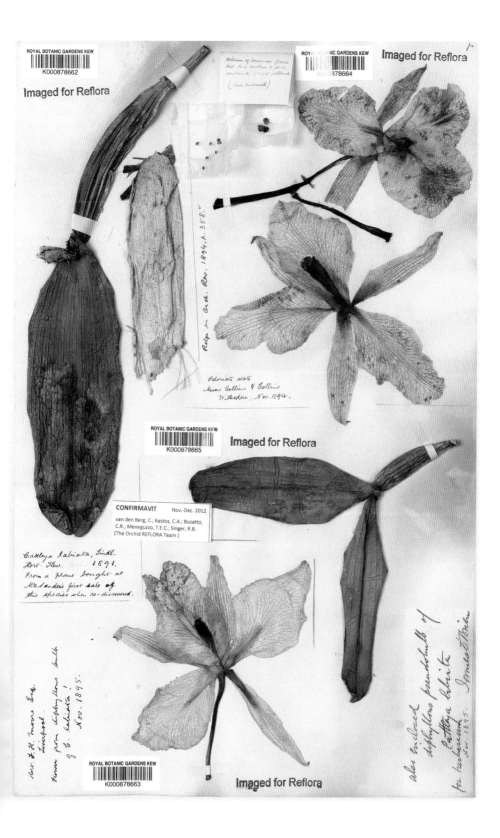

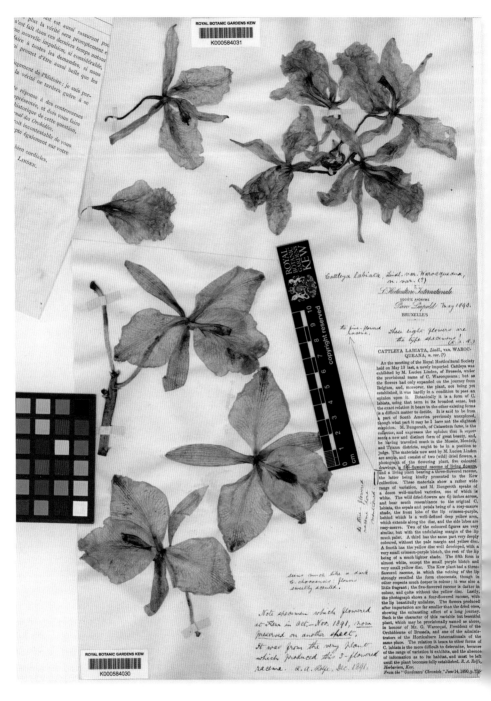

13 On this Kew Herbarium Specimen sheet, K000584031, a collage-like assemblage includes a clipping from the *Gardeners' Chronicle* by Rolfe with handwritten notes and a sticker from L'Horticulture Internationale. Board of Trustees of the Royal Botanic Garden, Kew.

10
"ORCHIDS FOR THE MILLION"

MOST OF THE GREAT early- to mid-nineteenth-century orchid collections were owned by the titled—the Duke of Bedford, Earl Fitzwilliam, the Duke of Devonshire at Chatsworth. Others were owned by rich men who made their money, like Cattley, in business. Still others used family money built on and in industry; Staffordshire resident James Bateman, for instance, presented himself as a member of the high elite, though his money came from the iron foundries his grandfather built and from his father's steam engines and cotton mills. Bateman plowed much of his income into orchids, and in 1843 he published a lavish and gargantuan tome entitled *The Orchidaceae of Mexico and Guatemala* based on plants sourced by a hunter in his employ, George Ure Skinner (after whom another purple exotic, *Cattleya skinneri*, would be named). Bateman was a respected collector of the century, his work a much-cited text; and in Bateman's opinion, orchid cultivation was—and should remain—an elite, wealthy man's pursuit.[1]

Bateman presented orchidomania as the nineteenth century's version of tulip fever in a ponderous introduction to the work. But tulips, he suggested—"that most strange infatuation"—are a lower-class interest. The far greater challenges of orchid cultivation reserved them for the "nobility, the clergy, those engaged in the learned profession or in the pursuits of commerce," and the royal family. Because orchids require such skill and care, "Orchidaceous culture will always continue in a (comparatively) few hands," he contended magnificently—but the "humbler sphere," "the great mass of mankind" would have its tulip, together with other, easy-to-grow plants like auriculas and carnations.

This, in Bateman's estimation, is the best outcome. He would not see orchid culture more "extensively diffused. Few will value what all may possess." Indeed, the pleasure of feeling oneself in an elite minority seems to be what draws Bateman to orchids, for their challenging cultivation allows those in the know—in what he calls, in pressing italics, the *"initiated"*—to experience the intense satisfaction of seeing what others do not. "So long, therefore, as each class has enjoyments suited to its circumstances and position, we are satisfied that the happiness of the community at large will be far more effectually promoted, than if all were interested in the same objects, or occupied with the same pursuits."[2]

Perhaps the Staffordshire collector was overcompensating; the most elite of the elite still tended to look askance at those whose money was made recently rather than inherited across many generations. His "introductory remarks" and the physical book itself—at a staggering seventeen kilograms, few could lift, still less afford it—worked hard to draw a firm line around orchid culture as an elite pastime, one that could and should be set off from the rest.

But orchids had been riding the wave of the new middle classes since 1818 at least. The key figures in the sourcing and naming of orchids were not traditional elites: Swainson was the son of a customs collector, Cattley a merchant, Hooker a brewer and son of a wool-stapler, Lindley the son of a nurseryman, and Gardner the son of a gardener. Meanwhile, dozens of botanical and horticultural publications were springing up to offer information and practical advice on how to get started on orchids, at a wide range of price points, suggesting a thirst for knowledge among even the kinds of collectors that Bateman disdained.[3]

A riposte to Bateman's claims about the elite nature of orchids came, appropriately enough, in Paxton and Lindley's resolutely middle-class, six-penny *Gardeners' Chronicle* in a series of articles entitled "Orchids for the Million," appearing between 1849 and 1851.[4]

The writer of Part I—who wrote as "Dodman" but was in fact Charles Bellenden Ker, a legal reformer and barrister—pointed out that, *au contraire*, the cultivation of rare plants was becoming accessible to all through modern technological and industrial advances. Orchids were no longer the preserve of the wealthy in an age when "hot-water pipes or tanks are now cheap, and a small boiler may be had for 2*l* 10*s*., or less; and glass (thanks to Sir Robert Peel), is also within the means of the humblest gardener."[5] The glass tax was repealed in 1845 under Peel's government, partly after pressure from Lindley, and advances in iron and glass production brought greenhouses within the reach of many middle-class families, while smaller, tabletop versions facilitated the importation of live plants. Doctor Nathaniel Bagshaw Ward had noticed, one balmy summer afternoon in the late 1820s, that fern and grass seedlings popped up in the wide-mouthed glass bottle he was using to protect the chrysalis of a sphinx moss caterpillar; he also noticed the way moisture condensed on the internal surface of the glass, keeping the insides damp. In this way, the "Wardian case" was born, a portable greenhouse that was used by plant hunters to bring back living specimens (this again helped drive down prices as more plants survived the long trips) while offering enthusiasts an affordable way to grow smaller, delicate plants at home, including dwarf orchids, by protecting them from the thick fugs of chimney smoke, the "smuts" and "blacks" (as they were known) of industrial pollution.[6]

Indeed, readers would learn over the next decades that orchids of some varieties and species could even be raised *outside* hothouses. Since some "grow at such elevations that hoar frost is found upon the leaves," they required less heat than early enthusiasts imagined.[7] Many could live happily in a parlor or sitting-room, at least for part of the year, and even those that required hotter temperatures could cohabit in a simple greenhouse with other tropicals and vines. They did not necessarily need vast, glittering orchid houses, of the sort erected on large estates and botanical gardens.

Dodman, taking direct aim at Bateman's claim that only aristocrats should raise orchids, that carnations and tulips were the plants of the "million," argued that it was not only becoming possible for those outside the elite to take up growing orchids, it was almost a matter of right: "we must have them for the million."[8] He agreed with Bateman that some exotic orchids were challenging to raise, but wasn't that all the more reason that they should be raised by the resilient, respectable working and middle classes making up the backbone of modern Britain? And while certain plants were undoubtedly still very expensive, Dodman argued, "like all matters of commerce, the supply will increase with the demand." Every day collectors were entering the jungles and scaling mountains, finding more choice plants, while nurserymen were learning new ways to propagate plants. Thus very soon indeed, he predicted, orchids would be available "at a piece within the reach of all."[9]

Taking up the charge in this effort to pitch orchidomania to the public as a middle-class pursuit, the next parts in the "Orchids for the Million" series were penned by a professional gardener, Benjamin Samuel Williams. He would soon take over nursery premises of his own, catering to the "million," in London's Holloway. Williams's columns were gathered into a successful how-to book, *The Orchid-Grower's Manual*, in 1852. Williams worked at the Hertfordshire property of City man Charles Warner at the time he penned these magazine articles, and his texts—the *Gardeners' Chronicle* sequence and the subsequent book—are intended for an amateur—not a rich, landed, gentlemanly figure but one with a relatively small house and tight budget. A second edition of *The Orchid-Grower's Manual* followed in 1862, and in a new preface Williams attested happily to the many responses he had received: "I have had ample assurances that some hundreds of cultivators have found my efforts useful." By the time the sixth edition was published, in 1885, Williams noted that those letters had swelled from the hundreds to the thousands, that they were pouring in from all over the world: "It is highly gratifying to us to know [that] . . . not only in this country, but on the Continent, in America, and the Colonies," readers had sought and found assistance from his manual, he explains. The work had even been translated into Russian.

The cornerstone of Williams's approach was to provide basic information on everything needed to get started with orchids, from watering to heating to potting materials and more. Williams attributed his success in each new edition to his use of plain language, to the fact that his material was written without the assumption of training on the part of the reader, omitting all "[c]uriosities of mere botanical interest." This was not a work of science or specialized knowledge, nor did it require much in the way of training or education to understand it. The goal was simply, or so each edition argued, to help an ordinary reader raise "handsome flowers."[10]

The pace of reprintings of Williams's book, which picked up in the 1860s and 1870s, with just three years between the third and fourth editions after a decade between the first and second, suggest he had read the market correctly.[11] The new middle-class demographic had money to spend on a leisure activity like gardening and raising house plants. But in a time of rapid rural depopulation and rising urban and suburban growth, the new middle-class horticulturalist was not necessarily raised on or with the land, so that the cultivation of any plant—still less exotics—posed challenges. A book like *The Orchid Grower's Manual* answered that call.

A whole genre of gardening books in the Victorian period, in fact, sprang up in response to exactly this problem. *The Suburban Gardener and Villa Companion* (1838), an influential book by editor, author, landscape gardener, and all-round workaholic John Claudius Loudon, was explicitly subtitled "for the instruction of those who know little of gardening and rural affairs." Loudon's text presumed that middle-class suburban residents, a new and growing demographic, were divorced from older, rural English traditions; they raised fruits and flowers for decoration and enjoyment, not sustenance or family support. They had markets and shops, he pointed out, in which to buy fruit, vegetables, and meats: the era of growing potatoes and carrots and raising chickens to feed a family was passing. Loudon—like Williams—saw this development not as a problem or failure of modern times but as pointing to progress in an industrializing, rapidly changing society. Gardening a small plot is positively productive of happiness, he contends, because it allows us to take on and succeed at smaller-scale, manageable challenges. To Loudon, the very confinement of the fenced or walled suburban garden was enlivening because it meant that the amateur gardener could realistically triumph in a smaller horticultural arena.[12]

Williams, too, in *The Orchid-Grower's Manual*, turned the shape of emerging middle-class life into a positive. Vast, aristocratic-sized greenhouses, for instance, he cast as actually *worse* for orchids than smaller ones, because it could be challenging to keep up consistent temperatures at night.[13] Moreover, many orchids positively throve if moved out of smaller greenhouses for at least part of the year into an ordinary sitting room or warm parlor—for all these reasons, middle-class gardeners were well positioned to enjoy the richest rewards of the age. "Many of these plants [orchids] are small growing, and do not take up much space to grow them in, so by having a small house a great number may be cultivated," Williams explained enthusiastically, while the large numbers of plants appearing on the market gave "all cultivators a chance of procuring plants at a cheaper rate, for many of the most beautiful kinds can be obtained for a few shillings at the present day, when at one time they were only within reach of the rich at exorbitant prices." His own

Benjamin Williams's 1882 advertisement in the *Gardeners' Chronicle* stressed: "No cards to view required. A hearty welcome will be given to all." Aiming to make orchid collection accessible to the "million," the advertisement detailed local rail connections.

nursery he proudly pitched to the world as open to all, to the "million": "no cards to view" were required for *his* exhibitions of rare orchids, he trumpeted, unlike some of the stuffier establishments—"A hearty welcome will be given to all who honour us with a visit." His advertisements explained in great detail which trains and cars on the metropolitan tramway could be used to access his nurseries, so that no personal carriage was needed.[14]

C. labiata, however, was not accessible to *all* the "million"—not quite yet. Though Williams celebrated *Cattleya* orchids as relatively easy to care for, they required more than passing attention: the leaves needed to be inspected and washed regularly with room temperature water, Williams conceded, to keep down white scale and insects. More important still, *Cattleya* needed a heated stove or greenhouse, with temperature systems regulated to drop no lower than 65 degrees Fahrenheit in the winter. And the buildings Williams suggested for *Cattleya* orchids were enormous. At eleven feet high, eighteen feet wide, and sixty feet long, they were far beyond the reach of suburban dwellers with a limited plot of ground. The first generations of suburbanites did, it is true, enjoy larger gardens than those who came later—Loudon's *Suburban Gardener* integrates designs for houses with up to seven acres, and in his classification of homes from first- to fourth rate (with fourth for the lowest classes and first for the upper) Loudon still presumes generous gardens at all levels: third-rate houses, for instance, he believes likely to have gardens up to seventy-two feet deep.[15] But those sizes could not be maintained as land became more desirable for building.

Moreover, *C. labiata* plants remained expensive. *C. mossiae* was advertised in the pages of the *Gardeners' Chronicle* in 1841 for 15 shillings (£80 / $102 today) by a Woolwich nurseryman, H. Youell; *C. forbesii* was listed for a relatively manageable 7 shillings 6 pence (roughly £43 / $55 today). Williams included notes for his readership in the "Orchids for the Million" articles on *C. labiata*—"One of the best of the Cattleyas"—yet the prices for plants with this name at the time were far out of the reach of the masses: Youell offered *C. labiata* for 31 shillings 6 pence (£181 / $231 today), while on June 15, 1850, Bateman sold a specimen of *C. labiata* at auction for a hefty £11 11 shillings, or £1,560 / $1,990. This put it far beyond the reach of a humble clerk at midcentury, earning in the range of £60–150, or $8,100–$20,260, each year.[16]

C. labiata thus remained a plant for the more affluent, fortunately situated, upper-middle-class resident. But more cattleya were being sourced by the week: nurseries were springing up across the country, many (like Williams's) touting their proximity to railway stations in advertisements. The bigger nurseries were also sending out collectors in pursuit of more plants. And so *C. labiata* could be presented, in advertisements, as a desirable commodity, a luscious collectible that would reach the grasp of the average middle-class collector sometime soon—just as the middle classes were looking about themselves, with interest, for something to collect.

THE BUSINESS OF COLLECTING was central to the nineteenth-century naturalist, who set about the job of amassing specimens with compass, trowels, pocketknives, hatchets, blotting paper (for drying specimens), vascula (on-the-go storage containers), books, microscopes, and more.[17] But collecting was at the heart of nineteenth-century culture more generally—almost key, scholars have argued, to the Victorian frame of mind. Naturalists collected abroad, but millions collected at home: on a table-top, in a bay-window, in the garden, greenhouse or stove. William Cattley's orchid collection in Barnet was not just one man's idiosyncratic interest, but a signpost to a century's obsession.

At the time Swainson's orchid arrived, in 1818, collecting was still a largely wealthy pastime. It was a good moment for the rich to collect, too: the upheavals of the French Revolution and the Napoleonic Wars had caused the dispersal of fine collections on the Continent, and the purchase of the Parthenon Sculptures by Lord Elgin helped inspire an interest in antiquities among those who could afford them.[18] Newly wealthy industrialists and businessmen turned to the kinds of hobbies that aristocrats had embraced for centuries, amassing collections of fine arts and other rare objects, including exotic plants, that designated connoisseurship and signaled leisure and wealth.

By the middle of the century, collecting was no longer the preserve of landed gentlemen.[19] The promise of the Linnaean system was that anybody could manage classification and, as the nineteenth century progressed, that classificatory urge extended into the broader population, from scientific specimens to just about anything. "This is a 'collecting' age," observed the weekly *Graphic* in 1869: "Never was the vocation of the gathererer of curiosities more followed than at present. Not only pictures, prints, coins, birds, insects, and fishes are collected, but there are amateurs who form cabinets of postage-stamps, first numbers of periodicals, playbills, and street ballads."[20]

Why was it a "collecting age"? Collecting—whether by the rich or the rest, in museums or private collections, of botanical specimens or tickets or playbills—is more than "just" a pastime, a collection more than a random assemblage. Apart from signaling leisure and the possession of income, collecting brings us together with other collectors: long before the internet, in an era of increasing migration, collection offered a means of forming connections, of battling the alienation and isolation endemic in a more mobile, urban age.[21] Collections gave people reasons to get together, at meetings, societies, and in competitions, while the fast-growing periodical press allowed for the vibrant exchanging of objects, ideas, and opinions through and on the print page. New magazines brought communities together in person, too—by touting the many garden shows, competitions, auction rooms, public garden displays, and nursery sales springing up all over the nation. In this way, magazines and journals helped provide readers a sense of location in (to adapt Benedict Anderson's famous term) an "imagined community," meaning a group that is not always physically together but that has a sense of itself, nonetheless, as a group.

To be a collector of plants, to be an orchidologist, offered entrée to a community of plant-lovers; to be a collector offered a sense of belonging. But the relationship of collection and community runs deeper still, for to collect is by definition *to bring together,* to unite. Collections are founded on the premise that the artifacts share affinities that the collection and collector stages. The collection, in this way, offers a sort of narrative, a story of kinship and belonging, with recovered objects supposedly reunited as if in the closing chapters of a popular novel. To collect is to care for, to gather, to sort into families. The rise of collection culture in the nineteenth century, every bit as much as the Victorian novel that restores the changeling to his or her parents, suggests the urge to counter what so many feared about a machine-driven age—isolation, loss, severance, alienation. Finding an intellectual community through collection, the collector told a story in material form that aspired to resist the driving norms of capitalist production.[22]

But what is gathered with what? Which families are preserved, which are sundered? Of course, collectors like Banks, Hooker, Lindley, Cattley and Bateman did not re-unite orchids with the soil or plants in which they first grew.[23] The stories unfolding in Victorian plant collections, the choices and selections collectors made continued to lean on and underwrite colonialist imperatives, to affirm Britain's power and right to take what it wanted. Victorian orchid collections did not connect plants with their habitats or local environments or people because collectors did not recognize those relationships. The collections they built in herbaria, in print, under glass, and in hothouses platformed relationships that required and provided the justification for more plant-hunting, more environmental damage, more colonial adventuring. Affinities of one sort were made invisible as affinities of another sort were staged.

The project of refining taxonomies, of building herbaria and plant collections thus became connected with, a further means of extending, Britain's hegemon.[24] Joseph Banks was dead, yet many of those taking up his mantle—like Lindley—were trained in his ways, his narratives, and his systems. Organizations like the Linnean Society and the Horticultural Society together with a growing network of nationwide, local horticultural societies increasingly took key roles in defining and setting standards of what was admirable and valuable—how specimens should be collected, preserved, displayed, named, what was to be celebrated and what judged of less account. These gatekeeping practices did not question older approaches, even tended to embrace them. This is part, too, of a larger phenomenon: the rising middle classes continued with gatekeeping practices (not just in horticulture, but in professional societies and organizations) we might have imagined they would reject.

As new professional groups and societies stepped into positions of authority over the course of the century, they put in place, Harold Perkins argues in *The Rise of Professional Society*, "the vertical career hierarchy rather than the horizontal connection of class."[25] That gradual yet seismic social reordering would help men like Hooker, Gardner, and Lindley rise from the provincial margins to positions of cultural and intellectual prominence. But the process was full of conflict, a deep fear of what might be lost in a new age shadowing advance. Birth is at least relatively easy to recognize; merit, rather less so. The new middle classes tended to attach themselves to older values and belief systems even as they invested hope in a future structured by new categories of value, ones that were (theoretically at least) available to all, not just those of a particular birth. The departure of the older way of life and its values *was* happening, but not, as Martin Wiener puts it, "before the aristocracy had succeeded in both prolonging its reign and educating its successors in its world view."[26]

What this meant in practice was that the new middle classes policed themselves rigorously. Their definition of merit looked (and, Daniel Markovits argues, still looks) quite a bit like what came before.[27] In the world of horticulture, societies and professional organizations had elaborate, complex systems to set off the worthy from the not-worthy, and these systems buttressed familiar gender, class, and race hierarchies. An 1849 guidebook on how to run village horticultural societies, for instance, included the lengthy "Rules of the Pytchley Horticultural Society" to give those interested in setting up a society insight in how to run it. Members were separated into classes of three different colors, depending on what they could pay, and membership was presumed male. "Members and their wives" were allowed one free entry per year, but no other family members were allowed entry. Specimen display was strictly regulated: Only a certain number of specimens could be displayed a limited number of times a year; the "Rules for Preventing Confusion" specified that drop-off must take place between 8 and 10:15AM; each specimen must be cataloged in the show book according to the color code of the specimen owner's membership; all specimens must be brought in at the same time; family members must be left at the gate while pick-up took place; any uncollected entries would be sold for the benefit of the society. These rules may appear as simply an effort to maintain order, yet in practice they surely advantaged the well educated over the illiterate, men over women, the wealthy over the poor, those with servants and nannies (who could take children while the whole operation unfolded) over those without. And of course, only those with income could participate in the category "For Greenhouse (or Tender) Flowers"; the "Open-Air Flowers" category sounds at first blush more open, yet it specifies that only "choice" wildflowers could be used—a term of art that presumably meant far more to some entrants than others.[28]

Still, some figures were beginning to rattle the gates. In spite of rules that championed openness in theory but firmed up hierarchies in fact, new kinds of collectors were finding ways to become part of the communities that were forming as new mechanisms of advance. By midcentury, some of the most formidable collectors of orchids were women.

11
BATTLES AT THE GATE

THE MUDDINESS OF GARDENING and the intellectual demands of botany may make both seem, on the face of it, arenas closed off to delicate Victorian ladies. Certainly, women's relationship to plants, as John Ruskin famously visualized it in his lecture "Of Queens' Gardens," was one of glancing relationship at best, the challenges of cultivation transformed from hard-won, earthbound skills to a sort of vague, airy test of morality: "She should revive," Ruskin insisted rhapsodically, of a woman's moving across the grass; "the harebells should bloom, not stoop as she passes." Such is the force of this image, so hard is it to imagine a Victorian woman filthy, sweaty, and calloused, historians long assumed women played little role in garden design or culture until the century's end; more recent scholars including Ann Shtier, Catherine Horwood, and Fiona Davison have argued that women were in fact making substantial inroads into gardening as a profession and botany as a discipline from early in the period.[1]

In December 1840, the *Athenaeum* reported a tremendous recent success in the august community of orchid collectors: a woman named Louisa Lawrence had won a prestigious Silver Knightian medal at a meeting of the Horticultural Society in London for a *Cattleya labiata* "with remarkably large and handsome flowers."[2]

Indeed, Lawrence had won fifty-three medals at the Horticultural Society by 1838 and was a fully credentialed member with the right to put *FHS* after her name. To consult the lists of prizewinners at the Horticultural Society and the Royal Botanic Society from the late 1830s through the 1840s is to find *Mrs. Lawrence* lodged, proudly and often, at the top. On the same lists, often further down, are leading Victorian nurserymen of the day, influential professionals whose businesses were growing rapidly to cater to a growing obsession. Low's, Veitch's, Bull's, and Rollisson's soon became household names, if they weren't already. These were large companies with hundreds of acres of land and employees. But in the growing and showing of orchids, Louisa Lawrence made a public name for herself as equal to all.

Lawrence was born in 1803 to a family in trade; her father made his money in haberdashery. But she climbed the social ladder rapidly when, in 1828, she married a wealthy surgeon twenty years her senior. The couple settled just outside Ealing in London, Drayton Green, and here, Louisa became a collector of exotics, known especially for orchids.

John Loudon, who featured her in *The Suburban Gardener* in a sort of celebrity spread, described her property as one of "the most remarkable of its size in the neighbourhood of London." The locale he considered rather unappealing, "flat, or nearly so," the area "straggling." But he was rapturous about her gardening: "The collection of green-house and hot-house plants may be characterised as among the most select and valuable in the neighbourhood of London." And the greenhouses were not just for show, either: "The numerous prizes which Mrs. Lawrence has received, for some years past, from the Horticultural Society of London, evince the excellence of the articles which she exhibits at their meetings."[3]

Lawrence possessed at Drayton fully 227 varieties of orchid. But she was known not only for the size of her collection. Lawrence designed her orchid house as a grotto, "with miniature rockworks and artificial hillocks, for terrestrial Orchidae; and small basins and fountains, formed of shellwork, for aquatics."[4] The 227 orchid species and varieties were not sited as individual specimens, in other words, nor was there an attempt to produce a naturalistic setting. Rather they were sited in a patently artificial scene of "miniature rockworks and artificial hillocks" and "shellwork."

This approach may sound kitschy to us. But the scene was received quite differently by its first viewers, who came in droves to admire it. William Hooker dedicated an entire volume of *Curtis's Botanical Magazine* to Lawrence, exclaiming that her many horticultural successes were "ONLY EQUALLED BY THE LIBERALITY WITH WHICH THEY ARE SHOWN TO ALL WHO ARE INTERESTED IN BOTANY AND HORTICULTURE."[5] Lawrence's gardens—especially those at Ealing Park—were often credited as rivals to those of the Horticultural Society of London (which later became the Royal Horticultural Society).[6]

The practice of assembling plants, shells, and rocks, of bringing together natural and marine objects from disparate places, points to a larger passion for craft and fancywork embraced by newly leisured, middle-class women. A common pastime of the moment involved making artificial flowers out of wax, knitting, paper, or shells; women also formed bracelets out of shells and other found natural objects or stuck them in a collage on top of boxes, tables, and other small household objects. Those with the space and resources might even construct a garden grotto of shells and rocks. Lawrence's 227 orchids in a shellwork grotto was, in other words, consonant with a larger crafting, collaging fever that raged at the middle of the century.

What did these grottos, collages, and other assemblages of natural objects signify? Were they just something for bored women to do, at a time when leisure signaled a family's high social status? Such created objects offer the crafter the enticing opportunity not only for tangible self-expression but also for taking a position on and in larger cultural, social, and political conversations. Art historians and literary scholars have pointed to the gendered and often political narratives these material objects convey. The nineteenth-century crafting fever for assemblage "made tangible the material accumulation of empire, naturalizing and familiarizing the scientific colonization of foreign specimens," Kathleen Davidson and Molly Duggins argue. To set natural objects in scenes and on objects was to stage their accumulation, reminding the viewer of the powerful networks that had brought them all together in the first place: the "extraction, removal and recontextualization of marine products, including seaweeds, shells, corals, and pebbles, enacted the colonial appropriation of natural objects on an individual and collective level, embedding nationalist ideologies of empire in everyday experience." Lawrence's curated orchid collection did not just reveal her skills and achievements as a great collector, then, it also communicated her involvement in a larger community of women working to "domesticate" the exotic, imported from afar through the auspices of Britain's expanding imperial networks.

The 227 orchids from all around the world sited with shells surely gestured to the oceans across which they had been brought, the waves Britain so proudly declared it "ruled." But Talia Schaffer's analysis of craft objects may suggest another possible meaning in the Lawrence scene as well: to *change* objects, to turn them into "personal emanations of their producer" was to make them "irreducibly and inalienably part of the private," she suggests. Their very oddness and curiosity was the point: Schaffer reads Victorian women's quirky crafted objects (wax coral, knitted flowers, hair jewelry) as part of an effort to rework the meaning of things in ways that expressed a growing unease over capitalism and the exchange economy. The woman who made knitted flowers and stuck shells on top of boxes was able to "imbue [the object] with personal significance," to resist or at least complicate the values of the marketplace that produced the wool or box or wax.[7] Read in this way, Lawrence's siting of her 227 expensive orchids among shells also reframed the meaning of these expensively imported commodities. Not just artifacts of a massive capitalist, colonialist machine, they are also presented to the observer as the products of a woman's care, collection, domestication. They are turned into characters in a different kind of story, one authored this time by a woman.

The Horticultural Society's decision to admit women in 1830 signaled a relatively forward-thinking approach in a country that that would deny women full suffrage until 1928.[8] But Lawrence fell afoul of the gatekeeping systems on a number of occasions in ways that suggest rising tensions nonetheless, a resistance to her successes and the narratives her collection practices enacted. Louisa Lawrence was moved to mention the different treatment meted out to her and her plants several times. Her complaints, recorded in the Horticultural Society's handwritten "Minutes," preserved at the Royal Horticultural Society in London, offer glimpses of the challenges she endured.

In June 1837, for instance, Lawrence and one of her gardeners wrote to protest the judges' decision at a recent garden exhibition. She'd won a coveted Gold Banksian for her "stove and green-house plants," but one Sir

Edmund Antrobus, displaying a very similar collection, had won the even more prestigious Gold Knightian.[9] Lawrence and her gardener felt that Antrobus had clearly been treated with an unfair generosity, though there is no record in the Horticultural Society's minutes of whether her complaint was upheld.

Then, a year later, on June 2, 1838, tensions resurfaced. The minutes record that members wrote to complain when *she* was awarded a prize and a similar display by Antrobus and his gardener, Mr. Green, was *not*. The assembled gentlemen decided that "it would be only an act of justice to give Mr. Green a medal of the same value as that awarded to Mrs. Lawrence." The *Gardeners' Magazine* reported without comment or explanation that both were awarded the Gold Knightian in late May. Surely Lawrence was pleased neither at the criticism of the judge's initial decision, in placing her above Antrobus, nor at the realization that the fellows of the Horticultural Society were quick to rally on his behalf in a way they had not been, a year earlier, on hers.

In 1853, Lawrence complained once again, the minutes reveal. This time, she was angry because the Horticultural Society had decided to reduce the value of medals awarded for orchids. She threatened to stop showing her exotics in retaliation. Lindley, who was the secretary at the time, took pains to explain that the early challenges of orchid growing in Britain "had been so much overcome," the plants "no longer demanded the encouragement they formerly received." Whatever Lindley intended by this, Louisa surely took the message that now a woman could do it, the society valued orchid-growing rather less; perhaps she also suspected that her own successes were the cause, that the Society did not want to give more money to *her*.

Six months later, a full-scale fight broke out on the grounds of the Horticultural Society when Lawrence arrived early to set up her display and was manhandled at the gate. The guards did not believe she was legitimately entitled to enter and display; the police were called. After a long and difficult meeting to unpack what had happened, apologies were extended to her, the minutes note, and she was given permission to arrive early in future. Evidently, she was not soothed: by March 1854 Lawrence had given up her entire orchid collection and eschewed future competition. Ill health may have played a part in her withdrawal; she died in August 1855, aged just fifty-two.[10]

Only glimpses are left to us of Louisa Lawrence's story; she did not publish a narrative in print. Still, the notes in the minutes, print descriptions of her collections, lists of her prizes, and published accounts of her achievements point to a woman who found, through orchids, a way of contesting and complicating familiar narratives of womanhood. The business of collecting, growing, and competing allowed Lawrence and other female horticulturalists to explore beyond the identity of the ideal Victorian wife, or "Angel in the House," as poet Coventry Patmore put it in a popular poem of 1854—"him to please," Patmore insisted, "Is woman's pleasure." The ideal wife should, conduct writer Sarah Ellis agreed, stay in the home so as to limit the polluting effects of the competitive public sphere on her menfolk. "How often has man returned to his home," Ellis demanded of her readers, "with a mind confused by the many voices, which in the mart, the exchange, or the public assembly, have addressed themselves to his inborn selfishness, or his worldly pride." A working man, in a capitalist economy, needed a domestic, home-based wife, "guarding the fireside comforts . . . clothed in moral beauty" in order to "[scatter] the clouds before his mental vision."[11] Rather than competing herself, a woman's job was to reward the men who entered the lists, as ladies of old did their knights: "[woman's] great function is Praise," Ruskin explained, "she enters into no contest, but infallibly adjudges the crown of contest."[12] In order to achieve the status of noble lady, the lifespan must be spent growing into goodness, Ruskin continued, and he recommended an entire education system arranged around learning such subjects as would allow a woman to understand the interests of her husband and grow her powers of sympathy with him.

Lawrence, striding into the Horticultural Society to scoop up her prizes, self-trained in horticulture, resisted that story, that feminine *bildung*. She participated in competition at a time when competition was framed as the preserve of men; through orchid-growing and raising exotics, Lawrence pursued her own interests energetically, across decades. While Ruskin argued women should aim "not for self-development, but for self-renunciation," Lawrence gained a reputation as an expert. In perhaps her greatest success of all, she flowered *Amherstia nobilis*, a luscious small tree from Myanmar, for the first time in Britain—well before the Duke of Devonshire, under whose auspices the *Amherstia* had been imported, and his gardener Joseph Paxton. Both were exasperated by her success.[13] Never one to hide her light under a bushel, Lawrence sent the first *Amherstia* spike to Queen Victoria and Prince Albert, the second to be engraved. She welcomed the royal couple into her home to admire her "night-blowing" cactus.[14] In 1852, the *Cottage Gardener* called her—in recognition of her successes with orchids, and with the *Amherstia*—"the best gardener in England."[15]

Orchid growing transformed Lawrence into the one doing the displaying, not the one on display; the collector, not the object; the author, not a passive heroine. Even into the twentieth century, a biographer characterized her as "a conspicuous beauty," adding "her looks may have been a shade insipid"; "lovely as she was, she had not the look of a great gardener, if the science of phrenology, which flourished in her day, means anything."[16] Lawrence likely confronted such prejudice head on as she was expelled from the gates of the Horticultural Society in spite of her fellowship, in spite of having opened her greenhouses to the public, in spite of her decades of achievements and awards.

Still, the *promise* of horticultural competitions, at least, was the equality of persons; theoretically, it was the plants that were judged, not the owners, and for all the gatekeeping mechanisms in place, plant competitions were noted for the wide social range of participants. "The principal part of the English aristocracy are present [in the London Society's exhibitions] and mix indiscriminately with the tradesman, the mechanic and the gardener," gushed the *Gardener's Magazine* in 1836.[17] And, while the presumption remained that the average horticultural society member was male, the business of cultivating and nurturing plants was carefully described by a number of authors as fully consistent with even the most conservative ideals of femininity: women practitioners were at particular pains to evoke horticulture as a means of tending to and beautifying the home. As I've argued elsewhere, women increasingly gained knowledge and experience of practical gardening and horticulture over the course of the century, such that gardening was one of the first professions women entered, even dominated. Training schools for women gardeners were established by the late nineteenth century, and one of the foremost Arts and Crafts gardeners was a woman, Gertrude Jekyll.[18]

Women played important roles in orchidology too. Apart from Hannah Moss, whose name is enshrined in *Cattleya mossiae*, a Mrs. Wells of Cowley House (outside Exeter) also owned, by 1842, a *C. labiata* as part of her extensive collection. Martha Wray, who lived near Cheltenham, was another noted orchidologist, as was Lady Dorothy Nevill. Jane Loudon, wife to John Claudius Loudon, edited the *Ladies' Companion, and Monthly Magazine*. She advised general readers on orchid cultivation and reported on the orchid craze with approbation.[19] The implications of these women's collection, cultivation, and competition practices were far-reaching. They showed women could develop expertise and participate in competition with men at the highest levels. Horticultural competitions were one of the few public arenas in which, as early as the 1830s and 1840s, women and men competed in mixed company, side by side. Even once women were allowed to enter institutions of higher education later in the century, they were prevented from being taught or from sitting exams along with men. The early generations of women orchidologists thus stood at the vanguard of a movement that would take nearly a century to yield, by stages, universal suffrage. In Lawrence's display practices, and in her battles at the gate, we gain glimpses of how gender shaped orchidology and the ways in which women struggled to resist conservative narratives of how a woman and a plant should interact.

THE RALLYING CRY OF "ORCHIDS for the million" and Louisa Lawrence's successes were among many signals of a subtly shifting world at the midpoint of the century.

The great Loddiges collection in Hackney was sold off when that company was unable to afford to renew its lease, given the high costs of buildable land close to the city. Auctions of Loddiges's formidable collections took place on the premises of auctioneer J. C. Stevens at Covent Garden. In a catalog of the sale, specimens of *Cattleya labiata atropurpurea* were listed, in May 1856, for £3 (£354 / $454 today). (A second auction that year, this time of plants from the Horticultural Society, included yet more *C. labiata atropurpurea* at £4 15 shillings (or £561 / $720).[20] It was felt to be a turning point, not just for the Loddiges family, but for London itself: the *Illustrated London News* mournfully pictured a great, fifty-foot palm tree, once owned by Empress Josephine, moved in a massive eight-foot-square box on wheels from Loddiges to its new home in the Crystal Palace by plumed horses.[21] "The progress of this stupendous plant through the metropolis," the paper pronounced heavily, "will not be easily forgotten."[22]

It was a time, too, for new figures. Hooker and Lindley were reaching the ends of their lives. At the Loddiges sale, in the Stevenson auction rooms, an anxious man named James Veitch stood determined to snatch up as many orchids as possible. A nurseryman, he was a figurehead of a new generation. He was deeply interested in *C. labiata,* and he was unconvinced by Gardner's plants.

Veitch's company was already hunting *C. labiata* once more through Brazil. But recovery of the orchid was proving surprisingly challenging, and Veitch realized—perhaps more than anyone in Europe at the time—that finding Swainson's orchid was not going to be easy. Veitch saw other problems ahead, too, in the establishment of exotic orchids as a thriving market in Europe and beyond. In spite of Dodman's assurances, back in 1850, that nurserymen were finding ways to propagate plants quickly, the truth was that many orchids, including *C. labiata,* remained slow to reproduce.

Focused on the growing mania for orchids, Veitch was turning over in his mind fresh means to meet, develop, and excite buyers. He was following scientists' conversations about varieties with particular interest. As more and more varieties were "discovered," as orchids of every color, size, hue, and fragrance piled up in nurseries and auction rooms, he was intrigued by the fact that no European, as yet, had worked out how to hybridize orchids—to create new varieties "to order," so to speak.

Perhaps there were other ways of producing remarkable orchids; perhaps orchids of the size and beauty of *C. labiata* did not have to be grabbed from a near-vertical precipice in low visibility and pounding rain. Perhaps a plant with a less desirable trait could be mated with a plant with a more desirable trait, "solving" its problems—producing more, bigger, beautiful, fast-growing plants.

After all, modern science, as Dr. Frankenstein discovered in 1818—the year the packing-case orchid first arrived in Suffolk—could achieve remarkable things.

PART III

THE DARKLING PLAIN

When I found so astonishing a power placed within my hands, I hesitated a long time concerning the manner in which I should employ it.

—MARY SHELLEY, FRANKENSTEIN (1818)

12

A NEW DYNASTY

IN 1832, JOHN VEITCH and his forty-year-old son James reached a painful parting of the ways. It was the year of the Great Reform Act, also known as the Representation of the People Act. In theory, birth was no longer destiny, and lower-class roots would present little obstacle to upward mobility: that, at least, was the promise of legislation that expanded suffrage deeper into the population. Small landowners, tenant farmers, and shopkeepers could vote, and cities gained more of a voice. The act did not give the vote to everyone—women were explicitly excluded—but it did point the way toward change.

In the shadow of the Great Reform Act, Veitch father and son argued over the right way to run a nursery in the small town of Killerton, eight miles outside Exeter in Devon. James was convinced they were losing business to companies located closer to the city. Even worse, the ground they operated on could only be held during the older man's lifetime, according to an arrangement made with the upper-class landowners.

James saw problems in his father's attitude to business. John Veitch, stocky and affable, was the sort of man who lived by the book—literally; he hung up illustrations from *Raphael's Almanac* and quoted it regularly. "Many a time he pulverized me with wise-saws from this Almanac," his grandson Thomas remembered ruefully.[1] John was interested in traditional horticulture, in trees, vegetables, and fruit trees. James—tall, angular, rather intense—had begun to try his hand at new trends in gardening, particularly the cultivation of exotics. The nursery, James believed, had the potential to be more than a place to buy plants. He wanted to make it a destination, a place to bathe the senses in the new, the remarkable, the strange, not just a practical place for the purchase of plants.

James, parting company with his father, bought twenty-five acres in a suburbanizing area of Exeter called Mount Radford. It was a forward-thinking move. "Villas"—as semi-detached and detached suburban houses were typically known—were popping up by the week. Streets and turnpikes were undergoing improvements to make travel between the city and the suburbs easier, while footpaths and pavements were constructed and enhanced. Mount Radford was, in other words, an ideal destination for those seeking a share in the excitement of networked modernity. The first generations of suburbanites were fleeing not the city but the declining countryside; to live in the suburbs as a Victorian was to move *toward* bustle, not away from it, and suburban living was from the first enabled by, and shaped around, transportation links.[2] James Veitch, grasping the possibilities of a new market, selected premises on Topsham Road that were well placed to attract mobile professionals, commuter families with evening leisure time and a little extra spending money.

After the purchase, James sent out a letter to customers. In it, he assured locals of the move's benefits not just to him, but to *them*. The new premises would have a shop for seeds and garden implements, he explained, even as "no expense has or will be spared in laying out, and ornamenting the Grounds." The new premises would lure in tourists, serving as "an attraction not only to the immediate neighbourhood, but to Persons from a distance visiting Exeter." And in the greenhouses, visitors would find truly extraordinary plants, including "extensive and *Choice collections of the following* Articles, Viz, Stove Plants including the various kinds of Orchidaceous, or Air Plants." Here, on a new twenty-five-acre spot in the cool climes of Devon, the nursery would dazzle with its "exhibition and sale of Exotics."[3]

James Veitch followed up his move into the new premises by sending his son, James Junior, to the capital city to learn more about the business. In 1833 James Junior began a job in Upper Tooting. The area—traditionally "a hamlet in the parish of Streatham"—was in the midst of

A New Dynasty

rapid development.[4] It was also the location of a large nursery, Messrs. Rollisson's, which was taking full advantage of its networked, suburban locale. "Being situated in the suburbs of London, just as it were beyond the smoke line, the nursery has the advantage of pure air," the *Journal of Horticulture and Cottage Gardener* enthused later in the century, even as "it is also convenient of access from the stations of Tooting on the South-Western Railway, and Balham on the Brighton line."[5] When Veitch Junior arrived, Rollisson's was what *Paxton's Magazine of Botany* termed one of the nation's "Principal Suburban Nurseries," rivaling Loddiges in Hackney and Low's in Clapton. James Veitch wanted his son to acquire more experience in the business, but he also hoped to learn more about what a "Principal" nursery was up to and what their consumers desired.

James Junior began his training at Rollisson's and soon saw for himself, firsthand, the rise and rise of orchidomania. Rollisson's was a key player, with plant hunters spread all over the globe: the company won a Gold Banksian for "Orchideous plants" at the Horticultural Society on June 6, 1835.[6] In the same year they won a Large Silver for *Epidendrum cuspidatum,* a fringed orchid originally imported by Loddiges.

James Junior seems to have picked up a thrill of winning, of being first, that never left him. He returned to Exeter bearing "a collection of the Orchid gems of those days" from Rollisson's, the family's published history would later record. These orchids became the beating heart of the Veitch collection, which was soon expanded with even more of Rollisson's orchids, in accordance with the terms of young James's hire (the Veitch nursery agreed to buy their orchids from Tooting as a sort of payment for his training). James Veitch proceeded full steam ahead with plans to make his Exeter nursery a destination for anyone who wanted to see, buy, and cultivate beautiful orchids.[7] But moving into the top tier of national nurseries required a plant hunter. Nurseries that imported rare plants were not only able to name them, they were themselves named and hymned in leading newspapers and periodicals. Rare discoveries ignited attention, attracted coverage, and drove business.

At around the same time, a very different sort of horticulturalist accelerated his own plans for a hunt. William Spencer Cavendish, 6th Duke of Devonshire, was a different kind of collector, too, owning eight homes and over 200,000 acres. He was an important Whig politician as well as president of the Horticultural Society. The duke and other illustrious patrons joined the Loddiges and Rollisson nurseries to fund two of the day's leading botanists on a trip through the Canadian Rockies. Robert Wallace and Peter Banks had been mentored and trained by Joseph Paxton, head gardener at Chatsworth; Hooker and Lindley were recruited to help identify the species they recovered, some of which were to be given to Kew and the Horticultural Society. This was a top-of-the-line cooperative effort, in other words, peopled by leading figures at every level. Its goal was to pick up the trail left by another illustrious plant hunter, David Douglas, who had been gored to death by a bull on a botanical expedition in 1834.

If James Veitch Senior was jealous that his Exeter nursery was left out of what Paxton rather smugly called the "select number of subscribers" invited to join in the venture, the annoyance was possibly short-lived. On October 22, 1838, Paxton's trainees were traveling on a boat with ten French factory recruits along the famously dangerous Columbia River. While navigating Dalles des Morts—Death's Rapids—the craft was swamped. Both men were killed, together with the ten others on board, including five children.[8]

Horrified, the duke retreated from funding future expeditions. But Veitch remained implacably determined to proceed with a plant-hunting trip of his own, though finding the right person to undertake such an expedition was difficult. A good collector, for Veitch's purposes, had to be a man who knew "what to collect for a nurseryman rather than one who only appraised plants with a Botanist's ego," as the family's history later put it.[9] A useful collector needed to understand, in other words, what his employer wanted. He needed to keep business interests, not just personal fame, at the forefront of his mind.

Not that Veitch was blind to fame; far from it. James Senior began a correspondence with William Hooker, now settled at Kew, shaped by the principle that scientists and businessmen could support and aid one another. Botanists needed plants to study, but plant-hunting expeditions were expensive, time-consuming, and obviously dangerous. Nurserymen could provide the plants, Veitch realized, and receive returns from this relationship with a knowledgeable professional scientist—a figure gaining increasing cultural caché. Such connections could help to separate the nursery from its competitors in a crowded market. Veitch would soon form a close relationship with Charles Darwin and be thanked gratefully in his work, in print. Victorian scientific discovery was substantially powered by nurserymen like Veitch, especially in a time of wavering and conditional support for science from government.[10]

James Veitch peppered Hooker with questions on plant hunting in his letters—how to find plants, how to remove them in sufficient quantities to make them commercially viable, how to prepare dried specimens, how to bribe local officials, how to store plants for a sea voyage, how to protect them from storms, salt, rats, human carelessness, and more. He also wanted inside information; he slyly asked Hooker how the Horticultural Society's own collector, Theodore Hartweg, was progressing. But Veitch was very careful to keep his own plans under wraps. As he moved toward making a decision about a first big venture, he warned Hooker: "Having decided on sending out a Collector, we would thank you not to name him."[11] James Veitch benefited from his collaboration with Hooker, but his letters indicate he was strongly motivated by competition, the urge to best other companies. That motivator was even stronger by the time the family's histories were more fully articulated, first in the 1888 *Manual of Orchidaceous Plants*, then in *Hortus Veitchii*, in 1906. Those texts were written in a time of open warfare over *Cattleya labiata*, and they suggest a fight for its recovery was unfolding energetically by 1840—though I have not found explicit mention of a hunt for *C. labiata* in Veitch's correspondence with Hooker from that time.

The letters *do* indicate that a man named William Lobb was sent on a plant-hunting trip to Brazil by Veitch and that he left Plymouth in early 1841. Lobb was the older brother of a man already working for the company. The working-class son of an estate carpenter, Lobb had experience with exotics from working at Carclew, a Palladian mansion in Cornwall owned by an orchid enthusiast named Sir Charles Lemon. He gained additional experience by working with the plant aficionado Stephen Davy of Redruth. Lobb wanted more from life—or, as *Hortus Veitchii* put it in more floral terms, he "cherished an ardent desire for travel and adventure": Lobb landed in Rio and headed for the orchid-rich mountains that Swainson journeyed through with other European naturalists, supported by the Russian consul Georg Heinrich von Langsdorff in the later months of his trip.[12]

James Veitch Senior reported Lobb's arrival in an excited letter to William Hooker, dated March 31, 1841. Veitch listed a range of plants

Lobb had already recovered, including a kind of orchid named *Oncidium pactum*. So far, so good: we are "pleased with him," Veitch said grandly. He was keen, he stressed, to maintain the shroud of secrecy: "of course we shall say nothing about" his achievements until the plants were safely home.[13] He made no mention of *C. labiata* in the letter; either he was not actually looking for it or, just possibly, he was concerned about the letter's potential interception.

Partly because of Veitch's unwillingness to put his plans to paper, it's hard to recover exactly what Lobb did, why, or when. Botanists like Swainson and Gardner had a motive to write up their experiences and publish them afterward; Lobb did not, for he was not seeking to establish a reputation with fellow scientists or to gain well-placed patrons. Lobb's income came from Veitch, and it was Veitch alone he had to please. Moreover, while Lobb clearly wrote to Veitch to recount his finds and challenges, unfortunately the Veitch correspondence is lost, burned by a descendant named Mildred.

Glimpses of what happened on the Lobb trip emerge only in fragments that speak to the challenges and monstrous losses of plant hunting in an era before steam, cable, and other infrastructural developments. A letter to Hooker, dated September 3, 1841, for instance, suggests Lobb's expedition in Brazil ran into a number of problems. First, there was competition over plants with George Gardner. Veitch was disturbed by Gardner's discoveries, and he pressed Hooker repeatedly for more information on what Gardner had unearthed. Worse was yet to come. Lobb's consignment of plants from the area were all but destroyed. Fourteen boxes of specimens were delayed at Rio de Janeiro when an agent failed to put them on board ship; a subsequent captain refused to accept them. By the time they arrived on a merchant vessel, after two months' delay, many of the plants were rotten. Lobb had already left Rio on a boat to Buenos Aires, so he could not return and replace them.

Years later, a descendant of Veitch's claimed that Lobb had been unable to recover *C. labiata* on the trip. "William Lobb, our own collector," Harry Veitch (son of James Junior) noted, in a section on "Cattleya" in the *Manual of Orchidaceous Plants*, "explored the [Organ] mountains in 1840" in pursuit of it, but was unable to find it, even though he found all sorts of beautiful orchids, including some "at that time extremely rare in English gardens." The *Manual* also stressed that Gardner had not recovered *C. labiata* either—as if motivated to explain, years later, that in the plant-hunting competition of Lobb and Gardner, the latter had not won: "the form [Gardner] met with [in the Organ Mountains] is, with a very high degree of probability, believed to be the variety Warnerii," Harry Veitch asserted. He suggested that the plant Gardner was unable to reach, the one on the Gavea, might have been the "true" *C. labiata* after all. But that area was soon devastated, turned into charcoal; *C. labiata* thus "is now known to have long since disappeared from that locality," the *Manual* concluded, stressing that it was almost certainly extinct.[14]

Lobb was to find many plants on his travels, both on this trip and others, that would satisfy Veitch in his desire for the new, the remarkable. He is particularly remembered for introducing the monkey-puzzle tree and the giant sequoia ("Wellingtonia") to Britain. *C. labiata* was not one of his successes, although he returned to the mountains outside Rio on at least one future trip.

James Veitch was, meanwhile, displaying *C. labiata* in competitions throughout the 1840s—possibly cloned descendants (from division) of Swainson's first plant, possibly plants from Gardner. At a November 1846 meeting of the London Horticultural Society, Veitch's magnificent specimen was singled out for special mention by the publication that had celebrated Rollisson's plant a decade earlier. "[A]although it had travelled all the way from Exeter," the *Floricultural Cabinet* marveled, Veitch's specimen of *C. labiata* "arrived in the finest condition. It had seven spikes, each with four or five large, handsome, purple flowers." Five other *C. labiata* were awarded prizes in the same exhibition, but Veitch's earned a Knightian medal.[15]

James Veitch Senior continued to invest in plant hunting. William Lobb's brother Thomas traveled on behalf of the company to Java and Singapore in 1843. At the company's height, it employed over twenty plant hunters, a job at once hugely wasteful, dangerous, and chaotic. William, taken seriously ill with dysentery in Peru, for instance, discovered on his recovery that a large number of the plants he thought had been sent to Britain were rotting in a warehouse. Collections sent on by Thomas froze on arrival at London in the very cold winter of 1844.

James Junior seems to have become particularly frustrated with the losses and impatient "for quicker rewards and returns," notes Sue Shephard, the company's contemporary biographer.[16] His early drive to win did not abate as he grew older. Irascible and famously short-tempered, the younger Veitch was increasingly convinced that the way to success in the world of rare orchids lay in a new approach: hybridization.

The boundaries of science were being pushed back daily; what seemed the food of fiction a generation ago was becoming possible. As a professor in Mary Shelley's famous novel explained, thrilling a young, impressionable Dr. Frankenstein, scientists "whose hands seem only made to dabble in dirt, and their eyes to pore over the microscope or crucible, have indeed performed miracles. They penetrate into the recesses of nature, and show how she works in her hiding-places . . . They have acquired new and almost unlimited powers."[17]

Could two glamorous orchids, mated together, produce an even more remarkable, gorgeous, prize-winning plant? "I will pioneer a new way, explore unknown powers," gloated Dr. Frankenstein, "and unfold to the world the deepest mysteries of creation."[18] The story of how orchids came to be hybridized would be narrated and renarrated in the second half of the century as a remarkable occurrence, in which science very nearly met its match.

13

HATCHING A PLAN

HYBRIDIZATION WAS A controversial practice with grave implications. It seemed, to many, to lead humans dangerously into the realm of the divine. Genesis 1 tells us that the Lord created species at the point of Creation: "See, I give you every seed-bearing plant that is upon all the earth, and every tree that has seed-bearing fruit . . . And God saw everything that he had made, and, behold, it was very good." Taken literally, this suggests that the species were created all at once, fixed from the start—so that to tinker with them, to change them, risks contravening the Almighty's Great Plan. The Bible also warns explicitly against hybridizing. "Do not sow your vineyard with two kinds of seed," we are told in Deuteronomy 22, "lest the fullness of the seed which you have sown be forfeited together with the increase of the vineyard." The blending of two different species may produce infertile offspring, in other words, and longer-term losses will follow.

Still, hybridizing became increasingly popular in the nineteenth century, at first used especially to improve crops.[1] It also gained valuable support from a clergyman, the Honorable and Very Reverend William Herbert, Dean of Manchester. Herbert would die before Veitch's put their hybridization plans into high gear, but the dean's scholarship, his narratives of possibility, were regularly referenced by Victorian authors as having laid the crucial groundwork.

Herbert was the son of an earl but, as a younger family member, he had to make his own way. He tried his hand at politics and the law first, distinguishing himself as an abolitionist and a supporter of the English downtrodden. "The first object of a good government," he argued, boldly, writing on the Game Laws, "is not that rich men should have their pleasures in perfection, but that all orders of men should be good and happy."[2] In 1814, Herbert changed careers and became an Anglican priest. For the rest of his life, he combined ecclesiastical labors and social justice with a passion for natural history, and for writing, painting, and translating from classical sources as well as from ancient Scandinavian texts. Herbert's translations of Icelandic poetry, his Greek Odes, and his twelve-volume epic poem on Attila the Hun were well known in his day.

Herbert rejected the idea that God made the species distinct, according to a Grand Design that was fixed at the point of the Creation. He argued that "it was preposterous to suppose all the existing form of vegetables . . . [had] been so specially created by the Almighty." Empirical study revealed that species were not immovably fixed or unchanging, Herbert observed, noting of amaryllis that "the species of this extensive genus slide into each other." "I am . . . unwilling to assent to the assertion, that every plant . . . or even a distinct species, or . . . genus, had a special creation," he concluded; as Charles Darwin himself later put it in a summary of the clergyman's work, Herbert saw that species "were created in an originally highly plastic condition."[3] To Herbert, hybridization was not a terrible taking-on of the Almighty's role but a continuation of an ongoing process. Herbert did not see himself as taking God out of Creation but rather as arguing that God made species (in both the plant, or "vegetable," and animal worlds) open to change.

Herbert's 1847 two-part article, "On Hybridization Amongst Vegetables," laid out his suppositions in detail. The dean began by referencing attacks made on him personally "as a person who was minishing from, instead of attributing infinity to, the power and wisdom of God" (*minish* is an archaic word for "diminish" or "make less"). Hopefully, he continued somewhat optimistically, "the progress of useful knowledge has nearly dissipated such absurd calumnies."[4]

Herbert pointed out that species obviously come and go: the mastodon is no more, while the rhododendron was not known to the dinosaurs.

Species were thus obviously not created all at the same time, as Genesis seems to suggest, and hybridization offered clues to a fresh understanding of how this could unfold. While some argued that the sterility that often followed cross-breeding was proof the Creator did not intend species mixing (as Deuteronomy and also Leviticus suggest), Herbert reminded readers that sterility was very far from an inevitable by-product of cross-breeding. He also noted that, when successful, hybridized plants could be *more* resilient than their parents. Herbert argued that this indicated some potential, as-yet-to-be-understood relationship between the plants that had successfully interbred. He posited a mechanism or shaping system that lay still beyond human comprehension—a system whose very complexity served to remind him of the greatness of what he reverently termed the "Allwise."

Orchids, to Herbert, offered an intriguing opportunity to push toward a fuller understanding of this still-baffling system. Like Darwin after him, Herbert thought that orchids and the very confusions that so frustrated Hooker and Lindley pointed to profound truths: "The Orchidaceous plants exhibit the most confusive diversities," he observed, suggesting that those diversities, the very challenges of distinguishing between multiple similar species and varieties, force us to the conclusion that changes are unfolding, and have unfolded, over time. After giving several examples of the kinds of perplexing questions orchids pose to the botanist, he demanded: "Can we, in face of those phenomena, assert that no vegetable since the period before the sun and moon gave it light, no bird or fish since the Almighty called them forth from the salt mud, no creature of the earth since it was evoked from the dust, can have departed from its precise original structure and appearance?"[5]

Orchids posed other, very practical problems for those interested in their propagation, and Herbert acknowledged those as well: "they are not easily raised by seed." Orchid seeds are extremely small, almost microscopic—what *Hortus Veitchii* would call "minute chaffy bodies of extreme lightness; so minute are they that an ordinary pocket lens is powerless to enable one to know whether the seeds are likely to contain a germ or are mere lifeless dust."[6] For a long time the belief persisted that they did not exist at all, and even once identified, the precise mechanisms of reproduction remained perplexing. Fanciful theories lingered for centuries, often connecting orchids to sex—for instance, that orchids developed from the semen of birds and animals scattered during copulation. Athanasius Kircher, a German Jesuit writing in the seventeenth century, suggested "these plants arise from the latent survival force in the cadavers of certain animals [and] animal semen."[7] This is not to say that no one had identified the plants' means of propagation, but the discovery didn't take root: a professor in Thuringia drew orchid ovaries and described the brown seeds in the late eighteenth century, while a German botanist named J. K. Wachter used the information to pollinate an orchid.[8] Numerous attempts were also made to pollinate various forms of the vanilla bean orchid in Indonesia.

In fact, perhaps the most important and successful of all such attempts was made by an enslaved child named Edmond Albius in the 1840s. Born in modern Réunion to a mother who died in childbirth, Albius discovered how to pollinate the plant by applying knowledge he had gained from the man enslaving him, a farmer named Ferreol Bellier-Beaumont. The latter possessed vanilla cuttings, but his plants would not bear fruit. To the farmer's astonishment, Albius applied a tactic he had been shown for propagating watermelon seeds to the vanilla plant, which soon began producing pods: as the farmer later reflected, in 1861: "This clever boy had realized that the vanilla flower also had male and female elements, and worked out for himself how to join them together." Just twelve years old at the time, Albius was soon marched around to show other enslaved people what to do and how to do it, and a global vanilla industry was born.[9] Albius did not reap the benefits of his own achievement; his reputation was repeatedly impugned by those who refused to credit his important discovery, and he was imprisoned for supposedly injuring a white woman during a robbery. He died destitute.

In spite of successful attempts like those of Albius, British botanists and horticulturalists remained frustrated in their efforts at midcentury. Reports of successes were few and far between, buried in rare books or spoken of by those whose words were ignored.[10] Orchidophiles wrung their hands over the matter in the pages of gardening magazines. "At the present time there are few subjects connected with plant growing on which there is less recorded information than that of growing Orchids from seeds," mourned David Moore, a Scottish botanist working in Dublin in the *Gardeners' Chronicle* in 1849. That paucity of information "appears the more remarkable when the great interest our ablest Cultivators have taken in growing this singular tribe is considered," he continued, because of "their intrinsic value" and "their tardiness of increase by division."[11]

Yet although Dean Herbert stressed the challenges of growing orchid seeds, he also indicated that he had made progress. Arguing that "Crossbreeding amongst Orchidaceous plants would perhaps lead to very startling results" if successful, he wrote:

> I have . . . raised Bletia, Cattleya, Orchis (Herminium) monorchis, and Ophrys ananifera from seed; and if I were not during the greater part of the year absent from the place where my plants are deposited, I think I could succeed in obtaining crosses in that order. I had well-formed pods last spring of Orchis by pollen of Ophrys, as well as other species of Orchis, which had been forced; and if I had remained on the spot I think I should have obtained some cross-bred Orchidaceous seed.[12]

Herbert's mention of the orchids he'd raised, including a cattleya, was enticing; but the dean died suddenly, before he could give details. In fact, in a strange coincidence, Charles Darwin was possibly the last man to see him alive. A shaken Darwin wrote to William Hooker's son Joseph on June 2, 1847: "I saw the poor old Dean of Manchester on Friday & he received me very kindly: he looked dreadfully ill & about an hour afterward died! I am most sincerely sorry for it."[13]

Two years later, David Moore's piece in the *Gardeners' Chronicle* offered details on how Moore himself had managed to sow orchid seeds:

> The manner of sowing the seeds, and treating the young seedlings, has been to allow the fine dust-like seed to fall from the ovaries as soon as they show symptoms of ripeness, which is readily known by the ovaries bursting open on one side. When this takes place, they are either taken from the plant and shaken gently over the surfaces of the other Orchid-pots, on the loose material used for growing them in, or on pots prepared for the purpose . . . In the course of eight or nine days after sowing, the seeds, which at first had the appearance of a fine white powder, begin to assume a darker colour to the naked eye, and if looked at with a Coddington [a microscope], or even a simple lens, evident signs of approaching vegetation may be perceived . . .[14]

But another piece published later the same month in the same magazine recorded that hybridized orchids remained out of reach:

> I believe with Mr. Moore that no cross has as yet been acknowledged to have been obtained in this class of plants; a few have been hybridised successfully here, so far as obtaining seed to all appearance perfect is concerned; and it has been sown, but it did not vegetate.[15]

One of the first plants this author attempted to hybridize was *C. labiata*: he attempted to cross it with another *Cattleya* species, *C. guttata*. It failed with all the rest.

Orchids were hard to propagate; orchids were astoundingly diverse: the two topics surface repeatedly in print and in letters about orchids, serving as energetic discussion points for debate in the 1850s. Could these phenomena be related? Could humans work out how to capitalize on orchids' capacity for diversity? The creation of a hybrid orchid—a plant that would not just rival but supersede a bloom like *C. labiata*—was rapidly becoming a new "horticultural holy grail."[16]

The passion for orchids was blazing ever fiercer, as the "Orchids for the Million" articles in the *Gardeners' Chronicle* made plain. The demand was prompted by a combination of more knowledge, more accessibility, more availability, dropping prices, expanding orchid-oriented communities, and an increasingly powerful conception—promoted and narrativized by advertisers and authors alike—of orchids as "exotics," exciting spectacles of the moment.

A man named John Harris, a surgeon living barely half a mile from Veitch's premises in Exeter, was himself one of the "million," a man with an ordinary day job who dabbled in plants in the evening and kept abreast of botanical developments in the pages of periodicals like the *Gardeners' Chronicle*. One day, according to Veitch's family history, Harris strolled up the road to Veitch's nursery and had a chat with their horticulturalist, John Dominy.

14

DOCTORING ORCHIDS

VEITCH'S BUSINESS WAS, by midcentury, flourishing. William Hooker toured the nursery in 1848 and published a laudatory account in *Curtis's Botanical Magazine*. The company received a further boost from reformist Charles Bellenden Ker, that "Dodman" of the *Gardeners' Chronicle*. Ker praised the premises and celebrated especially its "marvellous specimens of orchidaceous plants ... I never witnessed so many fine specimens, so much flower and such good treatment."[1] As far as orchids were concerned, Ker concluded, Veitch's was rapidly becoming second to none.

To cement the company's position as a leading national player, in 1853 Veitch's nursery acquired expansive premises on the King's Road, Chelsea, London. James Junior and his family moved—with vanloads of plants, and plenty of fanfare—to the city to operate it. Celebrating the arrival of the entourage, the *Gardeners' Chronicle* remarked with lofty urbanist condescension that the move to London was a sound one, as "everyone must have regretted the impossibility of seeing much of so important a collection, so long as it was buried in the heart of Devonshire."[2] James Senior stayed in Devon, plagued by health issues, though he remained formally in charge. James Junior set up the massive premises in Chelsea, running showrooms and managing offices, and quaking employees soon learned not to interrupt him because of his bad temper.

At around the same time, the company turned its attention to propagating hybrids under the aegis of their head horticulturalist, John Dominy (born in 1818, the year Swainson toured Brazil).[3]

"The chief aim of the hybridist is to improve those species and varieties already in cultivation by producing new and distinct forms," a later-Victorian writer explained in the magazine *British Gardening*, "with a greater brilliancy of colouring, an increased floriferousness, and an even more vigorous constitution." In other words, the hybridist could make up for the wants, the weaknesses of nature. Color could be rendered more brilliant; the number of flowers on the stem could be increased; the plant's lifespan could even be lengthened. Some plants, the author conceded, did not need scientific improvement, but the less resplendent orchids seemed to call out for human intervention. Looking back at decades of such labor in 1893, the *British Gardening* author remarked with satisfaction, "From several species hybrids of such excellent quality have been produced that the parent plant is altogether superseded by its progeny."[4]

Veitch's intended to be the first to "supersede" nature through the auspices of modern science. Just because cross-pollination *could* produce ill effects did not mean it *had* to; infertility did not always follow. Surely, sometimes, offspring in nature were not less but more vigorous (more beautiful, more colorful, more long-living) than their parents. Nurserymen around the country had experimented from the 1830s with hybridizing at a time of increased enthusiasm for the application of modern technology to horticulture.[5] The very aesthetic of horticulture at mid-century aimed to stage national and technological achievement. Mid-Victorian gardens were colorful places: the bright beds of single-colored geraniums and other exotic annuals set in geometric patterns showed off the plants that had been imported and were now able to survive through British industrial and colonial expansion.

In the history of orchid hybridization, a few details remain murky. John Dominy's role at Veitch's nursery is fairly clear, but quite who Harris was, exactly, remains obscure. He made, it seems, just one contribution to botany. James Herbert Veitch (James Junior's grandson) mentions a Dr. Harris from Exeter in the family's history, *Hortus Veitchii*, though both father John and son John William lived on a street named Southernhay in the city, according to the 1841 Census. Both were surgeons, and a local directory for 1850 indicates the two men were in business together.⁶ One of the few historians to research the relationship of Dominy and Harris identifies the father as the man interested in orchid propagation—and indeed, John Senior was a fellow of the Linnean Society, so this certainly seems possible. Yet John Senior, thirty-four years older than Dominy, died in 1855, while Veitch's employee was still experimenting.⁷ The son, who was the same age as Dominy, could also have spurred the latter's experiments in orchid propagation.⁸

Both John Harrises would have been trained in botany, which had been made a compulsory part of medical education in the 1815 Apothecaries Act. Plants were *materia medica,* meaning the stuff of which medicines were made. The Dr. Harris who met Dominy must have kept thoroughly up to date with new discoveries, for *Hortus Veitchii* explained that it was he who "pointed out to [Dominy] the reproductive organs" of the orchid.

Neither man left a written or printed account of what, exactly, took place in their collaboration. Harris, as a surgeon, was perhaps able to demonstrate the complexity of the column that housed the orchid's reproductive system, since he would likely have had the microscope—the Coddington—mentioned by the Scottish botanist David Moore in 1849 in *Gardeners' Chronicle. Hortus Veitchii* hypothesizes that such a scene took place, certainly, noting that nurserymen did not typically have access to this kind of information or apparatus at the time. Dominy, the work suggests, immediately saw the implications of what Harris had offered him: "As soon as an opportunity presented itself Dominy lost no time in turning [Harris's] suggestion to practical account."⁹

Hortus Veitchii, as a self-issued history of Veitch's business, predictably shapes its narrative of facts, dates, and events to suggest inexorable triumph. In fact, orchid propagation took time. Dominy began attempting to hybridize orchids around 1853, according to the *Manual of Orchidaceous Plants,* but it seems to have taken three years to produce a result. Just as with Dr. Frankenstein, achievement did not open "like a magic scene ... [all] at once; the information I had obtained was of a nature rather to direct my endeavours so soon as I should point them towards the object of my search, than to exhibit that object already accomplished."¹⁰ The *Manual* stresses that the business of raising seedlings in a climate and environment far from that of the parent plants remained fraught with peril. Even after germination, "The tender seedlings are brought into growth under circumstances so different ... that it is not at all surprising that multitudes of them perish in their earliest infancy."[11] Only the most skilled horticulturalist, the reader is prompted to understand, could overcome such serious challenges.

On October 28, 1856, James Veitch Junior took a trip to see John Lindley, by now a very eminent figure, resident in London's Acton Green. It was a memorable, even historic occasion, Lindley suggested in a piece describing the visit, published in the *Gardeners' Chronicle* in January 1858. Veitch arrived bearing a flower. Inspecting it minutely, Lindley saw that it

> combined the peculiar hairy forked spur and deeply lobed lip of the white *Calanthe furcate*, with the violet colour and broad middle lobe of the lip of *C. Masuca*. One might have said that the flowers were just intermediate between the two in all respects. A botanist could not have referred the plant either to the one or the other of these two species ... Had hybrids been suspected to occur among Orchids the plant would have been pronounced a cross. And such it was.[12]

The new orchid had first been created in 1854 by Dominy, Veitch explained, by crossing the two species Lindley suspected—*Calanthe furcata* from the Philippines and *Calanthe masuca* from the Himalayas. Amazed, Lindley is said to have exclaimed: "You will drive the botanists mad."[13] One of life's most prized secrets had been discovered—and by a nurseryman and his employee (with a little help from a doctor), not a botanist like Lindley or Hooker. The gorgeous new plant, *Calanthe × dominii*, was named to honor Veitch's horticulturalist—and, as Lindley explained, there were other hybrids in the works, "among which we understand are Cattleyas."[14]

Lindley's shock—"you will drive the botanists mad"—is much quoted in late Victorian and Edwardian stories of the hybrid (and, in fact, today).[15] The source of the great discovery evidently appealed to a Victorian middle class who believed that, theoretically, anyone could advance. The community of those interested in orchids were, indeed, much struck. "The raising of Calanthe Dominii is not such an occurrence to be passed lightly over, it forms an epoch from which to count a new era in gardening," declared an excited contributor to the *Gardeners' Chronicle*, a few weeks after the discovery was revealed: "I trust that that indefatigable gardener [Dominy] will lay before us the whole circumstances of the case, that others may follow in the track he has cut out." The plant's blooming so quickly was seized upon as especially interesting because "one of its parents is a shy flower"—meaning it took a long time to blossom. The new plant's quick blooming "points to one of the advantages which we may in future anticipate from hybrid Orchids."[16]

Veitch's displayed a selection of its "remarkable hybrid Orchids" at the Crystal Palace Horticultural Exhibition in September 1858, including *Calanthe × dominii* and four new *Cattleya* hybrids. The display caused quite a stir. *Cattleya labiata* was not one of the parents of the first crop of plants, but a year later, at a November 10 meeting of the Horticultural Society, *Cattleya × dominiana* was unveiled. It was the love child of *Cattleya labiata* and *Cattleya amethystina*, two orchids paired to produce new and spectacular offspring. Astonishingly hued—large and "blush coloured, with a delicately veined lip" and a "pale citron centre"—it was also botanically curious, "between a one-leaved and a two-leaved species." Awarded a first-class certificate, this "very remarkable plant" earned its own write-up in the *Gardeners' Chronicle* for the "new shock" it offered botanists and the plant world.[17] *Cattleya × dominiana* was on sale to the general public by 1862.

Veitch's received praise from many quarters for its achievement—including from the great Professor Reichenbach of Hamburg, now one of the world's leading orchid authorities, who published a fresh, admiring account of the Chelsea premises in October 1859. Veitch's had aimed to be a "destination" in Exeter; now it was a destination in the metropolis itself, and all were commended to visit and see firsthand how plants should be managed and propagated. "Among the many sights to be seen in the neighbourhood of London, few can exceed in interesting the intelligent observer more than the justly celebrated nursery of Mr. J. Veitch, jun., in the King's Road, Chelsea," Reichenbach cheered; "at first sight one scarcely knows which [orchids] to admire the most." He commended "a host of Vandas, Aerides, and other aerial Orchids":

> [H]ere are plants in every stage of growth all glowing with health, whilst the air is filled with a delicious perfume such as the beautiful wax-like flowers of these Orchids alone can give.... Mr. Veitch has

found out the exact manner in which these plants should be treated.... Leaving this house and crossing the centre path of the nursery to an opposite door, I found a house filled with Cattleyas, without speck or blemish, and looking as happy and as much at home as any of the afore-mentioned.¹⁸

Veitch's was arguably the leading nursery in the nation, with a collection that literally as well as metaphorically built on the Loddiges collection they'd swept up. Veitch's took on a second hybridist, John Seden, who went to Exeter to train with Dominy, now widely credited as the nation's leading authority on hybridization. In fact, he was literally peerless: Veitch's remained for fifteen years "the only nursery where orchid hybrids were produced." In spite of the hope so earnestly expressed in the *Gardeners' Chronicle*—that Dominy would "lay before us the whole circumstances of the case, that others may follow in the track he has cut out"—Veitch's kept resolutely mum about exactly what Dominy had done and how he'd done it. This secrecy about methods "gave them a huge lead in the highly competitive world of orchid collecting," Shephard notes in her study of the firm.¹⁹ More hybrids were churned out by Veitch's by the year if not the month. From 1856 to the late 1880s, "experiments have been carried on uninterruptedly in our horticultural establishments."

Veitch's nursery congratulated itself on a discovery that not only answered a scientific question but met rapidly growing consumer desire: the holy grail of orchids had been turned into something that could be quickly, easily, and cheaply reproduced. The Victorian urge for the biggest and the brightest seemed embodied in *Cattleya × dominiana;* who cared about *Cattleya labiata* when you could have a plant that was not just purple or pink but yellow as well? Who needed rare plants that required death-defying journeys all over the globe when you could make thousands at home under a microscope?

Except that somehow people did. Almost as soon as the first hybrids hit the shelves, a value-laden language began to creep into descriptions of orchids.

Home-grown in Britain, *Cattleya × dominiana* was on sale for £1 11s 6d in June 1862, at a time when middle-class incomes ranged from roughly £150 to £2,000–3,000 (a senior clerk might hope to earn in excess of £500, a bank manager £2,000).[20] While certainly not cheap, the price augured well for the future at a time when many imports still cost ten times as much or more. Hybrids opened the door to increasing accessibility of orchids to customers.

But even as many writers celebrated the ways in which orchids were, and looked set to become, available to a wider swathe of the population, others were horrified. Wealthy Staffordshire collector James Bateman was one of those who regarded the hybridization of orchids with horror, and he thundered his disapproval at the Royal Horticultural Society in a lecture of 1864. Improved propagation practices seemed worrying, too. "I am no advocate for multiplying specimens of the same species," sniffed "Serapius" in the *Gardeners' Chronicle*,—"quite an epidemic nowadays."[21] Even beautiful plants, he implied, if reproduced to excess, might spread insidiously, out of control. Hybrids were products of the moment, while plants like *Cattleya labiata* were evoked, with a sort of wistful sigh, as "established," "old."[22]

Indeed, *Cattleya labiata* regularly acquired the tag "old" around this time. On the one hand, the "old cattleya labiata" was a taxonomic-adjacent term, meaning that the plant in question agreed morphologically with the plant Hooker and Cattley first flowered, the one that Lindley named. But the adjective "old" was also a powerful descriptor in Victorian culture, laden with implications of tradition, stability, and endurance, and many writers and nurserymen began to apply it liberally, anchoring descriptions of their plants in the past.

On the face of it, this may seem strange. Early Victorians often celebrated the "newness" of their industrial age and, at midcentury, many were still proud of the new spending power and cultural visibility of the emerging "million." The new scientist and the new industrialist were powerful figures of advance, change, modernity. But, as the era progressed, newness became increasingly fraught. *New* suggested, at least to some, rapid change; *new* referenced new money and new values, the nouveaux riches. Many feared the loss of centuries of seemingly stable English identity. Wilkie Collins picked up on this dawning unease when in his novel *Basil* he described a modern suburban house in 1852 as "oppressively new. The brilliantly-varnished door cracked with a report like a pistol when it was opened; the paper on the walls . . . looked hardly dry yet."[23] A house as new as this telegraphed to anxious readers a strangely changed world dominated by a new group of people whose values dangerously oriented around acquisition, money, and the self.

Even the middle classes feared this, in spite of their own rapid growth and cultural takeover, not to mention their own domination of the architectural landscape. Angst about newness was not only articulated by aristocrats, it also expressed middle-class unease about the changes they'd unleashed. In a passage of painful self-mockery, for instance, satirist George Augustus Sala—friend to Charles Dickens—painted himself and his peers in 1859 as dominated by a sick urge for the new, for more, for "mere" accumulation:

> We change our dresses, our servants, our friends and foes . . . We tire of the old friend, and incline to the new; the old baby is deposed in favour of the new baby; the fat turnip silver watch our father gave us, gives place to a gold Geneva—we change, and swop, and barter, and give up, and take back, and long for, and get tired of, all and everything in life.[24]

Victorians deeply feared that theirs was an age of disposable pleasures. As factories and department stores made and offered wares at lower prices to the masses, readers were often exhorted to remember the past, to resist the lure of industrially produced modern goods.

New hybrid orchids were at once embodiments of the exciting opportunities of the moment *and* signposts to potentially larger cultural, as well as horticultural, decline. *Cattleya* hybrids were showy plants—that, after all, was the point of them, to be more beautiful, more remarkable than their parents. But their very showiness, their floral upward mobility made them potentially awful alter egos of the nouveaux riches themselves. New arrivals, they multiplied quickly, and they rapidly invaded homes of "established" plants—all behavior associated with the massed middle classes, with the stockbrokers and clerks peopling new suburbs across the nation. The fact that the consumers of cheaper orchids were often middle class themselves, that they were buying the plants to signal their new wealth and leisure, only helped cement the dangerous connection.

Veitch's nursery did not turn up their noses at customers without pedigree, and the success of their operation indicates that plenty of buyers were attracted to hybrids. James Veitch Junior knew the color of money when he saw it, and he seems to have regarded the growing veneration for the past as sentimental rubbish: when he bought a fine old nursery in Fulham, for instance, he cut down all the ancient apple and pear trees and promptly gave the land over to one-year-old fruit trees for transplanting (a remarkably apposite metaphor for his priorities).

But a yearning for the *old* was to shape the face of Victorian gardening for decades. Earlier Victorian gardeners centered their designs around brightly colored, massed annuals—plants imported from far-flung places and raised in England under glass. These designs reflected not just a love for bright colors, an aesthetic preference for order and geometry, but a celebration of the modern commercial, industrial, and colonial systems that enabled the arrival and flourishing of imported plants in the first place.

Late-century Victorian gardeners would reject the beds of bold annuals surrounded by box hedging. Horticulturalists turned to supposedly older ways of treating and managing the land. Annuals remained popular in suburban nurseries, but those who viewed themselves as tastemakers regarded plants grown under glass, like geraniums, begonias, and calceolarias, with emotions ranging from suspicion to horror. Many consumers still loved the quick bursts of color, but writers urged them to try native plants instead, pointing them to perennials above annuals, to gardens oriented around seasonal growth patterns, to large, unbroken areas of quiet green grass. Specimen plantings and neat formal rows were out; in came scattered, naturalized plants. The debate over wild versus formal gardens was to reach its height in the 1890s, but rumblings of it are in evidence much earlier in discussions of orchids in print texts.

Cattleya orchids were not, of course, native plants, and the roses-around-the-cottage-door aesthetic that dominated the "wild garden" craze at the end of the century did not quite stretch to accommodate them. But still, after 1850, as hybrids and other propagated plants poured out of Veitch's nursery and into the homes of consumers, a restlessness stirred all echelons of the horticultural world. The "old Cattleya labiata" was mentioned wistfully, as a plant whose beauty was not manmade, an orchid that embodied something that had been lost all too easily.[25] Though not native, it was an orchid with a long British history; when the first brash hybrid exploded onto the horticultural scene, *Cattleya labiata* had been blooming in England for a full forty years. To midcentury Victorians it was not a newfangled innovation but a reminder of a slower, Georgian past. It was an orchid with a heady whiff of romance.

Its resistance to propagation now looked less like an annoyance and more a signal of its authenticity. Unlike so much in a machine age, it could not easily be duplicated. And so it soon became a character in a gripping new narrative within the orchidomania of the last third of the century—the *vera*, or "true" species of *Cattleya* that was "lost."

Meanwhile, Veitch's successes—the ever-increasing consumer appetite for orchids—served to make plain to other businessmen that there was a lot of money to be made, one way or another. New competitors rapidly moved into the scene, filling the vacuum left by the breakup of Loddiges's business. William Bull, a former employee of the Rollisson nursery, purchased a business of his own on the King's Road, Chelsea, in 1861. Setting up in the same area as James Veitch, he too specialized in exotics, especially orchids. Williams opened the Victoria and Paradise nurseries in Holloway. And, in the 1860s, German-born orchidophile Henry Frederick Conrad Sander, working at a suburban nursery in Forest Hill, London, spied a seed business for sale, soon to become the great Sander & Co. in St. Albans, which Sander ran with Benedict Roezl, the collector who'd lost part of his arm in an accident in Cuba. Between them, these companies would import ever more orchids from all over the world, offering naturalists like Charles Darwin startling new insights into the complexity of the environment.

As consumers and scientists alike began to turn their attention to "old" orchids, *Cattleya labiata* re-emerged as a prize very possibly worth dying for. Veitch's and Sander's nurseries, and Hugh Low's in Clapton, were all especially motivated to discover it.

But British companies were not the only ones hot on the trail. Plant hunter Jean Jules Linden and his son Lucien were determined to win the great prize. Linden had developed a mighty horticultural empire, Horticulture Internationale, with premises in Paris, Brussels and Ghent. In addition, North American seed and orchid companies were also about to enter the fray. The most intense years of orchid hunting still lay ahead. The already-crowded competition was about to turn deadly.

In the meantime, the rising naturalist Charles Darwin was in the midst of a dramatic story of his own.

15
DARWIN'S CRISIS

THE SUMMER OF 1858 was intensely hot. Temperatures in London were over 86 degrees Fahrenheit (30°C) by the middle of June; a fierce sun blazed over the country for weeks. Farmers rejoiced as crops ripened early: "The prospects of the harvest, on which so much of the prosperity of the country for the next twelve months depends, continue to be everything that can be desired," crowed the *Liverpool Mercury* on July 24, but a growing drought soon began to threaten the bounty.[1] In the center of industrial London, the heat caused catastrophic problems in a city with outdated, overwhelmed sewers. The Thames stank so terribly that residents were overpowered, fainting and vomiting uncontrollably in the streets.[2] The Houses of Parliament, overlooking the river, had to be closed; Chancellor of the Exchequer Benjamin Disraeli was forced to depart the committee rooms clutching his papers, handkerchief pressed to his mouth, with William Gladstone right behind him.[3]

Down House, in the village of Downe, is fourteen miles south of London in Kent. It must have felt a world away from the metropolis and the horrors of "The Great Stink." Bromley would soon become a commuter suburb, tightly connected to the metropolis on a tangle of railway lines, but in the simmering summer of 1858 it was pleasantly remote. Here Charles Darwin lived with his wife, Emma, and seven children—the eldest, thirteen-year-old Henrietta, followed by George, Elizabeth, Francis,

Leonard, Horace, and the baby, Charles Waring, born in 1856. William Erasmus, the couple's oldest son, was not at Downe but at Christ's College in Cambridge, living in his father's old rooms.

Darwin himself was not particularly thinking about orchids that scorching June. He was enjoying family life outdoors while exchanging letters with his friends about a theory he had been developing and honing for decades. Or so, at least, suggested Henrietta, known as Etty, when she recalled the summers of the late 1850s. "I think of a sound we always associated with the summer," she wrote, employing a little storytelling of her own, "the rattle of the fly-wheel of the well, drawing water for the garden; the lawn burnt brown, the garden a blaze of color, the six oblong beds in front of the drawing-room windows, with phloxes, lilies, and larkspurs . . . the row of lime-trees humming with bees, my father lying on the grass under them; the children playing about, with probably a kitten and a dog." Her kind father played backgammon with her, Etty remembered, and her mother was always available to read and to talk. (Etty, too, knew her literary conceits; like the author of many a Victorian novel, she frames an Eden soon to be transformed.)[4]

One particularly close friend of Darwin's was Joseph Dalton Hooker, son of that William who first took receipt of the lost orchid in 1818. In 1855, Joseph had risen to become assistant director of the Royal Botanic Gardens at Kew. Darwin and Joseph communicated regularly and, in the early weeks of that hot June, Darwin eagerly awaited Joseph's reactions to a manuscript of his ideas that represented the accumulation of decades of thinking, reading, and research.[5]

Darwin knew he was walking into difficult terrain. Many considered the fundamental job of science, and of the scientist, as identifying God's Great Design, as understanding and categorizing, through careful observation, the species God made when he first filled the World.[6] Linnaeus had famously argued that "There are as many species as the Infinite Being produced diverse forms in the beginning." Species were largely fixed at the point of Creation, and the business of the scientist was to identify and delimit them (though, to be fair, Linnaeus's commitment to the idea of immutable species did wane over time).[7]

But Darwin was mapping out a theory to explain the plasticity of species that William Herbert and others, like the entomologist Thomas Wollaston, had noted. He believed species changed over the generations as part of *natural selection,* a process with competition for resources at its heart. Darwin, having seen the criticisms lobbed at ecclesiastical men like Herbert, knew a theory that suggested species were unfixed could be ill received, and his theory of raging competition did not accord at all with the idea of a beneficent God. Wary of publishing too quickly, Darwin solicited waves of private feedback on his work.

Joseph wrote positively, and Darwin expressed an initial torrent of relief. "You would laugh, if you could know how much your note pleased me. I had *firmest* conviction that you would say all my M.S. was bosh," he wrote on June 8; "Farewell—your Note has relieved me **immensely**."[8]

But just ten days later, Darwin received a cataclysmic letter. That remarkable and famous document—though it has not survived—was to change the course of his career; it would also direct him on a frantic personal hunt for *Cattleya* orchids. To understand Darwin's eventual obsession with orchids and his contribution to a larger story of orchids and orchidomania, we must first follow a few chaotic weeks in his household's life and risk retelling a well-known tale in order to catch it in a little fresh light.

The June 18 letter from Alfred Russel Wallace enclosed "On the Tendency of Varieties to Depart Indefinitely From the Original Type." Here Wallace outlined a theory startlingly like Darwin's theory of natural selection. Darwin described it in a panicked, desperate missive to Charles Lyell, a Scottish geologist. Enclosing Wallace's paper, Darwin spluttered:

> Your words have come true with a vengeance that I shd. be forestalled. You said this when I explained to you here very briefly my view of "Natural Selection" depending on the Struggle for existence.—I never saw a more striking coincidence. If Wallace had my M.S. sketch written out in 1842 he could not have made a better short abstract! Even his terms now stand as Heads of my Chapters.

If it occurred to Darwin to squelch Wallace's note and rush out his own theory into print, he did not say so. He firmly stressed his willingness to help get Wallace's ideas published: "I shall of course at once write & offer to send to any Journal." But he was devastated: "So all my originality, whatever it may amount to, will be smashed," he mourned, before, in the next breath, pulling himself together: "I hope you will approve of Wallace's sketch, that I may tell him what you say."[9]

It was to be only the first catastrophe of the day. Even as Darwin struggled to react to the news and choose his course, a second, more intimate horror hit the house. Etty fell ill with diphtheria, one of many stricken that summer as a deadly epidemic swept the nation. Down House may have seemed remote, but the bacteria pushed its way in regardless.[10] The bacteria infect the patient's respiratory system, resulting in a leathery membrane that swells the throat, leading in severe cases to suffocation. Diphtheria kills up to 10 percent of those infected and up to 20 percent of infected children.

Children were already dying in the local village, Downe, and Etty was very sick: "there was no actual choking, but immense discharge & much pain & inability to speak or swallow & very weak & rapid pulse, with a fearful tongue," a shaken Darwin reported to Joseph a week later, on June 23. His thoughts were turned from his own work to the pain experienced by his child: "She will, I fear, be some time in getting her strength & will require constant attention. We are both rather knocked up & I have not spirits to see anyone." At least the situation seemed to be improving, Darwin concluded hopefully—"The Dr. gives very good Report."[11] But his relief was short-lived. Charles Waring, Etty's youngest brother, fell ill within hours.

The baby, nineteen months old, had been fragile since birth.[12] Charles seems to have been slow walking and talking, though Etty's later claim that he was born "without the full share of intelligence" is at odds with Darwin's own description of the child as "intelligent and observant."[13] Quiet and placid, the child could crawl, Darwin recalled, and had clear, strong likes, for example for their butler, Parslow: "it was very pretty to see his extreme eagerness with outstretched arms, to get to him." He enjoyed teasing his father: once taught not to scratch, for instance, he loved to pretend to be naughty with a quick grab and "a wicked little smile." He was a much-beloved child—and his parents must have been acutely conscious, as fever struck, that the boy was vulnerable. The Darwins had already lost one baby, Mary Eleanor, and a ten-year-old daughter, Annie.

Diphtheria was not even the only disease stalking the country that blistering June. Scarlet fever was also epidemic: often following on the heels of strep throat, the disease produces, in severe cases, inflammation and rashes so bad the skin on the body and tongue peel. When Annie Darwin succumbed in 1851 to scarlet fever, possibly exacerbated by underlying tuberculosis, her father was bereft: "We have lost the joy of the household," a devastated Darwin wrote on April 30 that year, recalling a "buoyant" child "radiant with the pleasure of giving pleasure": "she seemed formed to live a life of happiness."[14]

Charles Waring, too, would die of scarlet fever on June 28, 1858. Darwin wrote to Hooker: "I hope to God he did not suffer so much as he appeared." In a note on the child's last moments, he added that "the last 36 hours were miserable beyond expression."[15] Darwin was exhausted by grief and caring for sick children, and his letters to friends and colleagues are scattered, full of mistakes: Darwin mentioned documents he had received from Hooker, but added "I cannot think now on subject," and referred to correspondence from Lyell before concluding somewhat incoherently: "You shall hear soon as soon as I can think."[16]

But Darwin had to turn his attention back to his work—heat, drought, exhaustion, and grief notwithstanding. The letter that arrived, hours before Etty fell ill, could not be ignored. Writing to Lyell just as Etty was declared out of danger, the day his son lapsed into what would prove his last illness, Darwin was plunged in ever more roiling distress. Could he even think of publishing his own work, now that he'd read Wallace's ideas? Wouldn't everyone think he was plagiarizing Wallace, even if he wasn't? "I would far rather burn my whole book than that he or any man shd. think that I had behaved in a paltry spirit," he declared; "Do you not think his having sent me this sketch ties my hands?" At least he had written evidence from Hooker and others that he had developed the theory in the early 1840s. "But I cannot tell whether to publish now would not be base & paltry," he repeated, unable to think how to proceed: "This is a trumpery affair to trouble you with; but you cannot tell how much obliged I shd. be for your advice."[17]

A further letter on June 26 suggests that Darwin continued to worry at the problem with a compulsive anxiety as his son's health deteriorated. He wrote to Lyell once more "to make the case as strong as possible against myself." Wallace was still out of the country and "in the field," he noted, thus unable to prepare his own ideas for print. Surely then, Darwin would betray his very honor by proceeding with publication. "First impressions are generally right & I at first thought it wd. be dishonorable in me now to publish." In a postscript he added that their young son was hot with fever, "but we hope not S. fever," though he admitted that three other children with the disease had just died in the village, "with terrible suffering."[18]

Three days later, his son's life lost, Darwin was full of self-hatred in a letter to the younger Hooker: "It is miserable in me to care at all about priority."[19] He declared himself unable to think of the matter—yet he could not bring himself to give up on his theory or his claim to have originated it. He parceled up a copy of his work from 1844, with handwritten notes on it by Hooker, to remind his friend that he had, indeed, read them earlier, that Darwin's thoughts on the matter were well underway in the years after his return from the Galapagos.

Darwin's good friends, Lyell and Joseph Hooker, had an impossible situation on their hands. The careers of two men seemed to hang in the balance; Darwin was sunk in grief and fear, unable to give up on his life's work yet unable to see a way forward. Meanwhile the younger Wallace was fully deserving of praise and recognition—and not present in the country to defend his interests. Whom should they support? In the end, the two men evolved a simple plan, penning a letter that aimed to give space to *both* accounts, sent on Wednesday, June 30, 1858, to the secretary of the Linnean Society, J. J. Bennet. Included with the letter from Joseph Hooker and Charles Lyell to the society's secretary were excerpts from Darwin's work. Some of the documents dated to as early as 1839; some were more developed scholarship from 1844 and 1857. Also included was Wallace's "On the Tendency of Varieties."[20]

The *Times* thundered again that day about the appalling condition of the Thames, noting that while the members of Parliament—and the prime minister himself—were no doubt planning on retreating from the capital for the summer recess, Londoners themselves "are bees who cannot fly off." "In September the steady Peer, who is now 'bearing the stench with patience,' the paper continued bitterly, "will be imbibing the invigorating waters at a German spa, or following up his pointers over a stubble field. But we—we and the working classes of the metropolis—shall still be in the line of fire of the pestilence; we, in the fiery focus of the miasma."[21]

Hooker and Lyell carefully and conscientiously explained that Darwin and Wallace "having, independently and unknown to one another, conceived the same very ingenious theory to account of the appearance and perpetuation of varieties and of specific forms on our planet, may both fairly claim the merit of being original thinkers." Wallace's work, they stressed, deserved a wide readership, and so did Darwin's, yet the latter was threatening to keep his work from the world of science in deference to his colleague. To prevent this from happening, they concluded, and to give Wallace his due, the Linnean Society should hear the theories of both scholars simultaneously.

The next day, the two revolutionary theories were read aloud by the secretary—in what is now the Reynolds Room, in the Royal Academy.[22] Apparently, the rafters did not ring with astonished shouts and rapturous applause. Darwin later recalled ruefully: "our joint productions excited very little attention, and the only published notice of them which I can remember was by Professor Haughton of Dublin, whose verdict was that all that was new in them was false, and what was true was old."[23]

Few questions were asked, and when Thomas Bell, president of the society, later summarized the year's research, he said, dampeningly: "The year which has passed . . . has not, indeed, been marked by any of those striking discoveries which at once revolutionise, so to speak, the department of science on which they bear."[24] So much, it seemed, for Darwin's anxieties about honor and "priority."

But Darwin and his friends understood the significance of the work and knew it was time, past time, to publish. Still, the family remained in crisis, and Darwin's letters are an expression of the intense coexistence of intellectual engagement, professional anxiety, and urgent domestic need, each threatening to derail the pursuit of the other. The children's nurse caught scarlet fever next, and the Darwins embarked on a series of seaside trips on their doctor's advice to inhale the bracing salty air and escape the heat and rising death toll—six children, by now, in the village of Downe alone.[25]

Perhaps Darwin reflected on the implications of his own theory, in the days and weeks after the baby's death: life is an ongoing, bitter struggle. Encumbered by packing and all the tasks of moving a large, sick family, Darwin nonetheless wrenched his attention back to work, and by the third week of July he was sketching an "Abstract" which he planned to publish in the *Journal of the Linnean Society*. This would be a more expansive version of the material Lyell and Hooker had presented on his behalf. In the end, the "Abstract" grew over the summer and autumn, through the winter, and into the following spring until it was finally published, in November 1859, as the book-length *On the Origin of Species*.[26]

Wallace's letter prompted Darwin's famous publication; without Wallace, Darwin might have sat on his ideas even longer. But the arrival of Wallace's theory, at a time of personal distress, also pressed Darwin to publish before he felt fully ready, fully prepared to offer up his ideas to the world, and this too had consequences. The *Origin* was conceived of as an abstract—as, essentially, an introduction to the theory. "This Abstract, which I now publish, must necessarily be imperfect," Darwin opened his work, "No one can feel more sensible than I do of the necessity of hereafter publishing in detail all the facts, with references, on which my conclusions have been grounded; and I hope in a future work to do this."[27] Even though the text was the product of decades of intensive, detailed study, that is, there were aspects of natural selection that Darwin covered only briefly in the text, corroborating facts and research he omitted, as he focused on communicating, in accessible terms, his theory's arc.[28]

If the talk given by Lyell and Hooker on Wallace's and Darwin's work produced comparatively little buzz, the longer work was a different story. A book that claimed, effectively, to explain the mystery of Creation itself—that signaled this, in a title that boldly referenced the very *origin* of species—generated fierce and wide debate. Much of it was negative.

Darwin knew he would inspire religious anger. But many reviewers came after him on the grounds that his work was not appropriately "scientific." More than one scholar charged, specifically, that Darwin lacked granular evidence of how evolution worked—that he'd jumped too quickly to loose, broad conclusions instead of following the strict rules of the so-called inductive method. (This required, essentially, that a scientist should begin with detailed and specific real-world observations, proceeding only slowly and cautiously from those toward general truths.) Richard Owen, anatomist and superintendent of the natural history department at the British Museum, expressed criticism of natural selection in particularly biting terms, writing that it was "just one of those obvious possibilities that might float through the imagination of any speculative naturalist; only the sober searcher after truth would prefer a blameless silence to sending the proposition forth as explanatory of the origin of species, without its inductive foundations."[29] Darwin—somewhat given to self-flagellation—wrote to another of his critics: "What you hint at generally is very, very true: that my work will be grievously hypothetical, and large parts by no means worthy of being called induction, my commonest error being probably induction from too few facts."[30]

In the early years of the 1860s, then, Darwin was staring down the barrel of a new problem. His great idea was out before the world, but the world was not yet convinced. It needed more; a new narrative, with evidence. He needed to show the complexities of his theory in fresh, arresting detail. He had to show that species changed, slightly, subtly, but definitely through the generations, he needed to show exactly and precisely how and why this happened; and he needed, somehow, to make the world care.

It was around this time that Joseph Hooker helpfully suggested Darwin might like to contact the nursery of Mr. Benjamin Williams and his business partner, Mr. Robert Parker, in Holloway, or James Veitch's celebrated premises in Chelsea, and take a look at their orchids.[31]

16
"I MUCH WANT A CATTLEYEA"

VEITCH'S ESTABLISHMENT was now well known for its successful hybridists. Still, James Junior understood that the market for imported exotic orchids, "older" orchids, was strong and growing too, and he invested in sourcing plants from other countries to pique and satisfy consumers. When Darwin wrote to Veitch's in the first week of July 1861, he tapped into a network that offered access to species from all around the world.[1]

William and Thomas Lobb still collected for Veitch's and had by now brought home many storied plants, including another much-desired orchid, the "blue orchid," or *Vanda caerulea,* in 1850 from the Khasia Hills of Assam.[2] Veitch's supplemented the brothers' discoveries with those of many other plant hunters, including Jean Linden; Colonel Benson, a military man in Rangoon, Burma (Myanmar); the missionary, horticulturalist, writer, and historian Reverend William Ellis in Madagascar; a young gardener from Plymouth named Richard Pearce, who was brought in and trained by Thomas Lobb, then sent to South America in 1859; and the Newcastle-born George Ure-Skinner, a close friend of James Junior and former employee of James Bateman, who'd developed a successful plant business in Guatemala. Ure-Skinner's collections became so extensive that an entire Veitch hothouse was devoted to his plants.[3]

Hooker's second recommendation to Darwin, the nursery owned by Parker and Williams, was something of a dead end, however. The partnership was dissolved abruptly on April 8, 1861. A small insert on the matter in the advertisement pages of the *Gardeners' Chronicle*, placed on April 13, offers only a glimpse of what happened: the public was assured that it was all a matter of "mutual consent," that Parker would continue another "business elsewhere as soon as arrangements are made for the purpose," that Williams was henceforth operating the Paradise Nursery alone and was good for all debts.[4] It sounds suspiciously hurried: perhaps Williams edged Parker out after some sort of disagreement—certainly, the business seems to have been operating normally enough on April 6, when the two men were hawking seeds together in the pages of the paper. Williams was obviously an ambitious man and within eighteen months had taken over a second property, the Victoria Nursery, on Junction Road, Upper Holloway. Robert Parker, meanwhile, went south; by the end of August 1861, he was established at the Exotic Nursery in Tooting, Surrey, offering "Choice Flower Seeds." But he seems to have left the Paradise Nursery before his new position was established, which suggests that the two men could no longer, for whatever reason, do business together.

Hooker seems not to have known any of this. The letter from Darwin asking for exotics must have arrived at the Paradise Nursery in a time of transition, if not chaos. Williams was, that same month, promoting the production of a brand-new catalog in the *Gardeners' Chronicle* and encouraging visitors to come to his property by dangling, in his advertisement, a new variegated phalaenopsis with over a hundred blooms; he had, in other words, other things to think about when Darwin's letter arrived—getting a new business on its feet, encouraging the public to keep coming, finding a way to brand his company. Whether he even opened Darwin's letter is unclear, and by July 13 Darwin, in Torquay, was frustrated and impatient. "I much fear I shall get nothing," he fretted to Hooker, noting that no one had even replied to him yet. Still, he persisted: "I am got profoundly interested in Orchids." Indeed, he expressed himself positively fascinated: "I much want a Cattleyea," he explained; he always struggled to spell it. "Orchids have interested me more than almost anything in my life," he wrote to fellow orchid-lover John Lindley: "Your work shows that you carefully understand this feeling."[5] In fact, at times, he feared he could hardly *stop* himself from thinking about them: "I am convinced that orchids have a wicked power of witchcraft."[6]

It wasn't only *Cattleya* orchids he hoped to find: "I shd. be very glad of a Catlya or Epidendron" [emphasis Darwin's]. He also wanted *Arethusa*, or "dragon's mouth" orchids, and those of the strange *Catasetum*, which appeared to produce flowers of three separate genera. He tried requesting these in another letter to Parker and Williams. Perhaps that letter too got misplaced in the turmoil surrounding the dissolution of the partnership. It seems likely that at some point Parker dropped by Upper Holloway to pick up the mail, though, because Darwin would later note that he got some orchids (type unknown) from Parker.[7]

The source of Darwin's attraction to orchids is evident in a word he would attach to them in the title of his next book, *On the Various Contrivances by Which British and Foreign Orchids are Fertilised by Insects* (1862). "Contrivances" is an odd term, a sort of puzzle in itself; it speaks to the ways in which orchids and their mechanics of reproduction had long been an enigma to botanists and plant enthusiasts, who struggled to account for what Herbert called their "most confusive diversities." Orchids raised particularly significant issues for Darwin and his theory of natural selection because most orchids have male and female parts and thus seem able to self-fertilize. Yet Darwin's theory was founded on the idea that the differences conferred by outbreeding, or allogamy, are important for vigorous, successful plants (and animals). Strength and adaptation comes from nature's capacity to select the phenotypes that confer advantage; *selection* indicates that there is a choice of sorts, between the phenotypes of two different parents, and this seems hardly possible if a plant self-fertilizes. How could Darwin's theory of natural selection be squared, then, with an apparently self-fertilizing plant—one that, moreover, grows luxuriantly and successfully all around the world? Orchids are found in many climates, not just in the lush forests of Brazil; they live on every continent on the globe except Antarctica. What was their secret?

Orchids also thrive in English meadows, as Darwin knew. Near Down House, native orchids grew in profusion in a lovely place the family nicknamed *Orchis Bank:* it was described by Emma Darwin as a "grassy terrace under one of the shaws of old beeches ... Here grew bee, fly, musk, and butterfly orchises."[8] Darwin examined the plants while pondering the lessons of Herbert, he reported. "The Dean believes that single species of each genus were created in an originally highly plastic condition, and that these have been produced, chiefly by intercrossing, but likewise by variation, all our existing species," Darwin recalled in the third edition of *Origin of Species* (1860), and he gave Herbert credit for helping him develop his theory of the struggle for existence in the first edition: "In regard to plants, no one has treated this subject with more spirit and ability than W. Herbert."[9] So much for the theory; but how, exactly, was intercrossing achieved? Darwin pondered this, watching the orchids on his hands and knees, his family contributing. Darwin was convinced of the answer by 1861: insects.

Orchids, Darwin noticed, actually seem structured to *avoid* self-fertilization. They possess a rostellum which functions, as Jim Endersby delightfully puts it, as a sort of "floral chaperone" to keep the male and female parts of the plant separate. The mere presence of the rostellum seems, in other words, to support Darwin's hypothesis—the plant is literally structured to discourage self-fertilization and encourage allogamy or out-breeding. So how did fertilization between different plants occur? How could the male and female parts come together? What Darwin grasped, through examination of British orchids, was the extraordinary ways in which the plants lure insects in and then "smear" on to them masses of sticky, gluelike pollen. The insects depart, enter another orchid, and leave behind the pollen, producing fertilization.

Or actually—not quite: if the insect comes into contact with the male part of the second flower, after it has picked up pollen from the male part of the first, no fertilization can take place. The pollen has to meet up with the female part of the flower, and this is set a little further back on the plant, harder to access. Darwin noticed that the gluelike substance from the first plant contracts as it hardens and dries. This changes the position of the sticky pollen-mass on the back of the insect, bending it slightly—Darwin called it "depression"—so that by the time the carrier of the pollen arrives at its next destination, the pollen mass is angled afresh, ready to miss the flower's own pollinia and connect with the female stigma instead.

So much for British orchids; Darwin told Hooker, in June of 1861, "I understand pretty well all the British species."[10] But this was only a part of the puzzle. The structure of many exotic orchids is disconcertingly different to that of British plants. Some have pollen masses that seem virtually inaccessible to insects. Indeed some, Darwin wrote, seem absolutely and "determinately contrived [so] that the plant should never be fertilised."[11]

If Darwin was to shine a light on the structure and traits of these extraordinary plants, he needed to be able to explain the many different processes of fertilization. And the stakes seemed high. As he put it in his introduction to *Contrivances*, he had been roundly criticized for failing to give sufficient detail and evidence in *The Origin of Species:* "Having been blamed for propounding this doctrine without giving ample facts, for which I had not, in that work, sufficient space, I wish to show that I have not spoken without having gone into details."[12] How could Darwin study exotic plants in action from a diverse range of ecosystems without leaving the cold shores of Great Britain? How could he examine the structure of these exotics, in addition to the arrival and departure of insects from them, while managing his own poor health and caring for his very considerable, and ailing, family?

Working with specimens from collectors like Joseph Hooker at Kew, communicating with readers in the *Gardeners' Chronicle,* and negotiating with nurserymen like Parker and James Veitch helped. Darwin would later thank Parker, Veitch, and many other gardeners, collectors, and botanists in the work on orchids that ensued, for the plants they'd sent him, the resources they'd provided. Whole pages, indeed, would be dominated by descriptions of the collaborations that inspired and assisted him. By listing these familiar names, by calling out to ordinary gardeners and referencing the pages of middle-class gardening magazines, Darwin's narrative of orchids not only identified and explained their "contrivances," then, but effectively positioned the plants too, Devin Griffiths argues, in a web of contemporary enthusiasts stretching across "well-entrenched divisions between naturalist and layman, between merchant, enthusiast, and expert."[13] The book's narrative depends upon, and calls attention to, the kinds of collaborations, in an ever-broadening community, that structured and enabled orchidomania.

Such gestures likely provoked a sort of nod of recognition in the reader; *ah yes, Veitch*. The reader, anticipated to know many of the names mentioned in the text, is encouraged to feel a part of this community of readers and enthusiasts. It is surely not accidental that the plant Darwin chose to interest and convince his peers was one that had already captured the

"*I Much Want a Cattleyea*"

imagination of the era, one at the center of energetic, lively, growing debate. As Griffiths explains, Darwin knew he had to "recruit interest" from "experts and hobbyists who could furnish advice and materials for study," then generate excitement in the marketplace. To do this, "Darwin sought to capitalize on [a] broad interest in 'orchidomania.'"[14]

Veitch had certainly answered Darwin's letter by Sunday, August 11, 1861; Darwin triumphantly wrote that day, from Torquay, to his good friend Joseph Hooker: "Veitch has sent me a lot of magnificent Cattleyas all in full bloom."[15] Veitch probably sent the plants from Exeter rather than from Chelsea—the Devon premises, after all, were just twenty-two miles from Darwin's summer residence in Torquay. It was, if the *Times*'s weather forecast is to be believed, a fresh if cloudy weekend, and the plants seem to have arrived in excellent condition. Darwin lost no time in examining his haul. In his letters he expressed himself thrilled to expand his understanding of the species and test out his hypotheses on exotics; though in his later published work, he stressed just how carefully and methodically he proceeded, how reflection and theory preceded action.

"I was anxious to ascertain," he explained in his chapter on *Cattleya*, the first chapter of *Contrivances* after his analysis of British orchids:

> whether the exotic forms . . . equally required insect-agency. I especially wished to ascertain whether the rule holds good that each flower is necessarily fertilised by pollen brought from a distinct flower; and in a secondary degree I was curious to know whether the pollinia underwent those curious movements of depression by which they are placed, after transportal by insects, in the proper position to strike the stigmatic surface.[16]

He wanted to know, in other words, whether he could prove that the genus was structured to be fertilized by insects, thus facilitating outbreeding. He further wanted to know whether *Cattleya* orchids, like British ones, were shaped so as to angle the precious pollen of the first flower at the stigma of the second. Darwin remained palpably concerned about his ability to persuade future readers as he began his study of *Cattleya*, and he asked Hooker, the same day, for yet more exotic plants to help him develop and back up his claims. He would go on to solicit exotic orchids from naturalists and orchid enthusiasts throughout the rest of 1861 and into the spring of 1862. He also committed himself to yet more reading, more study: "could you lend me Lindleys great work on Orchids," he asked Hooker, that August afternoon, "to which he refers in Veg. Kingdom, as explaining nature of Parts; to my sorrow I fear I ought not to publish, without seeing what is known."[17]

Once the Veitch *Cattleya* orchids were finally on the table, Darwin's "experimental strategy" was, Griffiths observes, while "elegant in its simplicity . . . ground-breaking." That is: Darwin "would play the part of the insect, mimicking its behavior in order to fertilize the orchids himself."[18] The naturalist delicately inserted a dead bee into one of the *Cattleya* flowers. At this point, "the tongue-shaped rostellum [was] depressed, and the object [was] smeared with viscid matter." So far, so good. He prepared carefully to withdraw the insect, at which point "the tongue-formed rostellum [was] upturned, and a surprising quantity of viscid matter [was] forced over its edges and sides, and at the same time into the lip of the anther [the part that contains the pollen], which is slightly raised by the upturning of the rostellum. Thus the protruding tips of the caudicles [were] instantly glued to the retreating body." The plant had perfectly prepared the bee, in other words, to carry its precious pollen away and angle it at the next flower.

Darwin, by identifying many different remarkable "contrivances," gained crucial evidence to support his theory. The *Cattleya* orchids were different in appearance, differently structured from the British orchids he knew best—the rostellum was tongue-shaped, for a start, while the anther was kept shut "by a sort of spring at the back."[19] The plant had selected for different characteristics, but the end result was similar. Insects in the wild would, Darwin theorized, press down the *Cattleya* labellum as they landed; this action would cause the insect to leave the rostellum undisturbed until it had enjoyed its fill of nectar and was ready to leave, at which point the gluing-on process would begin. Darwin tried the procedure on multiple *Cattleya* specimens to be sure his experience with the first bee was no fluke: "This [action] hardly ever failed to occur in my repeated trials," he reported triumphantly.

Darwin had struggled to catch fertilization live, in action, at Down House. "I have been in the habit for twenty years of watching Orchids, and have never seen an insect visit a flower," he explained (this was why he had to act the part of pollinator himself). But he was able to point more confidently to insect action in *Cattleya* orchids when by good fortune a colleague forwarded to him a bee from a hothouse where a specimen was in full flower. The dead bee arrived "with its whole back, between the wings, smeared with dried viscid matter, and with the four pollinia attached to it by their caudicles, and ready to be caught by the stigma of another flower if the bee had entered one."[20] It was exactly what he needed; he described the discovery with delight to Hooker. "I have examined a Humble-Bee with the pollinia of Cattlyea attached to its back. Really," he continued, "the contrivances of Orchids beat, I think, any animal."[21]

Insects fertilized orchids, a remarkable contrivance of nature that had far-reaching implications.[22] Darwin was increasingly convinced that species not only evolved in response to natural selection acting, over the generations, on chance differences, but *co*-evolved; meaning that they selected for modifications, over the generations, that gave each better access to the other.[23] Now Darwin felt empowered to make an audacious prediction about another exotic orchid: in January 1862, James Bateman sent Darwin a box of plants including the "star" orchid, *Angraecum sequipedale*.[24] First identified by a European in the eighteenth century, William Ellis had recently imported new plants in 1857: still, though, a single plant cost £20 (equal to as much as £2,300 / $2,900 today). It was, by any measure, an extraordinary gift, yet the Madagascan plant seems at first blush to test Darwin's theory of the fertilization of orchids by insects to breaking point. It was exactly the kind of plant that a detractor of natural selection could thrust in his face, for how could any bee or moth gain access to pollen set at the very end of a "whip-like" nectary almost a foot long, then retreat from it unscathed yet with pollen attached? Darwin hypothesized that a moth *must* exist—as yet unseen, unknown, uncollected—that had evolved with a foot-long proboscis to enable it to access the nectar. The prediction was met with derision in some quarters, but in 1903, twenty years after Darwin's death, the moth was finally discovered—*Xanthopan morganii praedicta,* or "the predicted one." It was filmed in action, for the first time, in 1992. Yet Darwin was musing to Hooker about the moth's existence on paper almost as soon as the orchid arrived in his house, before he'd even had the chance to properly inspect it—"good Heavens what insect can suck it"?[25]

Much of what Darwin's orchid book narrates, indeed the story it tells, is about the orchid's marvelousness, the truly astonishing range and variety of their puzzling contrivances. Darwin's book was at once a crucial intervention in, and part of a longer tradition of, Victorian writing about orchids, then, which so often began by noting their "inexplicable phaenomena" (as Lindley put it). "In my examination of Orchids, hardly any fact has so much struck me as the endless diversities of structure,—the prodigality of resources,—for gaining the very same end, namely, the fertilisation of one flower by the pollen from another plant," Darwin declared, suggesting that his theory brought about a meaning much longed for: "The fact to a certain extent is intelligible on the principle of natural selection."[26] His book also shows how a research agenda may be undertaken by just about anyone. These remarkable plants could be inspected with very ordinary instruments, at home. His gestures to his many collaborators, and to the familiar names of mainstream nurseries who helped source his specimens, all helped to cement the orchid as not only available to "the million" but as a collectible that could allow the average enthusiast, on an ordinary table, to experience the very great thrill of being first.[27]

Darwin's engagement with orchids would subsequently emplot the story of the lost orchid in obvious and less obvious ways. His study drew its life force from orchidomania and helped orchids disperse ever more widely into public knowledge. And his book encouraged the further use of orchids, by amateurs, for scientific inquiry; afterward, indeed, authors explicitly encouraged the public to cultivate orchids so as to follow Darwin's footsteps, to crowdsource as a way of continuing to help refine knowledge of the natural world.

But Darwin's theorization of the slipperiness of species also spawned ever-greater debates about authenticity and truth. What were the larger implications of Darwin's theories for society and the world? Later Victorians would become increasingly shaken by fears of "degeneration," anxiously seeing evidence in their people, their capital city, their empire, their environment, even themselves of pervasive decline, a *reverse* of evolution. And where, in all this, was God? Was it chance, throwing out the many variations from which members of the species, in whatever their circumstances or environment, generation upon generation, selected, or a beneficent Creator? Was there potential for a *contriver* behind the *contrivances*?[28]

Darwin had been careful, in his 1859 *Origin of Species*, to avoid rejecting God, even though his faith had been on the wane for years. Some of his more probing theological statements are left open-ended, half-questioning, while others seem orthodox even as they carry a whiff of vagueness and uncertainty—"On the ordinary view of the independent creation of each being," he wrote, "we can only say that so it is;—that it has so pleased the Creator to construct each animal and plant."[29] Darwin cited Genesis as source material in the book while presenting his theory in spiritually infused terms—"There is grandeur in this view of life, with its several powers, having been originally breathed into a few forms or into one," he concluded. Many took this to mean that Darwin saw the breath of God as the essential, animating force.[30]

But towards the end of *Contrivances* he demands:

> Can we, in truth, feel satisfied by saying that each Orchid was created, exactly as we now see it, on a certain "ideal type;" that the Omnipotent Creator, having fixed on one plan for the whole Order, did not please to depart from this plan; that He, therefore, made the same organ to perform diverse functions—often of trifling importance compared with their proper function—converted other organs into mere purposeless rudiments, and arranged all as if they had to stand separate, and then made them cohere? Is it not a more simple and intelligible view that all Orchids owe what they have in common to descent from some monocotyledonous plant . . . ?[31]

Is it not more simple, he asks effectively, at the end of this passage, *more likely, really, that species are created through the processes of natural selection, than that God did all at once?* Its implications were radical; one has a sense, reading it, that Darwin could no longer keep his views quite under wraps. Indeed, orchids often appeared to provoke Darwin into expressions of religious doubt. When the duke of Argyll remarked twenty years later that orchids seemed to him, in their marvelous and exquisite detail, quite obviously "the effect and the expression of mind," Darwin allegedly replied: "'Well, that often comes over me with overwhelming force, but at other times . . . it seems to go away.'"[32]

Certainly, some saw in the intimate, mutually beneficial relationship of insects and plants signs of an all-powerful, all-wise "contriver" or Designer at work. To many, "the tight fit between orchid and insect," as Jim Endersby puts it, "each supplying the other's needs (for fertilization or nectar), proved that the Creator was benevolent."[33] But Darwin, deeply conscious that each orchid displays signs of earlier iterations, of traits and features the species has progressively come to de-select, was less sure. To him, the lack of what one might call foresight in orchid evolution was a significant problem for theists: natural selection, as scholar Philip Appleman neatly puts it, "functions not by design but by incessant opportunism."[34]

Darwin would later write, in a letter to American botanist and University of Michigan professor Asa Gray, that his orchid book was effectively a "'flank movement' on the enemy," a stealthy yet inexorable attack on the idea of God as a shaping, creating force. Darwin even called this, by 1863, "my chief interest" in the work.[35] Gray himself was a strongly religious man, a Presbyterian: he would ultimately pen a collection of essays arguing for theistic natural evolution, or the principle that natural selection is consistent with Christian belief. Still, when forced to confront the full implications of Darwin's work, Gray wrote, on July 7, 1863: "When you bring me up to this point, I feel the *cold chill*."[36]

Millions of Victorians would come to feel the cold chill of a godless universe after Darwin. As Matthew Arnold famously put it in "Dover Beach," reflecting on the "melancholy, long, withdrawing roar" of the Sea of Faith from Britain at midcentury:

> ... the world, which seems
> To lie before us like a land of dreams,
> So various, so beautiful, so new,
> Hath really neither joy, nor love, nor light,
> Nor certitude, nor peace, nor help for pain;
> And we are here as on a darkling plain
> Swept with confused alarms of struggle and flight,
> Where ignorant armies clash by night.[37]

PART IV

SURVIVAL OF THE FITTEST

[A]s more individuals are produced than can possibly survive,
there must in every case be a struggle for existence . . .
Although some species may be now increasing, more or less
  rapidly,
in numbers, all cannot do so, for the world would not hold them.

—CHARLES DARWIN, ON THE ORIGIN OF SPECIES (1859)

The growth of a large business is merely a survival of the fittest.

—JOHN D. ROCKEFELLER (C. 1900)

17
"THE WHOLE WORLD'S GONE CATTLEYA CRAZY"

THE "COLD CHILL" WAS just a breath as yet. For several decades, optimism surged, and orchidomania was borne along on a story of scientific possibility and increasingly open access to information and botanical wonders from around the world.

Select Orchidaceous Plants (1862–1865) by Robert Warner celebrated the age in ringing tones. (A businessman, Warner was, appropriately, head of a bell foundry.) This was an era, Warner cheered, of opportunity, plenty, availability. Orchids were beneficiaries of the moment: in the past, they were rare and expensive. Now orchids were plentiful and accessible to all. With every passing year, ever more, ever lovelier orchids were becoming available to consumers: Shirley Hibberd's *Garden Oracle* for 1865 comprised a list of orchids in bloom every day, stressing collectors could now enjoy year-round color.

Ferried on fast steamers (first with paddles, later with screw-propellers) and railways, the prices of plants were dropping ever further. Warner singled out *Cattleya mossiae* as a gorgeous, once-rare orchid available for as little as 5 shillings (around £29 / $37 today), and he gave detailed information on how to grow, propagate, and protect this and cattleya of all kinds. Insects and white scale he identified as particular problems; he recommended application of Chase's Beetle Poison. Yesterday's orchid problems were becoming just that—yesterday's.

> Fifty years of ever-broadening Commerce!
> Fifty years of ever-brightening Science!
> Fifty years of ever-widening Empire!

Tennyson cheered on Victoria's Golden Jubilee in 1887—that very same event for which Sander prepared his monstrous display of orchids celebrating the queen as Empress.[1] Orchidelirium was framed as a happy product if not poster child of fifty years of commercial, scientific, and imperial expansion. It was not just the British who loved their orchids, either. Periodicals around the world proclaimed that a passion for the plants was raging—among those influenced by British and European tastes and values, at least. In Asia, the English-speaking world was exhorted to try its hand at orchid growing by the *Indian Gardener*, a magazine published in Calcutta that was framed as an aid to horticulture for colonialists. The magazine warmly celebrated the "autumn-flowering Cattleya labiata" to its readers in 1886 and published numerous articles on how to cultivate orchids and how to hybridize, including from *C. labiata* (the average time for the seed to ripen, readers learned, was thirteen months). The root of "colony" is *colonus*, meaning "farmer"; to colonize was not only to control, to dominate, but to transfer population to the new territory and, through settling, intensify the subjugated country's allegiance to the controlling nation. Horticulture, the introduction of nonnative species and their cultivation to India, enacted and enforced the colonial project. Orchid cultivation was positively patriotic.

In the United States, too, the magazine *New Remedies* announced that an "Orchid-Craze" was raging in the 1880s. "Talk about the tulip mania!" it exclaimed; those bulbs sank "into something like insignificance compared with the passion for orchids developed in England" and now crossing the Atlantic.[2] In the decades after the Civil War, fascination for orchids began to spread—slowly at first, then more quickly—emerging out of an interest in gardening and horticulture that had been establishing itself across the nation for decades. European gardeners who arrived as immigrants brought with them European knowledge, training, tastes, and connections. Horticultural societies—built along the same lines as those in the United Kingdom and established in Pennsylvania, Massachusetts, and New York in 1827, 1829, and 1855 respectively—fostered knowledge of plants and gardening in the middle classes. The falling price of greenhouses also brought hothouse cultivation within reach of an ordinary enthusiast: one Cornell-based horticulturalist wrote, at the end of the century, "Even the humblest gardener, if he is thrifty, can afford a green-house."[3]

The growing passion for orchids in the United States was also, as elsewhere, fostered by books and periodicals like those of Warner and Williams, now supplemented by works aimed specifically at American enthusiasts. Edward Sprague Rand Jr.'s *Orchids: A Description of the Species and Varieties Grown at Glen Ridge, Near Boston* (1876) aspired to fill a gap in the US market. Rand wrote that he was inspired by British writers like Williams, even as he saw the need for a

> trustworthy manual of culture adapted to the United States. English publications on this subject are not to be relied on, as the climate is so different from our own that the rules they give are not applicable to us; they make no provision for the brightness of our sun, the heat of our summer, the dryness of our atmosphere, and the cold of our winter.[4]

Columns in magazines like the *American Garden* and the Philadelphia-published *Gardener's Monthly* also advised readers on how to get started with orchids, noting that "the number [of collectors] is increasing very fast in this country," though the *American Garden* encouraged amateurs to steer clear of expensive and scarce varieties like *Cattleya labiata*, at least at first, and start with "good, common sorts" instead.[5]

Orchidophiles in the United States often used quasi-democratic language to evoke orchids and their culture, working to frame orchidology as consistent with American ideals. The plants' awkward, lingering association with aristocratic hothouses and elite collectors fit ill with the young nation's self-conception. Many authors worked to "correct" readers' supposed misapprehensions and to repackage orchids as plants for and of the self-made man and woman. Readers were told that orchids were not, in fact, expensive—that they were "as cheap and cheaper than many other desirable plants." Cheap orchids were also, W. A. Manda of the Harvard Botanic Gardens opined in the *American Garden* in 1887, arguably the most beautiful, so that the nation's ordinary purchasers were truly able to enjoy the hobby's greatest charms. Moreover orchids were painted as, in their very growth habit, consonant with America's democratic sense of self. Their trick of lustily cohabiting with other species, even as transplants, was a sort of marvelous botanical re-enactment of the United States itself: as Manda put it, "Orchids are not a stuck-up class of plants that would not associate with others not belonging to the order. On the contrary, they are most cosmopolitan; they enjoy growing among other plants in our greenhouses."[6]

US authors were also keen to stress that orchids were more than mere decorative objects, more than fripperies for a leisured old elite. They were plants that required, fulfilled, and repaid modern, hard-working American citizens. Manda stressed that orchidology was a way for ordinary collectors to become a part of the era's scientific discoveries, for instance: "Darwin could never have made progress in working out his important conclusions" without the help of ordinary folk, he explained—"conclusions which, however, may need revision," for Darwin had an "unfortunate tendency" to "mere general conclusions and rash, hasty explanations of the causes of the everything." In other words, the average amateur could hope to show the famous Darwin a thing or too. And, Manda explained, orchids also sold well as cut flowers, from 25 cents to $1.50 wholesale, and "quite twice these amounts at retail," so that growers could hope to increase their household incomes by raising orchids. (The *American Garden* reported, in an article on floral fashions in February 1886, that orchids were hugely popular as cut flowers that year.)[7] New hybrids, produced by amateurs at home, were also starting to generate large sums, so that orchids could and should be understood as both a means of participating in the pushing back of the boundaries of knowledge *and* as good investments and sound means of increasing personal wealth. An article on Frederick Sander in the *Illustrated American,* entitled "A Fortune from Orchids," not only presented orchids as a sound investment, but also stressed that Sander himself was a smart immigrant, a model for the paper's readers. While British interviewers tended to skip over the nurseryman's early years in Prussia, the American piece opened with details about Sander's decision to move to a new country and the fortune he accrued in just a decade, with these later successes framed as the product of an energetic willingness to leave home and start a business with other smart, risk-taking immigrants (like Benedict Roezl).[8]

Rand, too, anticipated and sought to soothe readers' concerns that orchids were only beautiful objects for the gilded rich. He conceded in a chapter on the history of orchids in the United States that, since their arrival in the 1830s, exotic orchids had largely existed in the collections of a few wealthy specialists and were imported through the auspices of English nurseries (especially Low's and Veitch's). But his effort to reframe the orchid business in the United States also turned on the claim that they were becoming good investments for the ordinary purchaser, "their money value increasing every day." Because orchid plants live so long, Rand suggested, they were, as investments, approximately equivalent to investments in fine art. The *Gardener's Monthly* in 1877 similarly told its readers, in an editorial column, that orchids "grow in value with age" and that they are regularly sold in Europe, at the collector's death, for "more than he paid for them." This was not yet quite the case in the United States at the time, the columnist admitted, but he argued that as collectors found one another, as networks of enthusiasts expanded, this would change.[9]

American women too were advised to take up and learn about orchids. *Godey's Lady's Book* reported on the growing fascination with the plants in Europe as early as 1852, explaining they might easily be grown in the United States in conservatories. The *Lady's Friend*, published in Philadelphia, informed readers in 1865 that many orchids could thrive as "room plants," without extensive accoutrements, and needed little care.[10] "Orchids are at present attracting a vast amount of attention," explained Nancy Mann Waddel excitedly in a richly illustrated article, "Flower of the Air," in the *Ladies' Home Journal* in 1893, while Tennessee-born journalist Martha McCulloch-Williams wrote of "The Flower of Paradox" in *Godey's* in 1897, explaining the "craze" for orchids in the United States. She listed key elite American collectors, including Helen Gould, daughter of magnate (and robber baron) Jay Gould, yet stressed that "you can now buy a choice single bloom for a half dollar" (about $17 today or £14).[11] Anna Nichols Goodno, writing in the Quaker *Friends' Intelligencer and Journal* in 1890, remarked, "Perhaps no other flowers are so much talked about at present as the orchids."[12] Meanwhile, women who did not collect and grow orchids themselves—who did not have the time, space, money, or inclination for greenhouses and heating systems—could enjoy cut orchid blooms in table decorations. Sprays could be bought from florists for the same price as a rose; in 1891, *Ingalls' Home and Art Magazine* recommended the thrifty housewife to place cut orchids near a mirror, to double the effect. Other advice included placing orchids in

Godey's Magazine pitched the "wonders of orchid culture" to a US audience.

simple vases with water and a little moss, or, for a more elaborate display, on a bamboo foundation with moss and asparagus grass.

The number of US seedsmen and nurseries grew rapidly in the second half of the century. Prince's nursery in New York City, known for its fruits, was already well established, operating since the mid-eighteenth century. New companies specializing in orchids included Siebrecht & Wadley on Fifth Avenue, John Saul in Washington, DC, and Brackenridge & Co / Rosebank Nurseries in Baltimore Co. Maryland: their sales pitch was orchids "cheap as good roses." Young and Elliott, seedsmen and auctioneers, operated extensive premises on New York's Cortlandt Street and nearby Dey Street: Elliott, born in Scotland, had sailed to the United States in 1851 and set up business with I. J. Young three years later. The area of their operation was, by the 1880s, a bustling mecca for the horticulturally curious, home also to August Rölker & Sons, at 44 Dey Street, and Peter Henderson & Co., at 35–37 Cortlandt Street. W. A. Manda operated premises just outside New York, in Short Hills, New Jersey.

Nearby, in Summit, New Jersey, Sander opened his second branch in 1889. The St. Albans businessman was especially keen to capitalize on the

SPECIAL OFFER.
ESTABLISHED -:- ORCHIDS !
No Risk. No Loss.

SIEBRECHT & WADLEY offer the the following named ch ice Orchids, all established plants, either in pots, in cribs, or on block, and all flowering plants, in fine order and condition:

Acineta Barkeri.	Cypripedium Lawrenceanum.	Odontoglossum Bictonense.
Ada Aurantiaca.	" Ven stum.	" Cervantesii Majus.
Barke la Elegans.	Dendrobium Nobile.	" Insleayi.
Brassia Vericosa.	" Thyr Iflorum,	" Russii Majus.
Brassavola Glauca.	Epidendrum Fragrans.	" Schliperianum.
Catasetum assorted.	" Aurantiacum.	" Cordatum.
Cattleya Citrina.	" Vitellinum Majus.	Oncidium Incurvum.
Cattleya Mossiæ.	Laelia Albida.	" Luridum.
Chysis Aurea.	" Anceps.	" Ornithorynchium.
Chysis Bractescens.	" Autumnalis,	Sobralia Macrantha.
Cypripedium Barbatum.	Lycaste Aromatica.	Stanhopea Grandiflora.
" insigne,	" Skinneri.	" Tigrina.

12 fine plants of the above varieties, our select on, $10.00. 24 fine plants of the above varieties, our selection, $18.00
12 " " " " your " 12.50. 24 " " " " " your " 30.00
or the whole collection of 35 for $35.00. With the whole collection a premium plant of a rare variety will be sent.

Also Largest Stock of Decorative Plants in the Trade.
—ADDRESS—
SIEBRECHT & WADLEY, - NEW ROCHELLE, N. Y.
Or 409 5th Avenue, New York City.

"No Risk. No Loss": this advertisement in the *American Florist* for New York firm Siebrecht & Wadley presents orchids as a sound investment.

possibilities of a growing American market at a time when, as he wrote excitedly to one of his employees, "the whole world's mad on Cattleya."[13] The US branch of his orchid business opened to fanfare in the press: the new establishment boasted eight growing houses, each 150 feet long and twenty-one feet wide, and connected to a central show house. One of Sander's former collectors, J. Fösterman, was established as manager of the branch.

Why should American consumers choose Sander for their orchids, above other, already established firms? In an article reporting on the opening of the Summit branch, Sander was commended especially for his experience. The business of importing orchids, of bringing them across oceans, was fraught with difficulties, the author wrote, and "Comparatively few gardeners in the United States have sufficient experience to warrant success in taking charge of Orchids when in an enfeebled condition; and when plants are lost dissatisfaction must follow." In Sander's growing houses, exhausted plants would be ably protected and nurtured after their long journeys, so consumers could rest assured that they were getting value for their money. Sander's nursery was also described as a place where young men who wished to learn the business could go for training, and these would "supply the skilled labor which is now difficult to obtain in this country."[14] The opening of the nursery was described as a benefit to the American enthusiast, in other words, and as a means of

The vast greenhouses of Sander are made visible to "American Eyes" in the *American Florist* in 1895.

aiding the wider flourishing of horticulture, and orchidology, in the nation.

Sander's competitors in the United States were, by the time the New Jersey premises opened, already sending out collectors of their own in pursuit of botanical novelties. In fact, Sander's correspondence at Kew reveals that at least one American nursery was actively seeking the lost orchid in the early 1880s. Continental European, British, and American collectors were entering a power struggle for the position of leading provider of orchids in a global craze, and the acquisition of the rarest blooms was one of the most important ways firms tried to generate buzz, boost brand awareness, increase sales, and ultimately dominate the market.[15]

18

FAKES

ON THE ONE HAND, the final decades of the century saw the rise of the orchid obsession on new continents; on the other, the queen of the *Cattleya* genus, the most beautiful of all orchids, was apparently missing, lost. All efforts to recover it so far had been unsuccessful. Even Gardner's apparent successes were judged, decades after the fact, a failure. The area around Rio de Janeiro in the Organ Mountains was, according to James O'Brien of the RHS, the site where Swainson found the lost orchid, but efforts in the region by hunters like Gardner, Lobb, and Digance had turned up nothing. Many specialists now asserted that the second *Cattleya labiata* Gardner found, from Sapucaya, was not *C. labiata* at all but *C. warneri*. Almost every serious enthusiast seems to have been convinced, by 1880, that the *C. labiata* found by William Swainson, cultivated in Suffolk by William Hooker, was lost forever.

Frederick Boyle, publishing in *Longman's,* presented the absence as a terrible, ghostly loss. "All importers are haunted by the spectral image of Cattleya labiata," he mourned, "which, in its true form, has been brought to Europe only once, forty years ago." That haunting sense of loss unfurled, too, in Harry Veitch's evocation of it in the *Manual* as unlikely to be recovered, ever, because of logging of the timber on which so many orchids twined.[1]

Robert Warner, in an 1882 book co-authored with Benjamin Williams, also argued that not a single genuine *C. labiata* had been recovered in decades: "it is this that makes it so scarce, and causes it to fetch the high prices that it does." The existing plants in the nation were mostly the products of division of Swainson's first plant, they explained, so that the old problems of access and high prices, which were supposed to be resolved by more hunting, more propagation, more transportation, and hybridization, had endured the passing of the years. "It is greatly to be regretted," they lament, "that our collectors do not again discover its habitat." W. H. Gower agreed, grieving in *The Garden*, "This fine species of Lindley's, upon which the genus was founded, was introduced seventy years ago, and it is said that no traveller has ever been fortunate enough to find it since . . . there are none which flower at this particular season to approach it." In the United States, *Garden and Forest* agreed that this most desirable cattleya was both the subject of intensive search and "now believed to be extinct."[2] Yet plants called *C. labiata* sold regularly, even quite cheaply, on both sides of the Atlantic.

Sprague's 1876 manual on orchid culture put *C. labiata* on its "List of Thirty Cheap Orchids for General Cultivation."[3] In 1887, *L'orchidophile: Journal des amateurs d'orchidées* reported that a recent selection of orchids at Veitch's seemed at first to include *C. skinneri*, but on closer examination this was, in their opinion, *C. autumnalis*. In the same issue, a report on a Massachusetts collection referenced an orchid as *"C. labiata vrai"* or "true"—meaning, apparently, Swainson's plant.[4]

How are we to understand the apparent presence of the lost orchid in collections and sales around the world in the years *before* Claes Ericsson's frantic journey to Pernambuco, hot on the heels of the mysterious Bungeroth? As we prepare to resume the hunt once more—to follow the Swedish plant hunter again on his plunge into South America on behalf of Frederick Sander—it seems worth asking: if the "lost orchid" was *not* lost, if it could be purchased, why were men sent in its pursuit?

It seems likely that many of the orchids described and sold as *C. labiata*, the type species, were actually something else. Certainly William Gower believed "the plants I have frequently seen lately in collections are either not true or they are inferior varieties."[5] A language of truth and veracity became more and more pronounced in the later years of the century—the hunt was on for *C. labiata vera*, the "true" lost orchid—but this expressed a *desire* for fixedness and clarity in the face of baffling opacity, of seemingly endless questions after Darwin and Alfred Wallace. Writers and enthusiasts remained deeply unclear about what was a variety or subvariety of what. There are dozens of what is today termed "the Labiata Group of cattleyas" (or the "Cattleya labiata group") and debates about species as a category endure.[6] In the mid-nineteenth century, some argued that varieties and species were fundamentally different—that varieties were unstable and would "in a state of domesticity" tend to revert to the "normal form of the parent species." Wallace explicitly refuted the claim of the stability of species, arguing that what happens in nature and what happens in cultivation are not at all the same thing, that the "struggle for existence" has a profound effect on organisms, that "natural means of selection" are constantly "[perpetuating] varieties" whose differences will, over time, result in new species. Darwin pointed out in his chapter on hybridism in *The Origin of Species* that there are plenty of examples of fertile offspring produced by interbreeding between different species, of sterile offspring produced from breeding between members of different varieties of the same species, so that fertility and sterility can't be taken confidently as markers of difference either: "neither sterility nor fertility," he proclaimed, "affords a clear distinction between species and varieties"; "there is no fundamental distinction."[7] Indeed, he privately doubted in a letter to Joseph Hooker whether "species" was a category of meaning at all:

> It is really laughable to see what different ideas are prominent in various naturalists minds, when they speak of "species" in some, resemblance is everything & descent of little weight—in some resemblance seems to go for nothing, and Creation the reigning idea—in some, descent the key—in some sterility an unfailing test, with others it is not worth a farthing. It all comes from trying to define the indefinable.[8]

Perhaps the very notion of species is little more than a feature of our desire to categorize, to define, he (and others) suggested; if what we see in nature is produced by processes unfolding generations ago, there will be some changes we can see, but many we can't. "The patterns of similarity that we recognize are the remnants of former evolutionary groups that might have long since shifted and splintered," explains evolutionary biologist Jody Hey.[9]

Some collectors and nurserymen seem to have cheerfully embraced the uncertainty, the slipperiness of species, to upsell their orchids. The labels "Cattleya labiata" and "Cattleya labiata vera" were applied to orchids as a sort of gesture, an indication of likeness. To say that an orchid was "Cattleya labiata" signified that the plant in question was a large, likely purple-to-pink *Cattleya* (or *Epidendrum*) from somewhere in South America that was judged as possessing a resemblance—in color, shape, and size—to what the owner or vendor knew of *C. labiata*. It did not mean that the plant was from the same population or descended from what Swainson had collected in 1818.

Victorians were far from unconscious of the problems of fraud and puffery in the marketplace; it was a much-discussed, much-debated issue in an era of emporia, bustling bazaars, department stores, and conspicuous consumption enabled by and embodied in all-caps, blaring advertisements. Many saw predatory, duplicitous practices at work, aimed squarely at the weak and vulnerable, but robust consumer protection laws lay in the future. For the time being, a commitment to free trade and low, low prices meant that there was little political will for enacting laws that might protect unwary consumers but could also drive up the cost of goods.[10]

In an intensely competitive marketplace, businessmen skated the edges of the truth, exploiting naming and identity confusions in order to pitch their plants as more desirable species and varieties. Sander would presently prove himself a master at this game. But he was also as confused as his predecessors. For many years, Sander expressed himself, in private letters, frustrated by the challenges of identifying this particular orchid.[11] The blooming season, color, shape, and size vary considerably; questions were also raised by the coloring in Lindley's illustrations from the 1820s and 1830s. Were the hues of deep purple, say, in Drake's illustration a feature of an inexact coloring process, or did they reflect the actual colors of the orchid?[12] Lou Menezes, Brazilian forestry engineer and orchidologist, notes that to this day, Lindley's coloring from the type plant, the orchid on which he established the genus, remains "practically unknown to Brazilian collectors of the species."[13] Moreover, the types of men who were willing to risk life and limb in the later nineteenth century, to plunder the environment for a little money and a chance at fame, were often ill educated; few had received the kinds of education enjoyed by men like Hooker and Lindley, still less Joseph Banks. The finer points of an individual specimen were hard for them to grasp, and this led them to grab plants *like* the ones they were seeking with abandon. Before Ericsson reached Pernambuco in hot pursuit of Bungeroth and the lost orchid, waves of hunters plowed across South America tearing at orchids, hacking at the trees on which they grew, sending hundreds of ill-packed cases of jumbled and quite different plants to London and New York. Sander and his peers named the plants and marketed whatever they thought would sell, then sent the hunters back for more, often armed with illustrations cut from botanical magazines.

The lost orchid was pursued through the 1880s with increasing, chaotic urgency.

FREDERICK SANDER'S VOLUMINOUS correspondence, preserved in manila files and boxes in the archives at the Royal Botanic Gardens, Kew, comprises thousands of hunters' and agents' letters, memos, and telegrams, together with several hundred documents by Sander. These attest to how plant hunting unfolded on the ground (and along the river, up in the trees) over decades. The hunters' writings, sent to St. Albans from all around the world, are not easy to read—written in multiple languages, many are on tissue-thin paper, with poor spelling and grammar and often indescribably bad handwriting. Sander's own letters are scrawled, colloquial, often in German—and there are fewer of them, for the obvious reason that, once received, they were generally lost or thrown away abroad.

Sander, having opened his St. Albans premises in 1876, had by 1880 at least two hunters seeking, among other plants, the lost orchid in South America. These were German-speakers J. D. Osmers in Brazil and twenty-year-old Fritz Arnold in Venezuela and Colombia. Both men's letters to Sander are among those preserved, while Arnold's correspondence *from* Sander constitutes the bulk of the Sander-authored material that has survived. From these, we learn just how frustrated and confused Sander was about the limits of species and varieties in the early 1880s, and how much the uncertainties about what *C. labiata* was, and where it was, galvanized him to send hunters in its pursuit. In German, for the most part, and in letters that conformed to few of the epistolary conventions that, say, Hooker, Lindley, and Garner struggled to master, Sander issued his private orders in rough, peremptory, sometimes brutal fashion.

"Cattleya labiata," Sander wrote to Arnold in December 1880, was a prize he had been pursuing "for quite a long time." "Now to the main thing and seriously and pay attention," he exhorted his young agent with urgent underlining, outlining Arnold's new mission by describing the orchid's color, history, and growing season. Sander also stressed the challenges of distinguishing the lost orchid from other, similar plants. The treacherously similar *Cattleya warneri* blooms in June, he wrote—"thus for God's sake, caution"; then there were rose and white cattleyas to worry about—"You must see that the C. labiata are indeed genuine," he scrawled, "they must not be Mossiae." Another similar South American cattleya had recently been named, by rival nursery Low's of Clapton, *speciosissima*.

Sander's frustration was built, it seemed, partly on recent errors. In early October, Osmers had sent seven crates of what he believed were old *C. labiata* from along the São Francisco River outside Rio. It had been an exhausting, expensive foray, Osmers wrote to explain, taking six and a half weeks. Some plants were damaged when the trees were felled, while others were mishandled on their journey to Rio. Sander was nonetheless extremely hopeful, and he wrote to Arnold in November, "I now know you will not find the Cattleya labiata in Caracas, she comes from Brazil."[14] He directed Arnold to turn his attention to finding white *C. mossiae* and *Masdevallia tovarensis* instead. (A collection of the latter froze in the Stevens auction rooms the night before their sale.)

But by December, Sander was demanding Arnold pick up the hunt again. The Brazilian plants were not what he sought: Osmer had not seen the orchids in bloom before he packaged them. "I do not know, for I am quite confused about it," Sander fretted on December 29, 1880; the businessman seems to have determined, in the later months of 1880, that the lost orchid could at least potentially be recovered from somewhere other than where Swainson originally located it.

Finding the orchid's habitat was far from the only problem confronting Sander and his team. The Sander correspondence suggests that multiple companies were pursuing the lost orchid with near-equal intensity at this point. Low's had two hunters in the area, first a man named Smith, who searched for a year, then a hunter named White, who arrived at the same time as, and seemed to be shadowing, Arnold. Sander suspected Low's was actively pursuing *C. labiata,* having learned that White traveled out on the same ship as Arnold—the *Moselle,* a steamship that deployed a propeller or "screw" rather than a paddle for faster, more efficient service.[15]

"[I]t is possible that White has also been sent for this Cattleya... Naturally I don't want White to get ahead of us," Sander wrote. To keep the mission secret, he told Arnold never to say in White's hearing what he sought: "you must never mention the name labiata, altogether act as if you don't know that name." To see off White, to win the prize, "you must be very smart and see what he is up to," Sander continued, urging Arnold to trail White as White was trailing him—"Do not lose the plant! And take your time, for if White will become aware what you are after, he will never leave Caracas." Buy a map as quickly as possible, Sander warned Arnold, watch out for illness, send back any and all information on Smith and White, be "sly as a fox"—and above all, he added with Germanic humor, "don't be scared, God will not abandon a Bavarian, especially if he wears such white trousers [*weisse Hosen*] as you do."

The competition with Low's evidently added extra spice to the hunt, at least as far as Sander was concerned: "It makes me very excited and gives me such pleasure for I love fencing very much."[16] The business of being "sly as a fox" was one Arnold had to learn, as Ericsson and Digance would learn after him: it seems to have required following the enemy (while trying to throw the enemy off one's own trail), bribing local people for information, putting fake stories around about what one had found, figuring out what plants the competitor had sent, then finding those plants and sending cases of them to one's employer as quickly as possible, shipping plants under fake names so that the enemy would not be able to do the same thing in return, all the while keeping the truth about what was being sent from everyone involved at the shipping company. Sander's biographer, Arthur Swinson, reports that hunters also sometimes even urinated on rivals' plants to damage them.[17]

Sander's warnings to Arnold were not enough. In spite of the increased speed of steamships, letters still took about a month to reach their destination. (Arnold's letters were annotated, presumably by local Colombian and Venezuelan mail officials, to indicate the date of receipt; comparison of this with the date of writing confirms the speed of travel.) This made it hard for Sander to get information quickly to South America until cable use became more widespread later in the decade.[18] Sander's most urgent warnings to Arnold about White, at the start of the trip, arrived too late; the young German was apparently tricked during the steamer voyage. The Low's agent befriended Sander's employee and, pretending to work for a Birmingham manufacturer, affected cheerful ignorance of orchids while they were journeying on the *Moselle*. When Arnold realized the truth, he locked White in a room and threatened him with a revolver, proclaiming "Either you or me must not leave this cabin alive"; or so, at least, claimed orchid fancier Frederick Boyle. In *About Orchids*, Boyle reported that a

"HE DREW FROM HIS POCKET A REVOLVER."

"Checkmate! The Romance of an Orchid." Illustrations accompanying this highly fictionalized account foreground Arnold's mastery over his rivals in *Wide World Magazine*.

startled White hid under the table and "discussed terms of abject surrender from that retreat"—though since Boyle was an unabashed groupie of Sander's, this story was almost certainly exaggerated. (A quarter of a century later, Sander renarrated Arnold's tricking and the scene with the revolver when writing of the hunt for a different orchid entirely, in the thrillingly illustrated "Checkmate! The Romance of an Orchid.")[19]

For months, Arnold's search remained unsuccessful, and Sander was impatient. "Afraid it is a hopeless job," Arnold told him, in spite of the support of a second hunter operating in the region, A. Ernst, whom Sander sent to support him. (Sander noted that Ernst was unusually well-educated and actually capable of identifying species.) Then in November 1880, Arnold reported progress—"At last, on the track of the Cattleya," he proclaimed, and Ernst wrote to Sander, on the sixteenth of that month, "Arnold believes he has found the old C. Labiata here in bloom and asked me to send you the enclosed flower." Evidently anticipating Sander's next, inevitable concern, Ernst continued: "White knows nothing of this, and I shall say nothing to him about it."[20]

A 1915 article about Arnold reported the young German recovered specimens eight feet in diameter on the trip, with 1,000 bulbs and 380 flowers, growing at an altitude of 6,000 feet. Certainly Sander, the letters indicate, was thrilled—"If the Cattleya is the old labiata, then you will hear more from me than you think," he wrote.[21] He sent the specimens to the great German orchidologist Heinrich Reichenbach, who was—since the deaths of Lindley and Hooker in 1865—the world's leading expert on orchids. Reichenbach would deliver his opinion in the *Gardeners' Chronicle* on June 17, 1882. The Arnold orchid was not *Cattleya labiata vera*, he decided, but a new variety to be named "percivaliana": "If a very young botanist had time and means to study a fine question, he might do grand work by travelling only for the purpose of studying Cattleya labiata," Reichenbach wrote thoughtfully. "Our plant is from a totally new place, yet it appears exceedingly near the oldest Cattleya of Lindley (*Collectanea*), and of Loddiges (*Botanical Cabinet*)."[22]

Meanwhile, the oldest cattleya—"*die alte labiata*," as Sander and his men called it in their letters—was in the sights of at least one other up-and-coming company as well.

Sander, who was often violently angry with the hunters, seems to have experienced fellow feeling with Arnold, at least some of the time—"we have to lead the Bavarians into battle" against Hugh Low, he wrote, and several times he commended Arnold's intelligence. But he could also be brutal: "If I'd got you here I would beat you until not a single patch of your skin was left unbruised," he wrote, complaining that his young employee had inadequately packed *Cattleya* bulbs. Arnold was certainly inexperienced; he was also, his letters suggest, desperately keen to please. He sent back immensely long, tightly written, often crossed epistles with catalogs of plants' names childishly underlined and decorated with curlicues.[23]

Perhaps Sander's roughness and miserliness, which would enrage so many of the other collectors, explains why Arnold came to send ten cases of potential *C. labiata* orchids, in the last days of 1880, not to Sander but to the New York firm, Young and Elliott.

Sander was deeply agitated about the progress of Young and Elliott in the region and wrote almost as much about them to Arnold as he did about Low's. The businessman would be reduced to near-apoplexy in a memo penned December 29. After forty years of searching, Sander spluttered in disbelief, "*40 Jahren suchen*," Arnold had given away the secret of a nation, the holy of holies, the grail. Sander could hardly even put what had happened into words, he raged, in case his letter was found, the truth revealed. Soon the American garden magazines would know what had happened. Soon Veitch's and Williams's travelers would find out, too—Sander suggests both companies had hunters in pursuit of *C. labiata* at the time. And all this because of Arnold! He had thrown away hundreds of pounds' worth of plants when he, Sander, would have paid a fortune for the smallest consignment of true labiata orchids![24] At a time, too, when "*Alle wellt ist Cattleyen verückt geht*"—the whole world's gone Cattleya crazy![25]

Either Arnold was fed up with Sander's berating or he took a bribe—or he made the mistake of telling too much to the shippers, who let crucial information slip to Young and Elliott's agent, who then figured out how to reroute a quarter of the shipment. Or it was simply a mistake—the archive reveals that mistakes were often made, that other crates went astray, and these could take weeks to recover.[26] Whatever happened, thirty cases of *Cattleya* were headed to St. Albans, but ten were steaming merrily to the United States—a terrible event, as far as Sander was concerned. "We have to cultivate the thing for 1 year in order to see if it is genuine," he fumed, and now Young and Elliott would be able to steal a march on them if it was indeed the lost orchid.[27] What happened when Young and Elliott received the crates is unclear; perhaps the plants were, as Sander expressly hoped, frozen by the time they docked in New York (it was, after all, January). Even if they arrived unscathed, they were almost certainly not *Cattleya labiata*, the "true" old plant, at least if these were the same orchids Reichenbach had named *C. percivaliana* in the *Gardeners' Chronicle*.

Fritz Arnold would continue searching in Sander's employ for several years, latterly with a colleague named Charles Palmer. He reached what a local banker called, in a letter to Sander, an "untimely end" in 1886, aged twenty-five, afflicted by a gastrointestinal disease near Maipures on a search along the Orinoco River. Palmer was evidently much affected by Arnold's loss, suggesting that, in the midst of the madness, real affections could still form.

A document, in Palmer's handwriting, describing Arnold's final hours, has survived at Kew. Dated "Port of Spain May 20, Trinidad," it offers a shaky, fragmented reconstruction of what happened, somewhere between an account and a list of facts and times. The young Bavarian, suffering a stomach ailment, tried pills, bananas, oranges with sugar. Matters took a decisive turn at 9 o'clock one evening; the next time listed—presumably of death—was the following day, at 5–6 o'clock. Arnold was buried in the San Fernando do Atapao churchyard. "*Totenschein*," or Death Certificate, is heavily underscored, possibly indicating that Palmer still needed to figure out the paperwork. Arnold's possessions included a revolver, suitcase, watch, papers, keys, money, bill. On the back of the paper, Palmer began to draft an awkward letter to Arnold's father (evidently Arnold, at least, retained connections to his family), mourning the "irreplaceable loss" of a young man with so much to live for, "the death of whom leaves a painful and hard-to-fill emptiness" behind.[28] The manager of the Colonial Bank in Trinidad, W. F. Kirton, tasked with advancing cash to Sander employees abroad, also described the "unfortunate matter" in a dolorous, black-edged letter to Sander—bursting out, in a less formal postscript, that Arnold could only have saved himself if he'd known a quick fix for such ailments, namely to drink hot water "as hot as it can be taken into the stomach."[29] Sander's reaction to Arnold's end was not recorded, though he was perfectly willing to renarrate deaths in more thrilling terms for the media. In the later "Checkmate! The Romance of an Orchid," for instance, he claimed Arnold was found mysteriously dead "in an open boat": "the cause of his death was never definitely known." (He once told an interviewer, in Kiplingesque form, that another of his agents was burned to death by priests for blasphemy.)[30]

Chaotic failures and loss of life did not affect the intensity of the hunt, whose progress we can also chart in the pages of the periodical press. Low's advertised *Cattleya labiata* for sale in the *Gardener's Monthly and Horticultural Advertiser* along with "unprecedented" numbers of other orchids in 1886—housed under a staggering (and proudly counted) 274,610 feet of glass. "The Stock is of such magnitude that, without seeing it, it is not easy to form an adequate conception of its unprecedented extent," the ad explained.[31] "Orchids everywhere!" was the cry at Sander's St Albans' premises, according to a round-eyed Frederick Boyle, while Brackenridge in Baltimore blared "Stock Immense," Siebrecht & Wadley promised the "largest collection of orchids and palms in America," and the ads of John Cowan, outside Liverpool, screamed "Orchids! Orchids!"

ORCHIDS! ORCHIDS! ORCHIDS! ORCHIDS!
ORCHIDS! ORCHIDS! ORCHIDS! ORCHIDS!
ORCHIDS!
ORCHIDS!   **M. GODEFROY LEBEUF,**
ORCHIDS!   Horticulturist,
ORCHIDS!   Argenteuil, Seine-et-Oise,
ORCHIDS!
ORCHIDS!   FRANCE,
ORCHIDS!   WISHES to purchase living samples of the
ORCHIDS!   most ornamental exotic ORCHIDS. He of-
ORCHIDS!   fers in exchange either French Vegetables, Trees,
               Rose-trees, Ornamental Plants, etc., or such other
ORCHIDS!   specimens as he may be able to procure.
ORCHIDS!
ORCHIDS! ORCHIDS! ORCHIDS! ORCHIDS!
ORCHIDS! ORCHIDS! ORCHIDS! ORCHIDS!

An advertisement for Godefroy Lebeuf screams "Orchids!" twenty-six times.

to consumers. Godefroy Lebeuf's advertisements, from Seine-et-Oise in France, repeated the word "Orchids!" twenty-six times, in All Caps.[32] William Bull imported, in 1878, a total of two million orchids in just two consignments.[33]

The dream of orchidomania, borne along on a tidal wave of globalizing energy, industrial growth, colonial advance, and capitalist enterprise, was that dearest wishes could be fulfilled, by anyone, at last—but it was *not* the case that every orchid could be had just for the wanting. The middle classes, average enthusiasts, had long been told that orchids were accessible to them, that the exciting possibilities of the modern age were fully available to all, the "million"; yet the stark reality, embodied in the "lost" orchid, was that some dreams remained stubbornly out of reach. By the time Warner published his next book-length work, the 1882 *Orchid Album* (in collaboration with the "Million's" Benjamin Williams), a chill was starting to creep into even the most frenzied British celebrations of the orchid craze.

The delight in new varieties, new species, new opportunities was still there, but it was tempered by the dismayed understanding that not every hope had been fulfilled. The lost orchid had not been recovered, the *Orchid Album* agreed. "True" examples of *Cattleya labiata* remained expensive and rare. No collectors had been able to find more. It was excessively beautiful, yes—but its true beauty could only be appreciated by those lucky enough to own one. Its loveliness thus remained elusive, a point many other authors would make, with metaphorical sighs, in the last decades of the century.

Was it truly lost—beyond the grasp of trains, steamships, cables, print media and all the rest of the complex infrastructure of enterprising modernity—forever?

19
DAMAGE

IN FACT, THE ORCHID was potentially lost *because* of modernity, Harry Veitch suggested in the company's *Manual*. The vast Brazilian landscape continued to be subjected daily to ever more intensive and sustained assault, that very assault burning up the kinds of trees, the canopy on which orchids like *Cattleya labiata* throve.

It had been central to the colonial view of Brazil that the vast forests of the nation were, effectively, limitless—that the remarkable array of resources on offer were not only so diverse but so extensive they could be extracted without significant cost. In the nineteenth century, in both the Empire and the Republic—which would be proclaimed in November 1889, when lackluster Emperor Pedro II was sent into exile—the botanical richness of Brazil remained a focal point for the creation of a stronger sense of nationhood in a still-young country. Brazil's natural resources seemed particularly valuable, particularly worthy of celebration in an era when interest in botany was spreading around the world. "The production of images of the empire adorned with plants and tropical products was common," José Augusto Pádua argues; "Museums sought to emphasize the diversity of the country's fauna and flora, disseminating the figure of spectacular, picturesque nature that held universal scientific value."[1]

Yet it wasn't hard for Brazilian communities, or outside journalists and writers, to see that in spite of the extensive natural resources with which the country had been blessed, human activities were causing damage. Tree burning was a common, cheap means of clearing vast swathes of land while fertilizing the soil to prepare it for agriculture. The production of coffee and sugar involved dynamic change to and destruction of the environment, as complex ecosystems were cleared in favor of monocultures; soil erosion was a serious problem, so severe that plantations were constantly abandoned and new ones created, setting the same destructive cycle in motion again—a practice Swainson noted in 1818. Gold was intensively mined, and increasingly by foreign companies, who deployed modern industrial equipment, including gunpowder and underground drilling rigs. Elaborate networks of roads, railways, steamships, and telegraphs crisscrossed and looped around South America, facilitating the movement of people, animals, plants and other export goods across nations, states, communities, rivers, and oceans, while the forests were rifled for timber, rubber, palm, animal pelts, feathers, Brazil nuts, and more.[2] All of this extraction palpably left its mark on the environment—and on the nation's economy, too. In 1876, Henry Wickham took 70,000 rubber tree seeds out of the rain forest and shipped them to England, then on to Malaysia and Sri Lanka, ultimately allowing Britain in the early twentieth century to flood the world rubber market with a cheaper product. The formerly thriving Brazilian rubber market collapsed.

When Dodman celebrated the age of "orchids for the million," back in 1850, he noted with a shade of unease that collectors "in all quarters" were "ransacking the forests" for the auction rooms; that a friend of his from Penang had observed, "Our jungles are nearly stripped of all the Orchidaceous plants, such has been the demand for them of late."[3] The unease did not lead to a change in approach. Nearly thirty years later, in 1878, the director of the botanic gardens in Zurich noted, with rather more emphasis: "These modern collectors spare nothing. This is no longer collecting; it is wanton robbery, and I wonder that public opinion is not stronger against it."[4] The practice continued, though the use of verbs like "ransacking" and "stripping" pointed to growing awareness of the vandalism in the act.

Public opinion did not demand the cessation of the extractive approach. But increasing numbers of people began to write about it, to argue that the other side of the superabundance of orchids in Europe and North America was their increasing absence from their habitats. The cry of alarm grew louder as the end of the century approached. Writers of plant-hunting narratives, now awake to the dangers of overcollection, increasingly articulated their dismay and even angst. *Gardening Illustrated*, edited by the Arts and Crafts gardening advocate William Robinson, called out the "great ransacking" to which Brazil was subject by orchid hunters in 1889; the *American Garden* pointed an accusing finger at Sander's and Siebrecht & Wadley's, noting acidly that their agents were bearing down on the "few orchids . . . still left trembling in their tropic homes."[5]

A wider conservation movement emerged out of such observations on the negative impact of humans on the environment. A growing fear "lest some rare species may be extinguished through the operations of man" motivated the founding of the Association for the Protection of Plants in 1883 in Geneva, Switzerland, for example: its agenda was not only to protect the plants of its nation but to encourage preservation in other countries also—including, explicitly, of the orchids in Brazil. The Selborne Society, founded in 1885 in Britain to protect and study nature, published magazines that reported on threats to species and conservation efforts around the world and at home. By the end of the century it had multiple branches across the country.[6]

Orchidomania was not the only phenomenon to provoke anxiety about the dangers of consumerism and capitalism, the impact of modern life on the riches of the natural world, the growing fear of extinctions and species loss. A tone of dismayed grief about the engines of modern life is a common feature of the Victorian fin de siècle in Britain and Europe; the cold chill crept into discourse in the United States also. In mourning the destruction of habitat and the loss of a flower, Victorians mourned a real landscape, a set of real practices, certainly. The growth of the conservation movement in the mid-1880s amply reveals that this was an era waking up to the impact of industry on the landscape, especially the potential extinctions of species through unrestrained collection practices and habitat loss, the latter produced through urban encroachments, through increased agriculture and industrial cultivation practices.

But grief about loss of species also tapped into a more pervasive angst at the end of the century about cultural and social decline, what Max Nordau infamously termed "degeneration," a sort of awful alter ego of evolution.[7] Literature of the period is full of articulations of a terrible fear that much that was simple, good, and pure was being obliterated, steamrolled out of existence by the forces of globalization and so-called civilization itself. The onward march of modern life, swept forward on the promise of bringing light to dark, was feared to have damaged, perhaps ruined forever something immeasurably precious—clean air, clean water, and also, so to speak, clean *us;* the simple, uncorrupted, innocent self embodied in preindustrial, medieval man. The modern self, by contrast— educated, urban, busy with capital and commerce, conversant with marketing and advertising—seemed a changed, terrifying, polluted thing.

And was there even a world left for the simple figure of yore to adventure in? If all the world had been discovered, if signs of human activity could now be discovered on every continent, then what? What remained? As Britain succeeded in throwing its net over a quarter of the globe, it became subject to a terrible fit of angst rooted in its own self-understanding as a modern, industrial, capitalist nation. This angst swept through other European nations as well, reeling from the consequences of rapid urbanization and social change. Angst about degeneration took longer to become dominant in the United States, a younger country where worries about urban growth, immigration, and the rhythms of modernity were often tempered in the decades after the Civil War by a sense that the benefits outweighed the costs. But as the cities grew and became more racially and ethnically mixed, as the dust and roar of a capitalist economy was felt to intrude on and pollute the great, wide landscape too, white upper- and middle-class Americans worried that the early promise of the nation would remain unfulfilled.

As a sort of counter to fears of the degenerate modern self, adventure and fantasy stories, tales of remarkable discoveries and bold heroism, acquired in many nations a rich, intense allure. American novels by Zane Grey and Jack London showed that opportunities for daring and heroism were still possible in a vigorous life outdoors. Late-century British adventure romances were often set in strange, fabulous locales that were named, presented as real, yet were in no sense realistic depictions of those places. Perhaps there *were* new places and things to be discovered still, writers suggested to a fascinated, half-desperate public: perhaps there were still remarkable spaces that could be accessed through séances, faith, telepathy, or just possibly through travel. "As the visible world is measured, mapped, tested, weighed," wrote Andrew Lang, the writer, collector of fairy tales, and anthropologist, "we seem to hope more and more that a world of invisible romance may not be far from us."[8] Novels and short fictions by writers like H. Rider Haggard, H. G. Wells, Bram Stoker, Sheridan Le Fanu, and Rudyard Kipling suggested to readers that dangerous adventures were still possible in mountains, forests, jungles, castles, and haunted manors. Some hinted that excitement might still invade and penetrate ordinary London through uncanny means. In the second half of *Dracula*, for instance, the eponymous vampire arrives from mysterious Transylvania on a ghost ship; he turns out to be very good at impersonating an ordinary Victorian gentleman in Piccadilly. Readers shivered at such tales, longed for safety and resolution—yet bought another novel, and another, that offered enticing glimpses of heroic confrontations still possible in a commodity-filled, material, urban world.

The assault on Brazilian natural resources provoked anxiety about environmental damage in that nation and played into larger, postindustrial, late-imperial angst about loss and degeneration. Brazil had been described up to the middle of the nineteenth century as one last Edenic idyll, a past gloriously—almost magically—preserved.[9] But if Brazil too was damaged, if even this vast landscape of rich timber and strange animals and myriad exotics was changed beyond repair—then where remained, on this ordinary earth of ours, for the projection of fantasies, for thrilling dreams of future adventuring? As the end of the century approached, Victorian millennial despair turned to Brazil and Brazilian plants. The loss of *C. labiata vera*—through the forces of modernity, no less, the destruction of timber for charcoal—seemed an achingly powerful pointer to the dark side of an advancing (or, dread thought, degenerating) age.

But perhaps there was a way forward, a glimpse of at least partial redemption or possibility. Perhaps, even if a nation could not be saved, a single flower could be rediscovered, found preserved? In the midst of a rising tide of fear, the *idea* of the hunt for the lost orchid emerged. The hunt was real, as we shall see, and it drew on deep concerns about the impact of human activity on species and the destruction of habitats. But it was also a narrative given imaginative wings by a moment that longed to believe in something purer, something untouched by modernity, by a fragile nature that endures. The hunt for the botanical grail seemed a lifeline to a better, different world, a dream of an authentic, uncorrupted past that still lingered. Susan Stewart argues that a desperate longing for an authentic object is almost constitutive of the modern self in a capitalist economy: "Within the development of culture under an exchange economy," she remarks, "the search for authentic experience and, correlatively, the search for the authentic object become critical." As the German Romantics (and many since) have dreamed of an unearthly "blue flower," so British writers, collectors, and enthusiasts pursued *C. labiata*, hoping that the Brazilian landscape was not so damaged after all; that some of its sweetest, most glorious flora remained—that a flower existed that was beyond, above, outside the corruptions of modernity itself.[10]

Finding it, recovering it, would—like the quest undertaken by a medieval knight—require and reward not just courage, but the values of endurance, commitment, and honor that industrializing, urbanizing nations needed to believe were still fundamentally at their core. Here, then, is the story that replaced the earlier tale of open access and bounty:

A lovely fragile flower is lost. To recover it, a brave, true knight must risk danger, death, and suffering. Finding it will prove his worth, but finding it will also require a hunt deep into a part of the world that remains unknown, unmapped, near mystical. Terrible risks will haunt our brave knight's every step, and there will be competitors. If the hunter is to earn his Galahad stripes, he will have to best these other, ignoble knaves. The prize for success, if won, will not be just money—mere *sign* of value—but a peerless flower secured for a grateful nation for all time.

This story, more or less, was told and retold in late-nineteenth-century periodicals, newspapers, and adventure fictions; it still tends to surface in stories of plant hunting and Victorian orchids to this day. It did not have much to do with what really happened, as we shall see. But to pause one last time before we return to Ericsson, plunging into Pernambuco after "Bung. and the Belgies," it is worth remembering how much the fantastic narrative of the quest, narrated in print texts, sidesteps other, violent truths. To pursue *C. labiata* in the later years of the century was to pursue consumer desires at the expense of the environment. To pursue it was to continue to use and reinforce colonial infrastructures into, throughout, and beyond Brazil. To pursue it was to continue to insist on the right to grab and remove resources, to exploit the peoples and communities of other nations. To pursue it was to continue to impose narratives and storytelling derived from British and European traditions on the world. The quest tale of pure motives, courageous derring-do, and a virtuous hunt was a story built to conceal other, darker truths. How did such a tale take hold? Where and by whom was this pervasive narrative constructed?

The pursuit of *C. labiata*, the hunt for the lost orchid, was at least partly a marketing strategy dreamed up by two big nurseries at the end of the century, each hoping to dominate the other by controlling the story in the media first.

20
ENTERING THE LISTS

SANDER & CO. AND L'Horticulture Internationale, run by the Linden family in Brussels, were by the last decades of the century competing to be the greatest orchid powerhouse in the world. Sander, whose business was still fairly young, was intensely ambitious. The Linden family was more established. The father, Jean, born in 1817, had traveled to Brazil on behalf of the Belgian government in 1835 and was probably the Belgian botanizer Gardner mentioned anxiously in a letter to Hooker. Opening his first company in 1846, Jean set up large greenhouses on expansive premises at Parc Léopold in 1868; he began publishing *L'illustration horticole* in 1869, the *Lindenia* Journal in 1885, and restructured his company into the successful L'Horticulture Internationale in 1887. Lucien Linden, Jules's son, was administrator and director by the 1880s. He launched a flourishing society of continental enthusiasts, L'Orchidéenne (1888), set up strong relationships with botanists around the world, and founded yet another new magazine, *Le journal des orchidées*, in 1890.

Both Sander's business and L'Horticulture Internationale sought, as the last decade of the century unfurled, to be the greatest orchid seller of them all. Both companies pursued this goal not only in and through imports and sales, but by fighting their battles openly in the press. Both struggled to prove that they knew more, had discovered more, were more widely respected—and both seem to have grasped that control of the public story, the company's presentation in the media, was a critical means of establishing a stronger position in the market.

As the quest for the lost orchid took its final shape, two other skirmishes inflamed their animosity, setting them on a path for particularly bitter conflict. The first concerned the World's Fair, the Exposition Universelle, held in Paris from May 5 to October 31, 1889.

The Paris World's Fair was an opportunity for Brazil to telegraph its position and standing to the world; forty countries participated, thirty-two million visitors would attend. This was the event at which the Eiffel Tower was unveiled to an admiring world, and it was also an event fraught with complex symbolism. The Paris Exposition commemorated the centenary of the storming of the Bastille and so the foundations of the French Revolution. European nations with kings or queens on the throne were not impressed. Most, including Britain, boycotted the event, keen to separate themselves as much as possible from what was, effectively, a celebration of the end of monarchy.

Brazil was in a tricky position. As the last American nation with an emperor, it could have chosen to boycott the event. Dom Pedro II would, it turned out, be exiled within days of the exposition's closing, and Brazil would be declared a republic—but when the event was being planned, the emperor was eager to participate, to set Brazil's achievements before the world and use the Paris Exposition to help showcase Brazil as a land of botanical riches *and* of manufacturing and entrepreneurial success. Cecília Resende Santos argues that, in the late 1880s, the nation was keen to present itself as not just a country of landscapes: this risked fixing it eternally in the European and American mind as a space unconscious of its own potential, as requiring outside intervention. Instead it worked to present itself as a land of innovation, productivity, and advancement. The very structure of the Brazilian pavilion enacted this, with the ground floor focused on natural resources, the higher floors staging manufactured products, from coffee to ceramics, furniture, machines, and pharmaceutical products. The pavilion itself had a glass dome and a tower over 130 feet (40 meters) high. Visitors could buy coffee at a kiosk.[1]

Orchids were to play an important role as well. As a handbook to the exhibition reported, the grotto and gardens outside the Brazilian pavilion were covered with orchids and with palms.[2] But the Brazilian committee charged with planning the display decided not to source them directly from Brazil, given the challenges of moving plants and keeping them in bloom for many months. Instead, they visited the big nurseries of Europe, auditioning them to see who could most effectively provide and stage Brazilian orchids. They sought a company with the requisite depth in orchid holdings, with the capacity to manage an event on a massive scale, and with a vision for the flowers' staging.

The company they chose for the curation of their display was L'Horticulture Internationale, which blazoned the achievement in its *Lindenia* in 1888. "We all know," the magazine explained, "that Brazilian flora are one of the great riches of the world," and the government had that year determined to give the plants especial importance (*"une importance inusitée"*). The committee was building a pavilion to be adorned, throughout the duration of the event, with flowering orchids and other samples of spectacular plants. Because of the challenges of transporting orchids especially (*"les Orchidées surtout"*), they had visited all the principal establishments of France, England, and Belgium and had, at last, chosen to entrust the care of their plants and the whole display to the august L'Horticulture Internationale.[3]

Select photographs of the event were taken. Sepia-toned images of machines, products, and perfumes offer glimpses of the ways Brazil worked to present itself visually to the world as a thriving economy. Images of the greenhouse give us access to Linden's accompanying vision also. Plants are offered to the exposition's viewers as managed, curated objects, not wild nature: a collection of blooms topped by a huge palm is positioned in a central urn, as if in a vase on a tabletop. Around the urn, dotted on the floor, stand individual blooms trained upright. This is not wild, luxuriant, untamed nature, plants as embodiment of untapped natural resources, but nature subject to the forces of reason and control. The very fact, too, that so many orchids were kept in bloom for so long implied successful management of the natural world, so that the whole communicated in an image what the Brazilian government sought to enact—Brazil as richly resourced yet conversant with the intellectual systems needed to understand how to manage its own assets. L'Horticulture Internationale had done its job: *Lindenia* delightedly reported that at a competition held during the Paris Exposition, from August 16–21, the company won numerous prizes. Brussels was now *"le centre orchidéen continental,"* it trumpeted, the center of orchids on the continent.[4]

This image of orchids from the official catalog of the "Exposição Universal de Paris: Exposição Brasiliera—Interior da Estufa" presents nature curated and staged.

Round one, in other words, had gone emphatically to Linden. His firm had been chosen by Brazil as its authorized specialist, as a nursery at once deeply skilled in and knowledgeable of the flora of the nation and one sensitively able to understand and curate the nation's vision of itself. Linden was able to communicate his successes in text and in images. Sander, not even able to participate, must have read about it, fuming, from the other side of the Channel.

Round two commenced. Linden's collector Bungeroth had imported into Europe in 1886 a particularly luscious Venezuelan catasetum which the company unveiled in the *Lindenia* that year. The catasetum produced great admiration in Europe, and its flowering at Kew was written up with éclat in the *Gardeners' Chronicle* in 1887. A further, very positive assessment by Joseph Hooker followed in *Curtis's Botanical Magazine* in 1888, while contributors at home wrote breathlessly to the *Gardeners' Chronicle* in 1888-9 to report on the progress of their *Catasetum bungerothi* blooms. A single specimen was sold in the Stevens's auction rooms for fifty guineas to Baron Schroder (£7,300 / $9,300 today; Schroder was known for splurging on botanical novelties).

The Lindens sent a rather smug postcard to Sander to make sure he was fully aware of (what they called) "the most remarkable introduction of the last time."[5] But although the plant and its sale were much-heralded, the Schroder specimen soon led the Lindens into difficulty. Schroder sent one of the blooms from his new plant to Reichenbach; the great German specialist decided that the *soi-disant Catasetum bungerothi* was in fact an orchid already imported by the Lindens in 1882 and named—by Reichenbach—*Catasetum pileatum*. Lucien Linden disputed this—the company had had no hunter in Venezuela in 1882, he protested, and the flower was not remotely the same. But to make matters worse, another possibly identical catasetum was found to have been lying at Kew for some thirty years, imported by a British collector. Was this the same plant or not? What was the relationship between the 1882 and 1886 plants? Reichenbach died before he could really clear up the matter, but he was on excellent terms with Frederick Sander, and indeed expressed his opinion on the catasetum's being no new thing in a letter to the St. Albans nurseryman in August 1887.[6]

Sander was surely thrilled to learn that "the most remarkable introduction of the last time" was nothing but a duplicate. He promptly wrote up "Catasetum Pileatum" in the second volume of a massive work on orchids he was starting to publish, *Reichenbachia*, named in homage to his friend; its colossal size and weight were surely attempts to signal that it was a worthy companion to earlier tomes like *Collectanea Botanica* and Bateman's *Orchids of Mexico and Guatemala*. The pileatum "of Reichenbach" and *Catasetum bungerothi* "are identical," Sander wrote firmly in the text; this was not a new discovery. The *Catasetum pileatum* had been sent to Reichenbach by "M. Linden, a nurseryman of Brussels," a literally correct yet studiedly offhand descriptor that must have driven the prominent Lindens, who perceived themselves to be *"le centre orchidéen continental,"* half-mad.[7] To make things worse, the plant was not even properly new, Sander continued, for it had been "collected over thirty years ago by [Richard] Spruce" [for Kew]. He managed to make the apparent sending of the plant twice, in 1882 and 1886, sound rather ridiculous, as if the bumbling foreign firm did not quite know what it was about.[8]

This round had gone certainly to Sander, and Lucien Linden must have been enraged. A first name always took priority, and the plant was known from that time forward as *Catasetum pileatum*. "Permit me to protest," Lucien wrote in a testy letter quoted in the *Gardeners' Chronicle* in May 1890. He complained bitterly that the latterly introduced catasetum could not possibly be the same as the 1882 plant, that Reichenbach and Sander's Reichenbachia were wrong: "This species is so extraordinary, that my father and myself must have recollected it."[9]

But others feared to gainsay Reichenbach, even though he was dead. And since Reichenbach had determined for no very clear reason that his own herbarium should be sealed for twenty-five years after his death, it was hard to compare the new plant with the old.

Since there was no way of proving matters, Linden had to move on. He had other orchids arriving from his plant hunters all the time, and he soon threw his energies into new discoveries, seeking new novelties to put Sander's nose firmly out of joint. One bloom in particular, an orchid recently arrived from Brazil, looked like a serious contender for the title of plant of the decade.

IN ORDER TO UNDERSTAND EXACTLY how and why Claes Ericsson ended up on the boat to Pernambuco in 1891, to unpack the meaning of his circling, incoherent letters, we need to go into the pages of botanical and horticultural magazines from the previous year, on both sides of the Channel. There too we find the glimmerings of what was soon to be an explosive battle between L'Internationale Horticulture and Sander's.

On June 14, 1890, Robert Allen Rolfe, a curator at Kew, reported—quietly—an interesting development buried in the middle of a column in the *Gardeners' Chronicle*. The previous month, at a May 13 meeting of the Royal Horticultural Society, an orchid provisionally named *Cattleya warocqueana* had been shown by L'Horticulture Internationale. Sourced in Brazil by collector Erich Bungeroth, again, the new cattleya had been named after the president of the company, the playboy industrialist, gambler, and orchidophile Georges Warocqué.

Two weeks earlier, the Lindens themselves touted *Cattleya warocqueana* in *Journal des orchidées*. The new cattleya had been a "sensation" at the last meeting of L'Orchidéenne, readers were informed, and it was surely one of the most exciting new arrivals of recent years.[10] Lucien Linden was secretary of L'Orchidéenne; the society met at the Central Pavilion of Parc Léopold on the second Sunday and Monday of every month. Linden adroitly used his own organizations and publications to help fan the flames of celebrity around his plants. On June 15, just as Rolfe's piece was published in Britain, the Lindens celebrated the new *Cattleya warocqueana* in a two-page ad in *Journal des orchidées*, complete with notes on its astonishing beauty by Bungeroth, the plants' arrival in Europe on April 18 (*"en excellentes conditions"*), and information about the prizes it had just won (conferred, it must be said, by Linden's own society). The advertisement also reported the decision of L'Orchidéenne—unanimous—that the plant was definitely a new orchid, hitherto unknown in Europe. Plants of the highest quality (*"plante choisi, parmi les plus fortes"*) were offered for sale at one for 100, six for 500 francs.[11] The orchid then went on to the Royal Horticultural Society in London.

When it arrived, it was visibly stressed from its multiple journeys. Only just coming into bloom, it was, as a specimen, underwhelming, Rolfe noted, "hardly in a condition to pass an opinion on it." But L'Horticulture Internationale had helpfully provided the Royal Horticultural Society with accompanying materials to bolster the plant's claims to fame, including dried flowers, a photograph of the orchid in

> SOCIÉTÉ ANONYME
>
> # L'HORTICULTURE INTERNATIONALE
>
> PARC LÉOPOLD
>
> ## A BRUXELLES
>
> Adresse télégraphique : LINDENIA, Bruxelles
>
> Administrateur-Directeur : LUCIEN LINDEN
>
> ## MISE AU COMMERCE
>
> A partir du 1er Juin 1890
>
> ### de la plus belle Orchidée introduite pendant ces dernières années
>
> ## Cattleya Warocqueana Lind.
>
> *DIPLOME D'HONNEUR de 1re classe, décerné à « l'unanimité » comme Orchidée nouvelle, au Meeting du 11 Mai dernier de L'ORCHIDÉENNE, la Société d'amateurs d'Orchidées établie à Bruxelles.*
>
> Le **Cattleya Warocqueana** est un type de Cattleya, formant une section spéciale comme les *Cattleya Mossiae, Trianae*, etc., parmi laquelle un grand nombre de variétés se sont déjà déclarées. C'est une brillante introduction, entre toutes, qui provient d'une localité de l'Amérique Méridionale, complètement inexplorée jusqu'ici, où l'on ne soupçonnait guère que des Cattleya pourraient être rencontrés. Sa station naturelle est à une énorme distance de celles d'où proviennent les Cattleya connus. Notre collecteur, M. Bungeroth, qui a été envoyé dans ces parages par M. J. Linden, écrit au sujet de cette grande introduction :
>
> « *Je suis émerveillé par la beauté de ce Cattleya. Je n'ai rien vu d'aussi beau ni en* « *Colombie, ni au Vénézuéla, parmi les* Cattleya Mendeli, Trianae *et* Mossiae. *L'épi floral est*

The first advertisement for "Cattleya Warocqueana" appeared in the June 1890 issue of *Journal des orchidées*.

bloom, and five drawings. Taken together, these revealed that, in the wild, in happier circumstances, the flowers of the orchid were large and luxuriant, while a hint of lingering perfume suggested the richness of its odor. Color drawings of specimens also suggested the orchid could exhibit a range of variations and hues. This impressive new orchid deserved its own, new name; or did it? "Botanically it is a form of C. labiata, using that term in its broadest sense," Rolfe said, hesitantly, "but the exact relation it bears to the other existing forms is a difficult matter to decide." It did, he noted, almost in passing, display a "resemblance" to "the original C. Labiata."[12]

Rolfe's article was translated the next month into French and published in *Journal des orchidées*. The Lindens probably knew of his views early, for Rolfe collaborated with the company on *Lindenia* and *Journal des orchidées*. However, the Lindens did not, if Rolfe is to be believed, share with him where the plant came from or exactly how Bungeroth found it.[13]

In fact the Lindens do not seem, at least at first, to have quite grasped the implications of Rolfe's words. In a few short lines at the end of their translation of Rolfe's text, a writer (probably Lucien Linden himself) argued against those who suggested that *Cattleya warocqueana* had already been imported. Obviously somewhat bruised from the *Catasetum bungerothi* contretemps, Linden energetically insisted that this orchid had never been recovered by *anyone* before; *this* plant was quite definitely new. To support this, he claimed it was found in an uncharted locale, "absolutely lost" in a place where "no European was established"—no chance of another turning up this time, in other words, in a dusty cabinet in Kew.[14] The Lindens continued to tout their orchid under the name *Cattleya warocqueana*, and to describe it magnificently as a *"grande introduction,"* deep into October of that year.

The plants they had imported, now in the greenhouses in Parc Léopold, were covered in buds and ready to flower abundantly.[15] By the third week of October 1890 they were sufficiently recovered from their exhausting trip from Brazil to achieve full bloom. Messrs. Linden proudly showed off the orchid again, first at the premises of L'Horticulture Internationale on October 25–26, then, a few days later, at the Royal Horticultural Society in London.

This time, the gorgeously blooming orchid was, according to a November 1 account in *Journal des orchidées,* a riotous success. And it wasn't just the Belgians who admired it: the journal was able to quote at length from multiple English-language journals (*The Garden*, the *Journal of Horticulture*, the *Gardeners' Chronicle*, the *Gardeners' Magazine*, and *Gardening World*) on the beauty of the plant and the impressive achievement of the Lindens on bringing it to market. It was true, they added, that among the blooming plants were a number that seemed to bear a resemblance to the famous *Cattleya labiata autumnalis*, missing for fifty years, despite all efforts to locate it; but the point was made at the very end of their report, almost as an afterthought.[16]

However, the orchid was starting to garner interest and attention on the pages of the press in Britain and outside, and not just in periodicals aimed at specialists. Was the Swainson orchid finally found? The implications of the question were soon to enfold *Cattleya warocqueana* and the Lindens in turmoil. Quite apart from whether the plant was found or not, if some of the new plants were indeed *Cattleya labiata autumnalis*, did this mean that the lost orchid was, properly speaking, a variety of *Cattleya warocqueana*? Or was it the other way around? Was *Cattleya warocqueana* a variety of *Cattleya labiata*? What was the "old labiata," the type species, on which the genus was founded? "Opinions are sure to differ upon this point, and will do so, so long as there is no definite line to be drawn between species and species, still less between species and variety," observed Rolfe fretfully in the *Gardeners' Chronicle* in December 1890—but the debate was only just heating up.[17] For this was about far more than taxonomy, it was about business.

Orchid specialists differed on the identity of the new orchid in the months after the Lindens' introduction of *Cattleya warocqueana*. Some referred to it as "Cattleya Warocqueana," some as "Labiata Var. Warocqueana." Rolfe tried the tangled "Cattleya Labiata, *Lindl.*, var. Warocqueana, *n. Var* (?)," suggesting that *Cattleya warocqueana* might be a new variety of *Cattleya labiata*. But an author with the initials H. G. C. (a gardener named Henry Chapman, most likely) wrote, of the Lindens' naming decision: "there must be some mistake." This was *not* a new cattleya, he wrote decisively, a *grande introduction*; closer inspection revealed that this was "the scarce and long-sought-for true 'old' autumn-flowering C. Labiata" at last.[18] But the paper itself wasn't certain; the very next week it dubbed the orchid "Cattleya Labiata Warocqueana" once more.

L'Horticulture Internationale seems at last to have grasped the significance of their potential discovery—to realize that, if the plant was indeed the Swainson orchid, the market for more could be considerable. But was it, or not? Understandably, if a little frantically, they tried to play it both ways—to sell it as both new *and* old. *L'Illustration Horticole,* another magazine run by the Lindens, suggested that the company had made a remarkable discovery in *Cattleya warocqueana,* that the plants would "eclipse" all other recently introduced orchids—while also noting among the collection a number of types of "the famous *C. labiata autumnalis.*" Swainson's original plant, "so sought after for fifty years, has therefore been found," they suggested, "enriched with a large number of new forms and superior varieties."[19]

But William Hugh Gower, a botanist who had worked in his career at a number of leading nurseries as well as at Kew, was having none of it. Gower argued in *The Garden* on November 15 that the Linden plants were absolutely *not* Swainson's *Cattleya labiata.* In fact, "they had nothing in common with the species. Their growth did not resemble that of the typical plant, neither did the flowers," he opined—and Gower should know, he said, because he flowered an old *Cattleya labiata* at Kew back in 1862. Gower hinted that the lost orchid was still out there, somewhere: "I hope from the extended search which is being made for C. labiata that it will end in the plant being again found and some magnificent varieties introduced." Gower claimed that the Linden plants were either a new species, *Cattleya warocqueana,* or a variety of another long-known orchid, *Cattleya gaskelliana.*[20] Either way, he made it perfectly clear he was underwhelmed by Linden's plants.

Lucien Linden was irate. The fact that he had just lost out to Sander on the catasetum must have made him only more determined to win this time. In a public letter to Gower published in *Journal des orchidées* on December 1, dated the day after Gower's article appeared (and presumably penned the day the issue of *The Garden* landed on the newsstands in Brussels), Linden fumed about Gower's dismissal of his plant. He listed the authorities who, at the last meeting of L'Orchidéenne, unanimously determined the plant was *Cattleya labiata autumnalis.* These gentlemen included not only Belgians but also James O'Brien, member of the august Orchid Committee of the Royal Horticultural Society. Linden further referred to the judgment of his own father, Jean, the man in Europe who has "without doubt, known *Cattleya labiata autumnalis* longer than any." Those who resisted Messrs. Linden's judgment were influenced by money, Lucien raged, because the value of older *Cattleya labiata* plants, descended from Swainson's, would now be reduced. "But such considerations obviously cannot influence the opinion of judges in good faith," he added, rather righteously; "who are only concerned with establishing the scientific classification and nomenclature of new species." And he warned, in strong terms, about the likely imminent arrival of fakes— "*contrefaçons*"—from collectors who intended to rifle known habitats of *Cattleya labiata* in pursuit of more plants: "I want to establish that the *Cattleya Warocqueana* comes from a fresh, entirely distinct locality, and to warn amateurs against false *C. Warocqueana*."[21]

Matters apparently quieted over Christmas and in the New Year; the *Journal* referred to *Cattleya warocqueana* only to praise it, to tell the world that everyone else was praising it, and occasionally to refer to it as "identical" to *Cattleya labiata autumnalis*. But Linden knew something was afoot. When Gower said that he hoped "the *extended search which is being made* for C. labiata . . . will end in the plant being again found," he presumably knew or guessed that others were actively planning a search. Linden's piece seems designed to undermine any claims of a future discovery, to undercut any businessman seeking to bring in similar plants.

Of course, Sander was the most likely candidate. And of course, deep in competition with the Lindens, Sander was hatching a plan of his own. Rather than hailing the Lindens' discovery, rather than cheering that the orchid was found at long last, Sander was in pursuit of *Cattleya warocqueana* of his own.

21

ERICSSON AND BUNGEROTH

CLAES ERICSSON HAD BEEN LIVING in rather cramped circumstances at 17 Orkney Street, Battersea, when, in January 1886, he put pen to paper and wrote to ask Sander for a job. He was working at Battersea Park, he explained; he'd also lived in Germany and France. His dream was "to cultivate if possible Orchis in Sweden" and he hoped to learn more about their cultivation first. Something about this slightly odd letter—perhaps the author's picaresque lifestyle—piqued the Orchid King's interest, and soon Ericsson was off hunting on Sander's behalf around the world.[1]

The delight Ericsson felt at the offer of a job in 1886 seems to have waned a good deal by 1890, as the hunt for the lost orchid reached fever pitch.[2] Sander demanded long letters; Ericsson obliged from Singapore, sending epistles in which the required information about plants and routes was grumpily supplemented by complaints about payment, the terrible exchange rate, the challenges of life on the road. He returned to Sweden in August, when his mother got sick, and hated it; he complained bitterly into the winter months about the climate.[3]

Then, at Christmas, Sander summoned him to London. By this point, the businessman had inspected *Cattleya warocqueana* closely, and he must have determined that it was at least potentially from the same population as the lost orchid. But the uncertainty encircling the plant gave him a crucial opportunity. Whatever the truth of the matter, the plant was not yet fixed in the public's mind as found. There was space, in other words, to be crowned its finder, if he could craft an alluring tale of discovery to entice the media and grab consumer attention.

The nineteenth century saw a boom in printed texts produced to cater to the first generations of a mass-reading public. "Nearly everyone was exposed to print," notes Matthew Rubery, "during an era offering over 25,000 different journals to the growing reading public."[4] By the end of the century text was everywhere, no longer locked away on the hushed shelves of an aristocrat's country home or parceled to subscribers in lending libraries. It blared at millions of new readers on the very streets they walked: books and newspapers were offered for sale in tobacconist's shops and railway stands, advertisement posters—"bills"—lined omnibuses, walls, bridges, carts, trams, trains, and more. (See Plate 9.) Handbills were thrust into shoppers' hands as they passed, men with sandwich boards jostled for attention, while to open a newspaper or magazine was to be near-bludgeoned with page after page of advertising in very large text. Fully half of the pages of later-Victorian periodicals were given over to advertisements.[5] There was a thin line, too, between the advertisements and the articles, editorials, and fiction. Sneaky product placement was common: interior design writer Jane Ellen Panton, for instance, recommended paints, textiles, and other wares in her popular how-to design columns from stores that paid her to be their spokeswoman—and that, in turn, advertised her books and designs in their advertisements.[6]

When large amounts of information are on offer, when lots of text and media confronts us, we find it difficult to sift. We give our limited attention to what engages us, what grabs us, what we can quickly comprehend. It is the business of the advertiser to try to hook our attention in the midst of the maelstrom because "our attention is the portal to our consciousness," Vincent F. Hendrick explains of media today: this "makes our attention valuable to anyone with a message, a news story, or a product to sell."[7] Those trying to get us to believe or buy their ideas or wares attempt to capture our limited attention.

That hook, that allurement, is typically very simple and easy to understand, underpinned with dynamics of good versus evil, us versus them because these are easier, may even soothe us, when complex, hard-to-sort information baffles. Sander surely grasped that the British public was not trained in or able to comprehend the finer points of orchid species and varieties, of geographical locations in a nation thousands of miles away.

But a stirring tale of a British company's great triumph, on the other hand (and he was careful not to mention his German birth too often) might appeal.

Grasping the anxiousness of an age riven with self-doubt about the shape and rhythms of industrial modernity, yet simultaneously convinced of its inalienable right to take over the globe, Sander & Co. constructed a tale that apparently confirmed British greatness, that (even better) pointed to the failures of Belgium, a European nation with its own colonial ambitions in Africa at the time. The company offered the kind of simple, jingoistic, us-versus-them dynamic that could plausibly cut through the noise and focus consumer interest on orchids and their business.[8]

The Lindens had not revealed where the *warocqueana* was found. But Sander seems to have grasped he needed simply to figure out where Bungeroth was at the time of the orchid's collection in 1889 and infer the orchid's location from that. Of course, the world is large; finding Bungeroth might sound a near-impossible task, especially as Linden's hunter was known to be so good at evading notice. But Sander had his collectors fanned out around the globe, and part of their job was to spy on other collectors as they traveled. All he needed to do, in other words, was tap into the elaborate information network already in place and catch a whisper of Bungeroth's whereabouts from the previous year.

Multiple letters from the plant hunters attest to Sander's explicit pursuit of Linden's *warocqueana* over the next months and years.

ON NOVEMBER 10, Sander received a letter from Charles Palmer, in Port of Spain, Trinidad, that prompted him to recall Ericsson from Sweden. Ericsson journeyed to Sander via Brussels. A January 4, 1891, postcard, franked "Bruxelles," hints that he hoped to pick up information from Linden's establishment en route:

Dear Sir

I will be with you in St. Albans tomorrow. here is a beastly weather, raining etc. and I have got a beastly cold I'm just going to try to get in to Linds place now see how it will turn out
Truly yours

Ericsson.[9]

On the back of this innocuous-seeming postcard, Ericsson and Sander began their pursuit of Linden's *Cattleya warocqueana*.

Exactly what happened, what Ericsson sought to do—in stealthy fashion, since he notes he'll have to "try" to get in, will "see how it will turn out"—we can't know. Perhaps he wanted to get a closer look at samples of *Cattleya warocqueana* so he would recognize it; perhaps he hoped to bribe someone to tell him where Bungeroth was collecting when the plant was located and sent to the Lindens. He doesn't seem to have met with much success, however. Ericsson went on to St. Albans the next day and met with his boss while the latter was still processing Palmer's letter of November 10 and the most recent Linden statement in *Journal des orchidées* on the *Cattleya warocqueana*, dated December 1.

Palmer's letter suggested that the "new Cattl" actually hailed from "Columbia." Palmer explained he had just learned that Bungeroth had been in Colombia at the appropriate time, and thus the orchid "must be from there." Palmer was himself headed out in pursuit of it: "I shall start in a few days and will do my best to give you full satisfaction," he explained brightly, adding "I do not suppose that you will hear from me again till May next year."[10]

Sander had been willing to believe, the previous decade, that the lost orchid could be sourced in Venezuela, and he was perfectly willing to believe Palmer's claim it came from Colombia now. But, keeping his options open, he and Godseff sent a flurry of letters in October and November to William Digance, the young laborer's son, who was in Minas Gerais in Brazil, suggesting he should start looking as well. These letters have not survived, but Digance's miserable reply has. Still struggling to find much of anything, he sounded utterly defeated: "you have put me on a big job this cattleya business as I beleve it to be very scarce," Digance wrote unhappily. He was also fixed in place by days of thunder, lightning, and torrential rain that turned local streets to "young rivers" and made the mountainous roads deadly—"there is only just room for the mule's feet & the rocks rising up es [each] side . . . the water rushes down 2 feet deep at times during storms."[11]

Sander decided to send Ericsson out after Palmer, to cover as much ground as possible (and indeed, Palmer promptly vanished down a non-navigable river and was out of communication for months). But where, exactly, should Ericsson begin? Though few of the Swede's letters from the early months of 1891 have survived, it's obvious neither Sander nor Ericsson could quite work out where to start the quest—and Sander did not want to pay Ericsson to rattle around looking. Linden had said in his December 1 article in *Journal des orchidées* that his firm sourced the plant from a new and distinct locality. What did that mean? Sander seems to have kept Ericsson hanging around in St. Albans for quite some time, cooling his heels, while he tried to get more information.[12]

William Digance died on March 12, 1891, at four in the morning, at the United States of Brazil Hotel in Rio. The local registrar reported that a businessman, most likely the hotelier, brought a letter from the local doctor attesting to loss of life from "bilious fever."[13] News of the death may explain why Ericsson was sent off to South America, still without a clear plan. He was in Trinidad, off the coast of Venezuela, by the end of the month, and nearly wild with desire to best Bungeroth and prove himself: "I must find him somewhere so soon I get away from here, I will hunt him up, never mind where," Ericsson wrote feverishly: "I must find him, I have made friends with good many people in this place and they will be useful in the future."[14]

Ericsson tried to collaborate with Palmer in the information-gathering process—though this was difficult, given Palmer's vanishing. Ericsson was also competitive with Palmer: "what is he collecting for you has he got any information from you about the Cattleya Warocqeana," Ericsson asked rather jealously.

But on April 9 Ericsson wrote that he was making progress at last. Bungeroth was in Bolívar, Venezuela, in 1888, then journeyed along the Rio Meta, which flows from Venezuela to the east of Bogotá in Colombia. Ericsson explained both Bungeroth's route and the ongoing challenges of getting anyone to tell him anything, given Bungeroth's great personal magnetism:

> Two years ago he is supposed to have come overland from Maracaibo [in Venezuela] and brought some plants and shipped them from Bolivar [northern state in Colombia] with the Royal Mail, but I could not find out from the German there what time in the year they seems to be spellbound but I will know today but to late to tell you in this letter.

Following in Bungeroth's footsteps, Ericsson made enquiries on a steamship in Bolívar, and from there he learned of another possible collection site, along the Orinoco River in Venezuela:

> In the beginning of 90—or end of 89. two fellows went up the Orinoco collecting and shipped their plants from Bolivar, but they could not tell me exactly when but they would look up the books today and let me know all.

"I hope I'm on the right track this time," Ericsson continued, warning of hunters from other companies in the region—though at least they were experiencing a range of dread outcomes: in Bolívar, "2 german died, one got married." Ericsson also seems to have bribed local customs officials to warn him of any plant shipments, so he could be sure no other hunter was stealing a march on them: "there will be no more plants shipped from Bolivar without me knowing it I have seen to that," he explained rather smugly.[15] There were threats to life and limb to be managed, too: "I believe I saw more snakes then people," he wrote of Venezuela, "sometimes the snakes hang up in the trees ready to drop down in the boat."[16]

Bungeroth's hiding places seemed endless. "Some say Bung. was last year in Congo (africa)," Ericsson explained in a postscript, "some say Columbien." Then, traveling onto Barbados, Ericsson met another collector named Vratz, whom he briefly suspected of being the man who sent the orchids to Linden. Ericsson took out a draft for seventy-five pounds, presumably to bribe Vratz for the information—"it might have been Vratz who sent the C. Waroq. and then I would have collected them," he explained. But Vratz insisted he had never sent plants of any species to Brussels, only to a company in Leipzig (and in fact Vratz, who seems to have been struggling to make ends meet himself, hopefully offered to send some *Cattleya gaskelliana* to Sander).[17]

Just as Ericsson seemed to be running aground again, he received a telegram. It was from Sander. He had learned that Bungeroth had been, after Colombia, to Pernambuco in Brazil. This, of course, made far more sense, since it was a place Swainson visited all those years ago on his specimen-hunting trip. The St. Albans businessman directed his employee to head to Brazil promptly and start searching in the state of Pernambuco—far away from Rio, in other words, the site of Swainson's later journeyings, and from the mountains Digance, Gardner, Arnold, and Lobb had already covered. The hunt now focused farther north, to the west of Recife, where Swainson first ventured after leaving the city in 1818.

Sander, typically, was impatient for immediate results, and in his letters expressed frustration with Ericsson, with the quest, with the time it was taking and the money it was costing him. "[G]oing collecting takes time," Ericsson retorted, although he admitted that he too wished the hunt for "C. Waroq" was going faster. He also reported a story he had recently learned concerning the finding of *Catasetum bungerothi*. Shortly before his death in 1886, Arnold had apparently found the plants himself; but Palmer, right after Arnold's death, was tricked out of them by Bungeroth. (Jean Jules Linden also mentioned in an interview in July that Bungeroth was dogged along the Orinoco by two English collectors—"*par deux collecteurs de maisons anglaises: ARNOLD et PALMER*"—though he made no mention of Arnold's death or the exchange of catasetum).[18] The possibility that the star plant could have been called *Catasetum arnoldi*, or at the very least that Sander could have made money off it, not Linden, surely added fuel to the fire.

Ericsson was hopeful: for all his frustration at Sander's treatment of him, he longed to be the one to satisfy his master and prove himself the worthiest collector. "Hoping to receive good news in Pernambuco," he signed off the letter acknowledging receipt of Sander's telegram.[19] And indeed, almost as soon as he arrived in Brazil, Ericsson began to pick up clues he was in the right place at last. He was so excited he descended into even more than usual incoherence. On July 19 he cabled Sander (code: "Vanda London") the mystifying words "Bungeroth collected."[20] Quite what Sander made of this we can't know, but Ericsson followed the cable up next day with a letter. By these two words, Ericsson explained, "You will understand that he [Bungeroth] is the shipper"; "Bungeroth collected" meant that he'd caught definite information that Bungeroth shipped plants from the region at the right time for them to be *Cattleya warocqueana*. "I have been here very near a week but not before yesterday could I get anything to know for certain. Of course I do not know if that is Cattle Warocqueana he collected here," he went on, "but I say as you, what could it els be?" He also learned that Linden's agent had journeyed as far as Garanhuns, a small city to the southwest of Recife; since he had no news of venturings beyond this, Ericsson guessed the orchid grew in the neighborhood of Garanhuns. He already had local men following in Bungeroth's footsteps and would follow presently himself. He was just one day from Garanhuns. "I'm very glad to see a finish on this business it has taken a very long time," he wrote, adding in a postscript a clear expression of his investment in the process, at least for the moment: "with best wishes for our success."[21]

Ericsson reached the area Bungeroth had visited in late June and was able to make contact with a local man who'd helped Bungeroth the previous year. Taken to the source of the orchids, Ericsson inspected the plants and was convinced—almost, very nearly—that these were indeed what he had been sent to find. "I'm collecting the Cattleya it is Waroqeana," opened a letter to his employer on July 1, and whomever received the epistle in St. Albans underlined the words excitedly in blue. Ericsson started removing the orchids immediately—though since the plants were not in bloom, just a shade of doubt about their identity remained.

It was hard, miserable work. July is the coldest, wettest month of the year in Garanhuns, with average precipitation of ten inches. The local team Ericsson hired to help understandably did not want to slide around tall, slippery trees in mud and pouring rain to retrieve the contracted 3,000 plants. Also, Ericsson believed that Bungeroth had cut up his plants small—too small, so the Sander agent contracted for larger plants with no fewer than ten fresh bulbs. Careful, sifting work of this sort would take even longer, Ericsson admitted, and he estimated that putting together the first consignment could take as much as a month. Communicating with the workers and explaining what he wanted was difficult, too, because Ericsson did not speak Portuguese. Tensions with the men rose; they were suspicious of him and of the whole endeavor and angry at the thought of all the money he would make. Ericsson attempted to point out that *he* was not the one who would make the money; this did not make anybody happier.

The bulbs were also moist and full of sap; it was very far from the ideal time of year to pack and send orchids. Ericsson was deeply anxious about the soaking flowers. He feared they would not survive the trip, and indeed his letters from this period do not communicate triumph or bold confidence. Rather, they express ongoing confusion about what was what, lament the difficulties of removing plants in the rain and managing righteously angry work teams, and detail labyrinthine travel and shipping arrangements. Ericsson also posed an innocent-sounding question to his boss, which must have worried Sander and Godseff: "The natives told me that the Cattleya usually flowers at Easter time, I send you here 2 flowers dried, what is the difference between this C. and the old Labiata?"[22] After all that, were these orchids the same as the Swainson orchid after all?

This did not derail the extraction of the plants. In a first consignment, sent on the *Thames,* Ericsson sent thirty-one cases of over 2,000 plants. (The *Thames,* commissioned two years before, was one of the first Royal Mail steamers built with a sleek steel, not iron, hull.) Collection of a second consignment continued apace: Ericsson became more confident that he truly had the right orchid when he received fresh intelligence that Bungeroth was in the region as recently as February and intended to come again, possibly because he'd realized there were more, different colored varieties of *Cattleya labiata* in the region (in particular, there were a lot of white varieties). Ericsson sent 5,000 orchids next in fifteen cases on the German steamer *Cintra* to Lisbon, then, via the auspices of Royal Mail agents in Portugal, Knowles & Co., on to Southampton, England.[23] Ericsson checked anxiously on the progress of his shipments as the weather got worse: "By this time you will soon know in what condition the first arrived I am very anxious to know I went to see them off at the Mail but such high sea I could not get on board."[24]

At least, Ericsson clarified, the orchids were fall-blooming, and there were other *Cattleya* orchids in the mix. Meanwhile, he was obsessively tracking detailed information about Bungeroth and his shipments—dates, numbers of cases, even the names of shipping agents used to move the freight—in order to try to locate more collection sites. There was another Brazilian agent working against them, too, presumably in the hire of Bungeroth and Linden: Ericsson was forced to pay for a substantial stock of orchids that were of less than ideal quality to prevent local people with whom he was working from selling them to "the Brazilian." The agent seemed to love playing cards and to be unfortunate in his losses; Ericsson tracked his foe's gaming difficulties with ghoulish relish.

Rain paused the collection. In between packing cases, Ericsson reflected on what he had learned of the growth habit of the cattleya:

> The cattleya flowers from November to April also in our winter month, grows not bye the rivers, grows on smaller trees ^on the hills surrounding Garanhuns about 2,500 feet high the temperature in this season being in the night 16–18 celsius but it has gone down as low as 11 but seldom, in the day the Temp. being 22.

He sounded exhausted and confused and somewhat desperate as he listed the many different orchids and explained the range of colors Sander could expect to find in various crates; at one point, he even lapsed into German, the mother tongue of his employer: "you will find some who they with hellen bulben und blattern they mit zeug ungebunden sollen gelb sein," he wrote, presumably indicating that those with light bulbs and leaves, with other unbound material, would be yellow.[25] On the same day, he wrote to Sander's business manager, Joseph Godseff, to assure him with oaths that he would remove literally every white and yellow cattleya he could get his hands on and to assure him that he had teams of men still working hard.

Was he exhausted—or drunk? Ericsson became intensely agitated about the condition of the plants on their arrival, given the less-than-ideal conditions in which they had to be packed. Godseff wrote to assure him that the first consignment had arrived—but even knowing this, Ericsson wrote incoherently, "Mr. Sander is very anxious about the Cattleyas yes and so am I just like sitting on neadles I sincerely hope I do not get any more competition here for a month after that don't care." His next letters suggest his mind was circling: his urgent need for Sander's good opinion, the stress of fighting off competitors, and the logistical challenges of packing up the orchids is expressed in angry prose full of spelling and punctuation errors, with awkward additions and crossings out. He sent another 4,000 plants in thirty cases and admitted to destroying a further 3,400 because they were also in less than ideal condition, and he did not want them in the hands of Bungeroth's agent. "[T]hat Gent he is still up the country playing cards with Lindens money and when it is finished he will bye up all what I have not taken he ^will takes all what they bring him," he wrote.[26]

Ericsson's confusions in person and on the page triggered Sander to send what must have been eviscerating letters accusing him of drunkenness. A thoroughly rain-sodden Ericsson was deeply hurt. "Tell me why you write to me in every second letter to keep sober," he replied piteously; "do you take me for a drunkard you might take it were light to write but it is very hard to receive." In the aftermath, he redirected his focus back to the hunt for Bungeroth. He wanted to be sure he knew where his enemy was in case he was sourcing desirable plants, but "I can not get hold of his coming and going," he wrote, metaphorically wringing his hands because Bungeroth's name was not on the police list of passengers in the area. In fact, the police suggested he was traveling under an assumed name; Ericsson sent a spy off after him to Paraíba.[27]

Ericsson was able to send another forty cases—this time by the *Clyde*, another Royal Mail screw steamer—to Southampton: he wrote to alert Sander of the shipment on November 7, and the freight of living plants is listed in the *Journal do Recife* the following day, under "Exportação": "*C. Ericsson, 40 caizas com plantas vivas.*" "Plants are now getting scarce," he warned Sander in his letter, adding, without compunction, "but still there is enough to take from."[28]

Back in St. Albans, Sander was unpacking the early consignments of *Cattleya* orchids. The first plants arrived safely in early September. More plants in good condition were received in New York, too, and they were advertised for sale the following month. The nurseryman was overjoyed. Sander certainly knew this was the orchid from the same population Linden had introduced the previous year; in letters, he and Ericsson repeatedly referred to the orchid as "Waroq" or "Waroqueana." Indeed, Ericsson wrote proudly to his employer on October 9, 1891, that "the same man who brought Bungs plants sells them to me, Bung bought 2000 from him."[29]

The fact that he was now in receipt of an orchid, *Cattleya warocqueana*, that had already been "found" did not stop Sander for a moment. He determined to present himself as the finder of the "true" orchid, the *labiata vera*. He had been bested before by the Lindens; as he prepared to launch the new plants on the market, Sander did not intend to be wrong-footed again.

22

THE GREAT DISCOVERY

ON SEPTEMBER 12, 1891, Sander took out a full-page ad in the *Gardeners' Chronicle*. It trumpeted a major orchid sale on Friday, September 18, on the premises of the horticultural auctioneers Protheroe and Morris in Cheapside, East London. Sander announced a "RE-DISCOVERY" of "THE GRAND, OLD AUTUMN-FLOWERING CATTLEYA"—"supposed long-exterminated." "PLANTS IN MAGNIFICENT ORDER" of this "Queen of Cattleyas" would be made available to the public.

To make absolutely sure that readers did not miss the event, Sander took out a second ad in the same issue. Here he added historical detail, to fill in the bare bones of the story. "Dr. Lindley first described it from Swainson's plant introduced in the year 1818," the ad explained, "and named it in honour of W. Cattley, Esq, of Barnet." The new plants were sourced from "Swainson's original hunting ground," the advertisement continued, and Sander explained that this queen of plants was brought to market through "the indefatigable enterprise and persistent zeal of Mr. F. Sander." This triumphant find was termed in both of the Sander advertisements "Cattleya Labiata Autumnalis Vera," or the true autumn-flowering *Cattleya labiata*.[1]

The British press was exultant. Publications outside the worlds of horticulture and botany quickly took up the tale. "For nearly seventy years the search for *Cattleya labiata vera* has been proceeding without

The Great Discovery

> 298   *THE GARDENERS' CHRONICLE.*   [September 12, 1891.
>
> # RE-DISCOVERY
> of the supposed long-exterminated
> ### OLD AUTUMN-FLOWERING TYPICAL CATTLEYA LABIATA,
> AND
> # RE-INTRODUCTION
> BY
> # F. SANDER,
> OF THIS
> ### GRAND, OLD, AUTUMN - FLOWERING CATTLEYA.
>
> All plants have double sheaths, some green, others red; some plants have reddish leaves, others green.
>
> ## THIS IS THE OLD ORIGINAL LABIATA OF LINDLEY.
>
> The genus Cattleya was founded upon this species, after William Cattley, Esq., of Barnet.
>
> # CATTLEYA LABIATA AUTUMNALIS VERA.
> ### THE QUEEN OF CATTLEYAS.
> *(GUARANTEED TRUE.)*
>
> ---
>
> ### THE WHOLE IMPORTATION WILL BE OFFERED BY MESSRS. PROTHEROE & MORRIS,
> September 18th, 1891.
>
> PLANTS IN MAGNIFICENT ORDER AND CONDITION.   COULD NOT BE FINER.
>
> Collector writes:—"Plenty of whites are among them, and all shades of rose and red; flower seats, 4, 5, and 6."
>
> **The Easiest-Grown Cattleya.**
> **The Freest-Flowering Cattleya.**
> **The Varieties are endless.**
>
> FORM and COLOURING SUPERB — embracing, White, Rose-Purple, Red, Maroon-Crimson, and Yellow.
>
> ## By Order of F. SANDER & CO.

In this September 12 advertisement in the *Gardeners' Chronicle*, Sander blazoned his supposed "Re-Discovery" of the lost orchid.

intermission," the *St. James's Gazette* began breathlessly. This was not a fly-by-night discovery but the product of a longstanding, deeply fought battle, one that had swept up an entire industry. "During that time [the seventy years] every one of the shrewd and energetic men who import orchids has taken his part in it—Loddiges, Rollisson, Verchaffelt, Van Houtte, not to name moderns or amateurs. The money they have spent must be reckoned in tens of thousands." (Louis Van Houtte and Ambroise Verschaffelt were leading Belgian nurserymen earlier in the century, the latter selling his business, and *L'illustration horticole*, to Linden.)[2] Re-narrating the brave struggles of Swainson, Cattley, the Hookers, Lindley, and Gardner, the *Gazette* argued that the hunt for the lost orchid was a modern-day British epic in its form and scale: "If the story of their [the hunters'] sufferings were dug out of the faded old reports, where it lies mostly forgotten, and were 'written up' by practiced hand, it would furnish a small library of adventures. Briefly, the search for this plant is the romance of botany." Romantic it may be, but "this was no hunt for a mythical treasure," the *Gazette* continued, for the treasure in this case was real. Sander had won out, had found the hallowed prize, and all because of a "stubborn intelligence" that "deserved reward." The full details of how the prize was won remained to be told, the paper suggested excitedly; but truly the plant, "which cannot be mistaken" was found at last.[3]

The *Saturday Review* made similar claims of virtue rewarded on September 19, 1891. "Assuredly [Sander] deserves his luck—if the results of twenty years' labour should be so described," the anonymous author remarked.[4] The writer was actually the orchid enthusiast Frederick Boyle, who repeated his claims several years later in *About Orchids*, presenting Sander as the indefatigable, enterprising, and persistent finder of the orchid, as having justly earned the praise of the moment, together with the return on his investment. (There was no reference to Ericsson.)

Sander's calculation—developed in consultation, most likely, with his extremely involved business manager—quickly paid off.[5] The story quickly slid from the advertisements into the news columns of the same newspapers and periodicals, then into mainstream newspapers and periodicals, finally into book-length works. The story of the lost orchid, so to speak, went viral—and in America too, the sale of a thousand plants was greeted with enthusiasm. *C. labiata autumnalis vera* had been found by "our men . . . in full blossom on Christmas Day last," Sander reported to the trade magazine the *Florist's Exchange,* and the orchids were auctioned off, with one other species, on November 6 at 11 A.M. at 205 Greenwich Street, New York. The event was announced by the *Florist's Exchange* as "one of the most interesting orchid sales that this city has ever seen."[6]

Linden was horrified, enraged at Sander's breathtaking deceitfulness. Having literally copied Linden's approach, and sourced the same orchid from the same place, Sander was now passing off his orchid as a startling discovery, all while undermining Linden's and Bungeroth's achievements.

What came next was an almost ridiculous duel of full-page ads. On September 19, 1891, the Lindens took out a full-page advertisement, which offered for sale the "GRAND, OLD, AUTUMN-FLOWERING CATTLEYA"—"THE QUEEN OF CATTLEYAS," "Supposed long-exterminated," the "CATTLEYA LABIATA AUTUMNALIS VERA." This time, however, the plant was made available through Messrs. Watson & Scull, agents on East London's Lower Thames Street. Bristling with rage, the Linden ad assured readers and potential purchasers that the plant otherwise known as "Cattleya Warocqueana" was in fact—and had been judged by "All Connoisseurs"—the true plant, and had been since May 1890, and that *C. warocqueana* was a synonym.

Combative ads, interviews, and articles would play out in the media on both sides of the Channel for months, as both sides ferociously staked out their claims to be the true finder of the *vera* or true orchid. Both companies worked to assure consumers that they were worthy of the grail, not just that they *had* found it: worthiness was established, or was imagined to be established, by 1) indicating extensive understanding of the history of the plant; 2) indicating deep knowledge of hitherto unexplored locales; 3) indicating willingness to pursue the hunt relentlessly, risking danger and defeat; 4) stressing the purity of the sourced plants, their "true-ness"; 5) indicating the *in*authenticity of their rivals' plants; and 5) generally indicating that cerebral, scientific, and intellectual interests, namely the pursuit of knowledge, not money, had motivated their search.

To that end, Sander released a series of photographs that apparently showed him and his men unpacking crates of *C. labiata* orchids.[7] One, for instance, published in December in the *American Florist,* shows sober men examining paper records while a fascinated child reaches

> # RE-DISCOVERY
>
> OF THE SUPPOSED LONG-EXTERMINATED
>
> ## OLD, AUTUMN-FLOWERING, TYPICAL CATTLEYA LABIATA,
>
> AND
>
> # RE-INTRODUCTION
>
> By the TRUE RE-IMPORTERS,
>
> # Messrs. LINDEN, Brussels
>
> (In MAY, 1890), of this
>
> ## GRAND, OLD, AUTUMN-FLOWERING CATTLEYA.
>
> Plants have Double and Single Sheaths, some on the same plants, some Green, others Red; some plants have Reddish Leaves, others Green.
>
> All Connoisseurs have declared **CATTLEYA WAROCQUEANA** to be the Old ORIGINAL **LABIATA** of **LINDLEY.**
>
> # CATTLEYA LABIATA AUTUMNALIS VERA
>
> (Syn. C. WAROCQUEANA),
>
> ## THE QUEEN OF CATTLEYAS.
>
> **PLANTS MAGNIFICENTLY ESTABLISHED.—COULD NOT BE FINER.** Plenty of Whites are among them, and all Shades of Rose and Red; Flower Scats from 3, 4, 5, and 6.
>
> *Plant, in Sheaths, from* **30/-** *to* **80/-**. *Grand Varieties, and Extra Specimens, Price on Application.*
>
> ### THE EASIEST-GROWN CATTLEYA. THE FREEST-FLOWERING CATTLEYA.
>
> **THE VARIETIES ARE ENDLESS.**
>
> *Form and Colouring Superb, embracing White, Rose-Purple, Red, Maroon-Crimson, and Yellow.*
>
> **NEW IMPORTATION.**—Just to Hand, a GRAND AND MAGNIFICENT IMPORTATION of this AUTUMN-FLOWERING CATTLEYA, in the Finest possible Condition.
>
> **SENT CARRIAGE FREE to LONDON.** { PRICE . . . 10/-, 15/-, 20/-, and 40/- each. PER DOZEN . 80/-, 120/-, 160/-, and 320/-. SPECIAL QUOTATIONS PER HUNDRED.
>
> ORCHIDISTS are indebted to the "INDEFATIGABLE ENTERPRISE AND PERSISTENT ZEAL" of Messrs. LINDEN, since May, 1890, for this greatest desiderata.
>
> *All ORDERS to be Sent to—*
>
> **Messrs. LINDEN**, Directors of L'Horticulture Internationale,
> **BRUSSELS, BELGIUM.**
>
> *Agents in London*—**MESSRS. WATSON & SCULL, 90, LOWER THAMES STREET, E.C.**

Linden, a week later, countered Sander in the *Gardeners' Chronicle*, naming his own company "the true-reimporters."

hungrily toward orchids that spill from a crate stamped with Sander's initials. Photographs, with their "inherent accuracy and detail of rendering," aspire to offer objective proof of what "really" happened, though the medium of photography was used, Ellen Handy argues, as a mechanism for "propaganda and selective concealment" as early as the Crimean War.[8] In this case, the scene of unpacking the orchids is clearly staged to point the viewer to the same messages F. Sander & Co. pushed in print. The child's grasping for the flowers communicates the plants' desirability, even as the sober men signal, through their reading, that they are not prey to unbridled emotion. Intellectual engagement alone has brought the cattleya to London. In this way, Sander & Co. frame for the viewer how very much they "deserve" their discovery, how much superior values are the root cause of their success. The huge initials at the center of the image further communicate Sander's claim to be identified in the consumer's mind with these desirable and rare exotics.

At the same time, Sander worked energetically to cast the Lindens' plant as not true, not worthy, false. Sander repeatedly noted, for example, in September-October that his own orchid was "not Gardner's plant," as if there was a false *vera* that was; not that the Lindens had said anything of the kind—and anyway Sander switched to presenting his orchid as "Gardner's Labiata" by December.[9] Sander also rather superbly explained that *his* orchid "has no synonym"—banking, perhaps, on the fact that readers weren't exactly sure what a "synonym" was. Linden surely meant that "Warocqueana" was another name for the same plant, but Sander used it to signify something like "doppelgänger," the Hyde to his orchid's Jekyll. Then, in early October—the typeface in the ads getting ever larger—Sander began to describe his orchid's locale, too, conjuring up a lost Arcadian idyll as the site of discovery: "Found in a new locality," his orchid was "growing principally on rocks and low trees, over small streams . . . No Orchids have ever before been gathered in this part of Brazil." The orchid was "(Not guaranteed new)," he concluded rather unctuously, for indeed this orchid was supposedly as old as Eden, unlike, tacitly, *C. warocqueana*.[10]

Linden published a furious, five-page riposte to these kinds of tactics in an October 1 supplement to *Journal des orchidées*. He attempted to place the same piece in *Gardeners' Chronicle* too, but the British magazine published only a small portion of it, sniffing that it was too long.

ARRIVAL OF THE OLD PLANT OF CATTLEYA LABIATA AT ST. ALBANS, ENGLAND.

In this 1891 photograph from the *American Florist*, released to celebrate Sander's "discovery," a child reaches hungrily toward *C. labiata* while Sander and his employees bury themselves thoughtfully in paperwork.

Linden took Sander roundly to task for repeating that his plant had "no synonym." He repeated, in detail, the many times *C. warocqueana* and *C. labiata autumnalis* had been judged one and the same. He would have to await the judgment of history, Linden said rather piteously, for this would surely prove him correct; Linden also demanded of Sander to tell the world exactly from whence his orchid came. If the *warocqueana* was not the same as Sander's plant, then why exactly had Sander's men trailed Linden's agent Bungeroth through Brazil, dogging his every footstep?

Sander condescended to explain the supposed origin of his orchid in more detail in November. In an advertisement for the sale of "2000 Orchids," Sander told the world that the orchids came "from Rio Pinto." This, apparently, was "Swainson's hunting-ground, the only locality for the true old plant."[11] More crucial detail about an encounter with an entomologist named Moreau in Paris was added to the story to explain how Sander knew where to look, how he had stumbled on Arcadia. The *Gardeners' Chronicle* gave wings to the tale:

> It appears, that in the spring of 1890, Mr. Sander saw in the collection of Mr. E. MOREAU, at Paris, some Cattleyas which had been sent to him from Brazil. With the pardonable scepticism of one who had already spent thousands of pounds in searching for true autumn-flowering labiata, Mr. Sander refused at the time to acknowledge the claims of the plant to be the long-sought species, to be correct. However, he could not get the matter off his mind so easily, and soon afterwards thoroughly sifted the subject, and sent ERICKSON, one of his hardiest and most reliable Orchid collectors, to the spot indicated, and in due time received a great supply of which those referred to above are the last arrivals and the best.[12]

Needless to say, there is nothing in hardy, reliable Ericsson's letters to support this story whatsoever. There is no mention of Mr. Moreau, nor his astonishing plant. I have been unable to trace an entomologist in Paris in 1890 with the (even quite common) last name of Moreau; I can find no mention of such a specialist in the *Bulletin des séances et bulletin bibliographique de la Sociétié Entomologique de France* nor in the *Annals de la Sociétié Entomologique de France*. Sander gave no more clues as to the identity of Mr. Moreau, at once capable of sourcing rare plants yet unconscious of the significance of his discoveries, and the entomologist did not write of his find personally. Perhaps he did indeed exist, but it is hard to believe that a businessman like Sander—quick to threaten to beat his collectors, even willing to send them to their death—would sit quietly in Paris, with possible *C. labiata* spread out in front of him, hesitating to ask a question. Did Sander and Godseff love the idea that a Frenchman was sitting on the truth in Paris, under the very noses of the Lindens?

It is also hard to know where "Rio Pinto" was or is supposed to be. The Rio Pinto is in Colombia, not Brazil: it is where platinum was discovered (Spaniards termed the find *"platina del pinto"*). There *is* a Rio Tinto in Brazil, to the north of Recife, but that is not the locale Sander gave; indeed, he doubled down on the name *Rio Pinto,* repeating it as the habitat of the orchid many times. On November 20, while pitching *C. labiata* in the *Gardeners' Chronicle,* he offered other plants for sale from Ericsson's trip in "the Upper Rio Pinto, locality indicated in Swainson's correspondence."[13] He implied that he managed his great success with excellent close-reading skills and a robust willingness to explore, that by paying closer attention than anyone else to Swainson's documents, he'd found a shadowy place everyone else missed, a mysterious locale that escaped other specialists' attention.

There are some signs that readers scratched their heads, got out their atlases, and wondered. In the *Journal of Horticulture,* orchid enthusiast Lewis Castle repeated the first four lines of a little ditty he had recently heard: "The woods of Rio Pinto are now a gaudy show, The time is early winter, where old labiatas blow, And Sander of St. Albans, the Mighty Orchid King, Is filling all his boxes for shipping in the spring." Terming this sarcastic rhyme a "remarkable production," Castle adds thoughtfully: "whether 'the woods of Rio Pinto'" is really the home of the plants or merely a poetical fiction, I must leave to others to determine."[14] What the account of the rediscovery of the *C. labiata vera* lacks in veracity, as Castle guesses, it makes up for in romance and "poetical fiction."

Indeed, Sander & Co.'s marketing stories for their orchids mimic the plots, tropes, themes, and characters of popular contemporary adventure fictions by authors like Rudyard Kipling, H. G. Wells, and H. Rider Haggard. In these "imperial romances," as Patrick Brantlinger terms them, regions in India, Africa, and South America "offer brilliantly charismatic realms of adventure for white heroes, usually free from the complexities of relations with white women": such stories worked to make what was complicated, murky, and disturbing seem easy, triumphant, and assured.[15] Famous texts by explorers Henry Morton Stanley and Richard Burton, animated by the same urge, were excerpted in boys' papers and served to radicalize new generations into the cause of the British Empire. "Stanley's Adventures in Africa" were reproduced, for instance, in 1879, the first year of the *Boy's Own Paper*; in the magazine's columns, native people and animals run wild and are shown to be in dire need of sage, levelheaded British control. Weapons, courage, intelligence, and immense sangfroid are required from dawn to dusk, while the reward is fame, fortune, a passive woman, and the adoration of one's grateful nation.

Sander and Godseff surely saw that here was material they could use. Orchid hunting had already proved fertile territory for those seeking to celebrate British derring-do and supposed fitness for global domination. In Percy Ainslie's 1890 story for boys, *The Priceless Orchid*, for instance, the pursuit of exotics is presented as a way of proving one's mettle, making one's family proud, gaining a fortune, and even getting the girl one loves (the fair Dolly appears briefly at the end). The plot explicitly turns on the pursuit of a lost cattleya: in the opening pages, a Sanderian eccentric named Mr. Elan places a newspaper advertisement for a young man to help find an orchid that has eluded capture for decades. *Cattleya dolosa* is thought to grow near Minas Gerais, but all hunts in pursuit of it so far have failed. Mr. Elan has just learned it may be in Yucatán, Mexico; Jack, the young Everyman Victorian who answers the ad, doesn't hesitate for a moment to take up the challenge—once his mother has given him permission, at least. Astounding dangers plague Jack's every footstep, pits open beneath his very feet, but Jack is tougher than every local man and raving beast he encounters. Ultimately our hero finds the orchid, saves just about everyone he meets from hideous death, and wins Dolly when a grateful Mr. Elan obligingly dies and leaves him a million pounds.

Another boys' romance, *The Orchid Seekers* (1893) by Ashmore Russan and Frederick Boyle, is even more clearly inspired by Sander and his business (and was published after the lost orchid's supposed "discovery"). Sander is explicitly mentioned in the preface as an advisor on the story and appears lightly fictionalized as Ralph Rider. The lead orchid hunter, Ludwig Hertz, is modeled on Sander collector Benedict Roezl. (Rider, like Roezl, has a hook for a hand.) Two young brothers accompany Hertz in pursuit of orchids, a decision again thoroughly endorsed by the narrator as a sign they will do Britain proud. These stirring chaps are not, we learn when another long-suffering mother gives permission for her offspring to head out, "of the *genus* 'Molly Coddle.'" It apparently further proves the older brother's fitness for his orchid-hunting task that he "could throw a cricket ball one hundred and ten yards."[16] The plant-hunting adventures continue in Russan and Boyle's 1895 sequel, *Through Forest and Plain,* the brothers now seeking a lost cattleya with a crimson lip. In each text, the moment of final discovery is depicted—and visually represented, in accompanying illustrations—as the achievement of a bold European hunter, alone. Women, like orchids, are mere objects of desire (see Plate 10), while local people are threats only, challenges to be overcome.

Sander's version of events deftly pulled threads from these kinds of tales and wove them into the marketing for *Cattleya labiata autumnalis vera.* That marketing was subsequently sucked up by press stories about Sander and his business. In interviews, the St. Albans businessman was represented as a living embodiment of a figure from the pages of adventure stories, a little eccentric but upstanding, resilient, self-disciplined, brave. Measured, devoted to knowledge and truth, the Sander of the stories follows up on his lead sagely, after giving the matter sincere thought. There is little of the quick temper, carelessness of human life, or rapaciousness for profit we see spilling from his letters; the version of Sander presented in magazines and newspapers sees the plant he wants and waits judiciously before sending off a steady, "reliable" collector in its pursuit. Meanwhile the European specialist, Moreau, is figured as happily unconscious of the significance of the plant sitting on his tables, a move that positions Sander by contrast as the insightful, learned figure in the room.

But Lucien Linden was not one to admit defeat, and he too had access to the printed page. In January 1892, he assured the world (or at least his readers) that no, he did *not* believe the ridiculous story that at Christmas the English collectors had found *C. labiata autumnalis.* The Swedish Ericsson was too "corpulent" to achieve much of anything, he raged, still less to journey deep into the interior of Brazil and achieve a success worthy of the brilliantly talented Bungeroth.[17]

In March 1892, Linden brought out his own story of remarkable coincidence in a fresh article entitled "History of the Second Discovery" (*Histoire de la seconde découverte*). Again the astonishing event took place in Paris; again, the tale reveals determination, knowledge of the past, brilliance at searching, and adoration of the plant. Again the story seems a

As Hertz watched, he saw a Chinaman approach the orchids.

p. 264.

In *The Orchid Seekers: A Story of Adventure in Borneo*, the brilliant Hertz, with a hook for a hand, is a thinly veiled version of Sander's hunter Benedict Roezl. Orchid hunting is derailed, in the novel, by fierce local conflict.

"poetical fiction," but this time, Linden was the genius discoverer, the hero of the age.

And *this* time, there is at least some evidence in support of a truly remarkable narrative produced to explain how Bungeroth was sent, by L'Horticulture, to Pernambuco in the first place.

23
THE SECOND GREAT DISCOVERY

LINDEN CLEARLY FELT he had to act; many other tales were swirling to account for how the lost orchid had been found. A writer in *L'orchidophile* reported some of these many rumors: it was said, for instance (*"mon Dieu, ce ne sont que des on dit"*—"my God, what stories they say") that *Cattleya labiata* was introduced several years back by the Botanical Garden in St. Petersburg, sent by a M. Lietze, based in Brazil.[1] Others claimed it been introduced by a French importer from Bahia "in circumstances worthy of One Hundred and One Nights."[2]

Linden had to explain, then, how Bungeroth came to be searching in Brazil. Bungeroth published several pieces on orchid hunting in South America in *Journal des orchidées;* these went into detail about the environments in which cattleya and other orchids grew, but Bungeroth said nothing about what sent him to Pernambuco. Now it was Linden's turn to fill the gap, which he did in an article entitled in "Histoire de la seconde découverte."[3]

During the preparations for the Brazilian display for the 1889 World's Fair in Paris, the Exposition Universelle, he explained, Linden and his father Jean were intrigued by an oil painting of *Cattleya* orchids submitted by a Brazilian artist (Linden hastened to remind his readers that L'Horticulture Internationale had been hired to curate the Brazilian exhibit). The painting's blooms were colored somewhat fancifully, in violets and blues; still, the form of the orchid struck both men immediately. This was not the cattleya that would become known as *C. warocqueana*, but it was an orchid that nonetheless suggested to both men—Jean Jules especially, who had traveled extensively in South America—the presence of still-unknown (to Europeans, at least) *Cattleya* orchids in Brazil.

To find out more, the Lindens asked one of their employees at the exhibition, Florent Claes, for news of the artist, and the latter learned that though the oil was painted in Pernambuco, the painter himself, by great good fortune, was presently in Paris. Moreover, the artist had brought a specimen of the orchid with him; this he had given to a lady as a present. However, the lady was struggling to cultivate it, the Lindens learned, and would be happy to exchange it for a palm tree.

"The deal," Linden explained, "was done immediately." The flower was purchased and brought to the premises of L'Horticulture, while the painter was able to give directions to the orchid's habitat. The Lindens arranged for zealous Erich Bungeroth to go in pursuit of it and others like it, assisted by a local Brazilian man, known to the painter, who helped out as a guide. When Bungeroth reached Pernambuco, he sent back a few thousand plants, among them what "proved to be the old *Cattleya labiata* of LINDLEY."

It was true, Linden conceded, the first orchids to arrive in Europe were in poor condition because of a decades-long drought in Brazil; this explained their appearance on arrival. Still he, unlike Sander, could give a precise habitat for the plant. It only remained for Linden to add that the whole notion of "Swainson's hunting ground" was fake news—"intended to capture the imagination of buyers, but absolutely meaningless"—and to stress that Sander's collectors were simply following in Linden's own hunter's footsteps.[4]

Some of this is true, of course. Sander's collectors were, without question, following Linden's hunter, Bungeroth; their letters amply confirm this (though Bungeroth's from the period have not survived). But the detail of the painting sounds, on first reading, utterly implausible, another marketing tale produced to remind us of the Lindens' prowess at the Paris Exposition. No mention of the painting appears in the plant hunter's letters, certainly.

Ericsson informed Sander in 1892 (the Swede reported twice, on January 30 and February 14) that Florent Claes told him Linden discovered

Ericson
Pernambuco
30/1/92

Dear Sir
Yours of the 24 December to hand, about the labiata, Claes told me it was first discovered at the Exhibition at Paris 89, and it is very likely that is the same thing about the Rex &c. &c.
Bungeroth was here in the year 89-90 there has been no other here except Claes who arrived here in November last and left last Saturday for the South, of course I have been told that he left for the North but within one hour I will know, and wire to you. Bung. is supposed to be in Ecuador but that may mean the opposite direction. I believe Lindens new things come from this port

In this January 1892 letter—heavily underlined by the Sander office—Ericsson reports what he has learned from Florent Claes about the origins of the *Cattleya labiata/warocqueana*.

the orchid during preparations for the 1889 exhibition.[5] But there is no reference to a mysterious canvas, nor to a painter who traveled over continents with an orchid in his luggage, still less to a lady willing to exchange her orchid for a palm tree. On the face of it, then, this looks like a fiction worthy of Sander—not least because the trope of a painting that just happens to give access to hidden truths was a well-established, even hackneyed literary convention in the period. In Charles Dickens's *Oliver Twist*, for instance, a haunting portrait of Oliver's dead mother offers pointers to Oliver's true identity; in Mary Elizabeth Braddon's *Lady Audley's Secret*, a flaunting painting of the heroine offers clues to the seemingly angelic heroine's debased character. In Oscar Wilde's *The Picture of Dorian Grey*, a portrait of the hero registers moral decline and degradation even as the subject's living face remains untouched. Linden's claim that he saw a fantastical painting that directed him to the hidden site of the true *C. labiata* sounds as if it draws from the box of most clichéd Victorian tropes—one that Sander booster Frederick Boyle would also dip into, when he later wrote that Sander sent off a hunter in pursuit of *C. labiata* to Mount Roraima (at the apex of Brazil, Venezuela, and Guyana) after seeing a drawing of a similar plant in the British Museum.

But there *was* a painting at the exhibition that matches Linden's description. An 1888 canvas listed in the exhibition catalog, titled "Orchidées" by Estêvão Silva, represents lustrous white *Cattleya* orchids whose fantastical blooms, winding around a trunk, are rendered in weird, unearthly tints. This is surely *"un tableau a l'huile représentant un Cattleya,"* a painting of a cattleya, in tones of strange blues and purples, *"des teintes absolument fantaisistes, du bleu, du violet etc."* (See Plate 11.)

Did it indeed prompt the Lindens to approach the painter? The details about the lady and the present are surely fictional, but Silva was very real, an important Brazilian painter and teacher. The *Cattleya* painting was one of twenty-seven by the artist on display at the Paris Exposition, the largest number of any Brazilian artist. Linden does not give the painter's name in his account, however, which seems odd; one might imagine the information would serve to back up his story, if true.

Silva was Black, the son of an enslaved couple and the first Black artist to graduate from the Imperial Academy of Fine Arts. Silva specialized in still lifes, luscious evocations of fruits in particular, and he worked at the Lyceum of Arts and Crafts in Rio de Janeiro. He was something of a rebel, known for protesting before Emperor Pedro II when his art was given second place in the 1879 General Exhibition of Fine Arts and not, as he and others believed he deserved, first. He was censured for his protest and suspended from the Imperial Academy for a year.

Silva's prominence at the exhibition may have been part of the nation's attempt to position itself to the world a decade later as a country that had shaken off its associations with slavery—the endpaper of the exhibition catalog highlights the abolition of slavery—"L'Abolition de L'Esclavage au Brésil"—in 1888. Silva's painting of an orchid and his still lifes of succulent fruits also served to hint at the richness of the country's natural offerings, even as its creator's neoclassical training pointed to the intellectual and artistic rigor of the nation's institutions. This, as we've seen, was how Brazil wanted to pitch itself to other nations. Silva's painting could well appear, at least when placed in the context of the remarkable pavilion, consistent with a certain kind of upbeat Brazilian national story.

Out of that context, though, his art may prompt more disquieting thoughts; as Edimilson de Almeida Pereira writes, in a poem "The Painter Estêvão Silva," the artist's subjects hint, on close examination, by those willing to look attentively at something else, something darker:

> . . . his work
> has an unsavoriness that
> binoculars from a distance
>
> would find plump and pleasant.[6]

A luscious surface, in other words, is all that will be seen by those looking at the canvas from far away, for entertainment. The more attentive may find something less comfortable in the paintings' dark decay and loss. Silva's skill in a genre so associated with European "Old Masters," and with aristocratic collectors, reminds the viewer that Black artists have been excluded from generations of art. Silva takes on a genre found in every major gallery in Europe and North America: to look at Silva's haunting canvases is to glimpse the unfathomable generations of artists and tastemakers the world has lost, the millions of enslaved men and women unable to pick up brush or pen, the denigration of art forms Europeans chose not to understand. Silva points inexorably to those who came before him as well as to those still fighting for access and respect.

Linden sidesteps all of this by not naming Silva. He surely saw, and remembered, Silva's painting. But he did not name the rebellious Black artist who produced the work. The complexities of Silva's art might, perhaps, have been a distraction from Linden's purpose—which was firmly to foreground the perspicacity of his own work and company. He only mentions, in passing, that the anonymous painter introduced him to a Black man ("*un nègre*") in Pernambuco who knew the locality well and who served as guide. In this way, a man of color appears in the tale, but in passing, in the form of one who acts as servant. Linden foregrounds himself and his father as the ones who see, act, and comprehend. If any part of this story is true, if indeed a Black guide showed Bungeroth the place where *C. warocqueana* grew, this may indicate that the lost orchid's habitat was well known by members of Pernambuco's Black community all along.

Silva died in November 1891, aged just forty-six; he presumably never heard or learned of the role given to his painting by Linden.

The British press, meanwhile, all but ignored Linden's revelations, still less the implications of Silva's art. They continued to fall into line with the tale Sander told, gleefully crowning him the discoverer of the lost orchid. Robert Allen Rolfe at Kew described Sander's discovery "one of the most interesting and important events to the horticultural world that has occurred for many years. Its continued existence in a wild state, notwithstanding all that has been written to the contrary, is now an assured fact." Rolfe also wrote that the Sander plants were introduced "from Swainson's original hunt-ground" (not that he knew where this was, he admitted), and he described the recent thrilling sale of the plants "to eager buyers," whipping up the importance of Sander's discovery. He did not press too closely on the story of the locale—even as Linden very reasonably pointed out that if the actions of Sander's men were to be believed, this so-called "Rio Pinto" seemed to extend over three Brazilian states at least (Paraíba, Pernambuco, and Alagoas). Linden also noted repeatedly that his men, including Florent Claes, had literally seen the Sander hunters following around after them—they "followed [our] trail with dogged obstinacy."[7]

But Sander's story of the lost orchid was too good to give up. "Messrs Sander are ... heroes," gasped Frederick Boyle in 1893. The tale of recovery is "pure romance" realized, Rolfe agreed. Rolfe's September 1891 article in the *Gardeners' Chronicle* worked to help Sander turn recent events in Brazil into a British-centered story of derring-do, with a beginning, middle, and ending shaped by (mostly) British actors. He renarrated the discovery of the plant by Swainson, centering the parts played by Hooker, Cattley, Lindley, Gardner, and finally Sander, concluding with a triumphant end in London; this was a *Boy's Own* romance of a tale, for grown-ups.[8]

Of course, he also had to redescribe his own part, in naming *C. warocqueana* in the first place. Rolfe held out the possibility that Linden's plant was indeed the same as *C. labiata autumnalis,* in which case "there will remain the remarkable fact that two horticultural firms should have independently re-discovered a plant, which has been searched for over and over again, and both within a short period of each other." But Rolfe placed Messrs. Linden at a disadvantage in the great competition, for not having realized that *C. warocqueana* and *C. labiata autumnalis* were one and the same. As for his own part—well: if only he, personally, had "pronounced it to be the genuine old C. labiata," he mused regretfully, if only—but "we may leave the reader to finish the sentence."[9] A veil of mystery was thrown, once more, across the tale.

Linden did not give up. Dueling ads throughout the autumn of 1891 laid out the two firms' claims to the public, with lots of recourse to capital letters and terms like "TRUE OLD AUTUMN-FLOWERING" and "supposed long-exterminated." Sander's and L'Horticulture continued to lob insults at one another and at their orchids, too—Sander pointed out that the *Cattleya* orchids imported by the Lindens had flowered that first summer of 1890, thus could not be "true." Linden argued that the flower had been disturbed by its journey and had consequently flowered at an odd time, but that it had now settled into its autumn bloom.

British journalists continued to champion Sander's part, and Linden and his colleagues in *L'illustration horticole* decided to stop mincing words. Yes, Sander was known as "King of the Orchids," one author fumed—because he had given *himself* that title. The great Belgian achievement in finding the "true" cattleya was being concealed, pushed aside by the British for reasons of rank prejudice: "it is chauvinism applied to commerce, and I imagine that our neighbors would be very surprised to hear themselves called chauvinists." The Belgians had won, the British couldn't take it—that the "Belgian horticultural industry . . . succeeded in defeating the English in their own country."[10] This was little more than a case of a greedy businessman unwilling to see the market flooded with new plants when the old ones fetched such high prices; one writer claimed, on Linden's behalf, that *C. labiata* had sold for 1,500 francs a bulb before the "rediscovery," a fully developed plant for six to seven thousand francs, and he suggested that what truly stood behind the conflict was an urge to keep a stronghold on the market. The author was at pains, in other words, to puncture Sander's claims—made by him and those defending him—to self-disciplined, worthy high-mindedness. With bitterness, he was framed by the Belgian press as motivated instead by a grotesque combination of chauvinism and rank avarice.

Sander and Godseff had no intention of changing course. In shades of Malory's *Morte d'Arthur*, the company sent two more seekers to quest with Ericsson, Frenchman Louis Forget and Cornelius Oversluys from Delft. The two men arrived by ship, the Royal Mail steamer *La Plata*, in the third week of November 1891.[11] All three were charged to fend off the Lindens' men by any means necessary; and to strip the area of every last *C. labiata vera* they could find.

24

ERRORS

IN 1897, LOUIS FORGET published an article in the French periodical *Le jardin* about his experience sourcing *Cattleya labiata* in Pernambuco, entitled "Erreurs géographiques concernant les orchidées," or "geographical errors concerning orchids."

The title may sound dull, the article abstruse. But the "errors" of which Forget writes are the ones that sent pursuers of *C. labiata* in circles for decades, the ones that so distressed William Digance, the ones that dogged Forget and his colleagues in their last, anxious days hunting the orchid to near-extinction in 1891–2. Forget's article is, at its heart, a cry in the dark about abuse of the natural world and the treatment of socially marginal men like him by the driving forces of business and capitalism. It is a rare moment in which a hunter's frustrations made it onto the printed page.

Forget takes aim, obliquely, at orchidomania itself. Interest in orchids, literature about them, focuses on the plants *once they arrive in Europe*, Forget notes. The focus of those describing and celebrating an exotic is on the plant's European afterlives—those who found it, who imported it, who bought it, who named it, who first wrote of it. The plant is treated as a material, aesthetic, fundamentally European object, astonishingly little attention paid to it as a constitutive part of its habitat, in relation to other native plants.

C. labiata vera, for instance, was "sent to Europe in 1818 by William Swainson, without indication of the place of origin." This posed a challenge in its recovery, and what may have looked merely like Swainson's carelessness or forgetfulness at the time (personal distress caused by "costive bowels," say, or political insurrection) appears, seventy-five years later, part of an entrenched phenomenon. Swainson viewed the specimens he collected as objects that acquired meaning under the microscope, beneath the eye, and under the pen of the European botanist. He had little interest in the environment of which the plant was a part, except insofar—again—as *he* gave it meaning: as unexplored space, virgin territory, or dangerous threat. Of the "hunting ground" as ecosystem, of the plant and its seasonal growing patterns or interactions and relationships with nearby plants, animals, insects, the climate, and the soil, Swainson knew little. He did not see the plant in those terms. He plucked the orchid with few referents.

Forget points out that men like him are among those who suffer the consequences. With little information about geography, climate, and population to go on, collectors are set a near-impossible task: "Given the extent of Brazil ... it's as if you were saying: *Such a plant lives in Europe—search.*" Digance made almost the same claim fifteen years earlier, when he wrote to Sander plaintively, "please state the localities they come from when possible as Brazil is a large country to search for plants in."[1] To Forget and his peers, writing in a decade waking up to the emerging discipline of plant ecology, the issue is not just a scholarly one. The study of the relationships among organisms and their interactions with the environment was an emerging science.[2] But for Forget, Digance, and their fellows, such information was a matter of life and death: "what disappointments does the collector experience," Forget writes: "what ordeals is he not subjected to in seeking according to such imprecise data! And that is why, some, launched on false tracks, exhausted by vain research, do not survive, while many others, for the same reasons, are accused of ineptitude, lack of flair and lose even their reputation as intelligent collectors." If only a fuller, more "accurate history" of each plant could be given to them, how much would this help the collector and, later, the horticulturalist, struggling to acclimate the plant to its new home.

Forget touches on habitat loss, too, in his account of "errors." Approaching Brazil as a resource ripe for exploitation has done so much damage that the area in which *C. labiata* once grew luxuriantly is stripped. "The foreigner, in this environment, wonders where and how *C. labiata* and its allies can exist, because the mountains are deforested in all the regions accessible to the gaze," Forget notes. *C. labiata* as a species particularly battles to thrive on dead wood; it needs living, healthy plants to grow among. Its struggles, then, augur ill not just for its own survival, but for the area as a whole: it is the canary in the coal mine (an analogy scientists continue to make to this day, terming orchids "bioindicators and early warning systems ('pit canaries') of ecosystem health.")[3]

Forget makes an effort, rhetorically and imaginatively, to re-situate *C. labiata*, to rewrite it in relation to its geography, climate, native plants. The supposedly "lost" orchid may be found in the mountains of the interior in the state of Pernambuco and in two neighboring provinces to the north and south, Paraíba and Alagoas, he explains, and he evokes how the landscape changes over the course of a year:

> The flat land situated between the mountains is called *Certon* [Serton]; you have to see the vegetation that covers them to understand it: here are low, rickety shrubs called *Capeira*, there, a slightly higher vegetation called Catinga, where everything grows during the three months of the rainy season with incredible rapidity; after which, the leaves and the young shoots dry on the plants, and this, apart from the terrible heat, gives these regions at this time the wintry aspect of Europe.

By searching through "ravines and caves," *C. labiata* emerges in Forget's account entwined with its neighbors, interacting dynamically with the air, the landmass, the vegetation, and the microclimate:

> And then we find *C. labiata* under large trees whose trunks are at times decorated with *Pothos, Philodendron, Begonia, Anthurium, Davallia*, etc. It is mainly on the side branches of these trees that *C. labiata* grows, together with other Orchids, Bromeliads, Lichens, etc., its coiling roots extending right and left up to two meters, their *suckers* always intact drawing from the air the ingredients that suit them. It is there, in the penumbra produced by the evergreen leaves of the trees and receiving the fresh and invigorating air which always blows on these mountains, that the *C. labiata* thrive.[4]

Forget's vision of ecology is one that resituates him and his peers in the narrative of plants, too. The pervasive ignoring of geography connects, he suggests, to the ways in which men like him have been treated as expendable, their day-to-day struggles of little account. He tries to remind readers of the roles played by hunters in the face of cultural and literary narratives of plant "discoveries" that worked to conceal the conditions of their labor.[5]

To read the eccentric, rain-soaked, exhausted letters of plant hunters like Forget, Ericsson, and the grim Oversluys, resting after more than a century in their manila files at Kew, is to be vividly aware of this. Forget's account in *Le jardin*, frustrated as it is, remains a muted expression of the anger and confusion that hunters articulate in their private letters. It is also a tidied-up account of the plundering, the shockingly violent and haphazard collection practices that continued apace in the hunt's final months. Sander evidently doubted that Ericsson had stripped the region of *all* its *C. labiata* orchids; after all, the Swede still seemed to be finding new varieties, new colors into the last months of 1891. Sander and business manager Godseff had serious doubts too about Ericsson's sobriety and judgment. And they wanted to be sure they had found all of Bungeroth's orchids so Linden's man could not return and source more; they wanted to reach the very farthest boundaries of the mythical "Rio Pinto."

And so the hunt continued, violent, disorganized, and destructive, deep into 1892–3.

OVERSLUYS REACHED Pernambuco in November 1891, to Ericsson's relief: he spoke Spanish, at least, and "Spanish is very much like Portuguese," the latter wrote hopefully to Sander, weary of ongoing communication problems. A local German-speaker had taken advantage of Ericsson's lack of Portuguese to convince him that there were two more "Belgies" in the area—aka agents working on Linden's behalf. A panicked Ericsson paid for days of help before he grasped that the German's story was a lie.[6] Forget arrived next, in late January 1892. In his first letter from the region, he reported traveling some sixty miles on horseback from Caruaru and finding, in the woods outside Brejo, "the cattleya." But he was swiftly sent off again, on what he felt was a fool's errand—to Rio Formoso this time, over a hundred miles away, toward the sea. He blundered on—to find "nothing," he wrote angrily to Sander, "for I will tell you a thing of which you ought to have been told before": the orchid does not grow much below 500 meters above sea level. Even at 500 meters, *C. labiata* is "still very weack and scarce"; the higher you go, the more they "are founds in more aboundance and healthy."[7] The trip toward the coast was long, arduous, dangerous, and frightening; perhaps he began mulling his article on "geographical errors" around this time.

Sander's poor understanding of the area was far from the only problem facing the three men. The letters from this period recount trickery, danger, illness, panic, and Gilbert and Sullivan–style chaos. Linden's agent Florence Claes—the man Linden approached, apparently, after seeing the Silva painting in Paris—appeared in person in late 1891 to pursue Linden's interests and disrupt Sander's plans. By this time, the battle between the two companies was raging violently in the European press, and the dueling hunters eyed one another with deep suspicion. Ericsson pondered whether Bungeroth and Florent Claes were actually one and the same (they were not); at least, he mused, he had priority with boots on the ground. He'd made it known to local people that he would pay more than Claes for anything they found. He had spies at the ready to notify him if the new arrival attempted to send anything to Linden.[8]

His cool soon passed, however. Claes was terrifying. "[I]f Class could, he would eat me at least he looks like it," Ericsson wrote frankly, and Oversluys agreed, writing in unusually clear and transparent terms that Claes was "mad like hell." Forget also reported a barrage of rage, insults, and lies: he is "terribly wicked" and "able to do anything," Forget reported, indicating from the first an urge to give his adversary a pair of black eyes. Inevitably the conflict did turn physical: Forget noted pugnaciously that he "made [Claes] very obliged to close his mouth."[9]

Oversluys began to suspect someone in the Sander establishment was giving information directly to Linden about their location—though Claes may have simply poured this poison into his ears to rattle him and Sander.[10] Claes seems to have given out information to his adversaries strategically: he put the story about, for instance, that there were seven hunters on the case, to make the Linden side seem more intimidating. Ericsson knew enough to distrust this piece of information, at least. He also told the Sander men explicitly that *C. warocqueana* surfaced during the Paris Exposition several times, and Forget and Ericsson repeated this in letters to their boss (though, as we've seen, without the detail of the canvas, the painter, and the lady). Claes must have realized the hunters were reporting to St. Albans, and he clearly wanted to rub in Sander's face that Linden, unlike Sander, had a plausible origin story.

Claes was headquartered at Canhotinho by December 1891, twenty-two miles to the east of Garanhuns. The Sander men were stationed three hours to the southwest of him, in Palmeira. They glowered at one another across the trenches for weeks, the Sander men exploring farther south (toward Correntes) and west (toward Bom Conselho).[11] Claes worked to disrupt the Sander men's collection in numerous ways—by bribing local people to stop them handing over orchids, by enticing the Sander men to waste money so as to exhaust their funds, by besmirching their reputation in the area ("everywhere he make us black," Oversluys wrote), and by generally distributing false information to anyone who would listen. Ericsson reported several times that his crates of plants had been held to ransom by a seemingly uninvolved third party in ways he suspected could be traced back to Linden. Once, a charge of misconduct was filed against him. "I had got into trouble with a native and if I had not paid up before a certain date he very likely he had sold my plants to the Brazilian," a chastened Ericsson reported to his boss. After receiving the letter, Godseff scrawled on it, "Curious letter I hope Lindens maneouvres have not defeated him." (It was a common practice of the office to comment on letters and underline key phrases, and anything involving *Bungeroth, Claes/Class* or *Linden* earned heavy underscoring.) Claes was also traveling with *Cattleya* plates from the *Lindenia*, trying to train local people to go off and search for them. Oversluys thought this an excellent approach and encouraged Sander to send him pictures, too.[12]

The Sander men also knew that Linden's agent was trying to intercept their correspondence with St. Albans. Telegrams were a necessary means of communication for urgent matters, yet interception was a problem, even as various legislative attempts were made to give telegraph lines the same protections as postal routes. Still, workers could be bribed to disclose the contents of telegrams they'd transcribed, or alter the content of cables going out, thus the system was known to be vulnerable. When August Strunz, commission merchant and agent of the Royal Mail Steam Packet Company, received a cable reading "Give Ceredit [Credit] Overslays," he wrote dubiously to Sander: "I do not feel sure about it owing to the manner in which telegrams get changed in this country."[13]

The Sander men elected to use "John" as code, to render the content of their messages safe from prying eyes. But what did John mean, exactly? The first rule of a code is that everyone has to agree on its meaning, only everyone seemed to think "John" meant something different. Ericsson thought "'John' stood for Linden not for Bung specially," but Oversluys reported no, that Sander said "John" meant Bungeroth—"what means Linden then??" Ericsson demanded. He had been using "John" all this time—"I have been wiring to you all the time 'John' send so and so many cases?!!"—and now "Mr. Oversluys just arrived tells me 'John' stands for Belgies. Also not Bung specially."[14] Either Sander did not give clear instructions, or the men misheard, or they were unable to agree.

Then Claes abruptly vanished, leaving the Sander hunters further baffled. Ericsson, hearing that he had gone to the north, immediately assumed he had gone south.[15] Making further enquiries at Claes's hotel, he learned that the agent was gone into Bahia, under an assumed name—"but how to trace this people if they travel under other names?" he fretted. He sent Forget after Claes, just to be sure they knew what he was up to. Forget located Claes successfully, but he walked into danger not once, but twice on the journey: first in a railway accident, then in local political unrest.[16]

Sander's entrenched unwillingness to pay what his collectors needed, meanwhile, stood in stark contrast with Linden's open-handedness with his men—a point Sander's employees made regularly in their letters. Sander was furious about what he considered the pointless expense of following Claes, and he excoriated Ericsson on paper. Those particular letters have not survived, but Ericsson's dismay and sense of betrayal spills from the pages he wrote back in answer: "I know you are not Rotschild," he wrote, "but I did think when I asked Mr. Forget to go to Bahia that it was in the interest of the Firm." Then, just as Claes vanished, a new collector, thirty-year-old Edward Kromer, appeared on Linden's behalf. Kromer was a colorful character who loved annoying his rivals—"He is a humbug I think in his white waistcoat and white cravatte," Ericsson wrote—and he was swimming in ready money, thus pushing up the local cost of plants. He tried to bribe the team to join him and literally walked around clinking the money in his pockets. The Sander men, remarkably, stayed loyal, though they did point out that Kromer understood their plight: "Kromer will lik [lick] you soon he must have a devilish lot of money," Oversluys informed Sander; he "has played hell and through every thing he could see you perhaps afraid that one of us is going to run away with your money." Now, Oversluys explained, they were in the ridiculous position of having collected orchids but of being unable to afford the packing cases to ship them in, still less the cost of the freight. Claes would scoop up what was left on his return, for sure, and they would all be forced to sit and watch. And Sander, it seemed, had more or less abandoned them: "you might at least send us three armchairs," Oversluys wrote acidly, "so that we can see at our ease what the others are doing."[17]

Ericsson, in desperation, threatened to throw the plants of "old Labiata" he had just collected on the "rubbish heap tomorrow." Indeed, the whole situation stretched him to breaking point: He was at war with the elusive Bungeroth, then the angry Claes, then the grinning Kromer. Insinuations had been made about his probity, and he faced the challenges of navigating rising costs and the local legal case against him. He made pointless and embarrassing trips to the bank to ask for money, only to be laughed at by the teller, all while reading in the newspaper about Sander's magnificent sales in London. (A local Englishman had shown Ericsson the blaring ads; the Swede was not amused.) On top of all, Sander demanded he track a new *Cattleya* orchid named *C. rex*. Ericsson dared Sander to come out himself, to see how they were situated: Sander dismissed this as "ridiculous."[18] Ericsson, ever more aggrieved, reported that they were all on the edge of stopping work, his anger disrupting his letter writing and very nearly his practice as a plant hunter. "Now I am so upset I can not write any more," he exploded, on March 3, 1892, expressing himself abandoned by the man thousands of miles away. In a postscript he added piteously: "Do not forget your Travellers in Brazil for the Bricks of St. Albans."[19]

What Sander wrote in response to this I could not discover. But Oversluys wrote, in March and April 1892, that he thought he had found yet another spot where Bungeroth collected quantities of *C. warocqueana*.[20] The site of the plant pictured in one of Linden's plates was Paquevira, near Canhotinho, he reported, close to the border of Pernambuco and Alagoas, though it was difficult to be certain when collecting plants not yet in bloom. Ericsson, jealous of his colleagues early in the hunt, seems to have softened toward them as they too suffered the ill-effects of Sander's close-fistedness. Sander's response to Forget, and his trip, irked Ericsson particularly. The Swedish plant hunter began to describe his colleagues more warmly, and the three worked collaboratively that spring, sending back hundreds of cases to St. Albans.

But the risk of disease returned. Yellow fever was in the region, introduced (or so Oversluys heard) by the boat that shuttled passengers between Pernambuco and Rio. Two men died on board, and since a quarantine had not yet been established, the disease spread rapidly, possibly from the doctor who treated the dying men. Oversluys, according to Ericsson, was "very, very much afraid" of yellow fever; Digance's still-recent loss can't have helped. After packing up *labiata* plants into fifty cases, and scribbling out roughly what was in each case, Oversluys left for St. Thomas. (Ericsson sent another fifty cases in the same consignment.[21])

It was the beginning of the end. Kromer, too, departed the region, though not before getting into a spat of his own with the local authorities when he refused to pay what amounted to a plant-hunting tax. He called them a "lot of canailles"—rascals, vulgar folk—and told them they could go to hell; not impressed by this, the authorities took away his horse and saddle. Despite this setback, he was still able to send out forty-three cases of "plantas vivas" in April and a further two in May, according to the local newspaper.[22] Forget headed out too, at least for a while, leaving Ericsson behind; he rather preferred it that way, the latter wrote sadly, with just plants for company. "I will go my own way," he wrote to Sander, "and after my own wishes." Perhaps rather lonely, in spite of this bold claim, he planned a future visit to St. Albans, to see "some of my old labiata friends."[23]

There is little indicator, from the letters or other documents, that the well-being of Ericsson or any of the hunters concerned Sander or his associates much. Right up to the last months, weeks, days of the hunt, their focus was on how to best Linden. Ericsson, in his final weeks in Pernambuco, was asked to research how much rain had fallen in the region over the last ten years—no doubt because Sander was working out how to answer Linden's claim that the only reason the first *C. labiata/warocqueana* plants were small was that the region was suffering from a drought. Ericsson explained to his boss that it had rained every year, certainly—if not, perhaps, as much as in previous decades. Interestingly, he explicitly connected this climactic change to "the cutting down of the big woods" which, he thought, had "a great deal to do with" it.[24]

All three men's plant-hunting narratives, as expressed in their letters, remained essentially the same as their lives diverged. Their scrawls tell of ongoing fights over money with Sander, battles with other collectors and agents, the threat of disease, and of hundreds, thousands of cases of plants of many species and varieties wrested from their homes and into packing cases, across the ocean, to feed consumer appetite at St. Albans, New Jersey, and in Belgium. Forget continued hunting *C. labiata* in the region, although he periodically disappeared, according to a hunter named Louis Perthius, whom Sander sent in pursuit of the vanished Forget and any last, lingering *C. labiata* orchids in 1893.[25] Oversluys got the fever he so feared and struggled to write his letters with a violently trembling hand—"the post will leave I am all shaking: and cannot finish."[26] The last document in his archive notes that because of ill health he was advised to stop collecting by his doctor: a handwritten addition from the Sander office comments unemotionally that efforts to reach him have gone "not answered." Ericsson—that supposedly "reliable" collector—was abruptly fired when Sander decided he was too expensive to keep on. He got stuck in Singapore when Sander refused even to send him the fare home, and only managed to return to Europe when another collector, Wilhelm Micholitz, lent him money.[27]

Perthius too was cut off by Sander with nothing, abandoned in the middle of his journey. His father's articulation of pulsating fury in the aftermath offers us a rare glimpse of a very different narrative—a full articulation of Sander's faults, that is, a *j'accuse* attempt to hold the businessman accountable. Léon Perthius wrote to Sander to tell him he was "vile," a "criminal" for leaving a young man in the "virgin forests of South America" to die alone. Léon even wrote out two copies of the letter and sent them to Bruges and St. Albans, to be sure Sander would receive it. These two carefully handwritten documents point us resolutely to the human costs of Sander's business, the pain and suffering that stood behind the tales of triumph Sander blazoned proudly in the press. How was it even possible, Léon asked, that a man with his own family, a father, could treat another man's son this way; could risk another man's life and sacrifice his youth to "enrich [his] establishment"? If it hadn't been for the help of strangers who took pity on Perthius, showing him more consideration and kindness than his own employer, Léon wrote, "I was condemned never to see my son again."[28] When Ericsson wrote to Sander "Do not forget your Travellers in Brazil for the Bricks of St. Albans," he surely recognized all too well that the edifice of orchidomania was built on the forgotten foundations of expendable hunters like him.

25

"SOME BRAZILIAN"

ORCHIDOMANIA WAS NOT ONLY exploitative of men like Forget and Ericsson, of course. The Sander & Co. successes depended on networks and teams of men Sander never met, whose names he barely knew, whose labor was rarely mentioned, even in passing, in European-authored literatures. Sander did make brief reference to his European-born plant hunters in interviews, articles, and advertisements, and sometimes hunters' names became attached to the plants they had collected. Even Digance had a variety of cattleya named after him, *Cattleya diganceana*, though that name was later replaced. Occasionally, too, the voices of the plant hunters (like Forget and Bungeroth) appeared in print, though the vast difference between the polished texts associated with their authorship and their ragged, ungrammatical letters suggests that a third party was heavily involved in the writing and revising of their manuscripts.

But published works on orchid hunting in the era almost never referenced, indeed carefully concealed, the labor of the local people whose expertise would prove so crucial. The moment of "finding" an orchid was described—and visually represented—as a moment involving the European orchid hunter, alone. In *The Priceless Orchid*, for instance, a stirring boys' novel, a hunter reaches out for the orchid he has sought deep in a jungle, which he has penetrated through sheer determination and

"There, growing in the marsh, was the priceless orchid he had sought for so long."
—*Page* 193.

Percy Ainslie's *The Priceless Orchid: A Story of Adventures in the Forests of Yucatan* shows a lone British hunter finding the orchid he seeks. Sander's archive reveals that local, on-the-ground knowledge, expertise, and labor was critical in the "discovery" of plants.

perspicacity alone. No one has helped him get here. No one else, apparently, merits a place in this scene of great success.

Albert Millican's *Travels and Adventures of an Orchid Hunter* (1891), a supposedly nonfictional work representing the author's search for *C. mendelii* in Columbia, is also typical for the ways in which it erases the agency and knowledge of those in whose communities orchids grow. The people who row boats and canoes, climb trees, carry provisions, fish, and provide food and drink, pack and move plants on mules appear only briefly, in passing. Their labor is depicted as a carrying out of the white narrator's orders; they seem unable to act with decision and hardly comprehend the challenges of living in their own landscape. A "native" foolishly attempts to remove the teeth of a living crocodile, for instance, and is unsurprisingly eaten for his pains. Claims of laziness, foolishness, drunkenness, and mendacity abound. Millican interprets his employees' unwillingness to follow his rules on one voyage as "a fit of laziness," even as it turns out they were fishing for food—and, by the next page, they are making fires to cook the fish and prepare the coffee. The narrator does not seem able to see their work, even when it is literally performed in front of him and benefits him personally.

As in the botanists' travels discussed earlier by writers like George Gardner, Millican's narrator presents himself as the one who not only works, subdues, and adventures in the terrain he visits, but also as the only one who interprets and understands. He strengthens his claims to both heroic and intellectual status by contrasting himself with the supposed passivity and ignorance of local people.[1] Accompanying images again reinforce the text's structuring pro-imperial hierarchy. An image entitled "Ready to Enter the Forest," for instance, shows Millican alone, dominating the visual field, with a big hat and gun, positioned to do battle and subdue. When local people appear in the text's illustrations, they are either small, faceless, dominated by the landscape, or they appear as a general, undifferentiated type (as in "a solider"; "native ladies"; "native boatman").

Published accounts of brave white men hacking their way through jungle undergrowth, dueling with "hasty-tempered" alligators and ready-to-strike snakes, carefully conceal the ways in which "plant-hunting" was, for the most part, a matter of finding knowledgeable local people, asking them where the orchids were, then paying them to source, collect, and

READY TO ENTER THE FOREST.

Albert Millican's *Travels and Adventures of an Orchid Hunter* encourages readers to empathize with the white protagonist in his violent pursuit of a pink cattleya.

move the plants.[2] European plant hunters did not so much hack their way through the jungle as rely on deep, local botanical knowledge and teams of laborers. Local people not only knew what was where, they also understood the growing seasons; this particularly helped in the identification of plants not yet in bloom. Linden reported in 1891 that Claes had learned there were two main varieties of *C. labiata* in the region, for instance—one, that bloomed around Christmas, known as *Flor Natal*, and a second, blooming March-May, known as *Flor de Guaresma*, or Lenten Flower. Understanding this fact alone might have helped earlier generations of European hunters. Either no one thought to ask, or local people did not choose to lead the vandals through the forests.

But for all that the published works sidestep the ways in which European adventurers were dependent on the expertise of Brazilian-born locals, the plant hunters' private letters make plain the labor and networks of knowledge on which they constantly relied. These offer us some glimpses of the identities of those involved. Sander was almost as keen for information about his rivals' business practices as he was for news of orchids, and Ericsson, Forget, and Oversluys in Pernambuco, Arnold in Colombia and Venezuela, Digance in Rio then Minas Gerais, all gave detailed information about Linden's men, like Bungeroth, Claes, and Kromer.

They also, though less clearly, gave accounts of "the Brazilians" with whom the Linden men, and they, worked closely.

Digance, learning the ropes on his first trip, for instance, gave quite a bit of detail about the local networks he was trying to tap into, even as his challenges with the language made it difficult for him to communicate. He was baseline suspicious of all those who offered to help, partly through a xenophobic suspicion of foreigners and partly through naked fear. As an English-speaking and working-class traveler, on his first trip out of the country, he struggled, on arriving in a new area, to grasp the arrangements and hierarchies of the communities through which he passed. Yet he needed to make sense of them and quickly locate those with expertise in order to make his voyage a success.

Digance's letters indicate that workers in hotels that catered to tourists and collectors like him acted as go-betweens for new arrivals. Digance was asked by the keeper of his first hotel, The Beresford in Petropolis, for a list of plants he wanted, for instance; the hotelier worked to help

Digance find people in the area who knew the habitats of the plants he sought.

Digance suspected that the hotelier planned to make him wait for the plants in order to keep him as a paying guest. Baffled by much of what went on around him, he was quick to reach for negative interpretations of his new associate's motivations: "I can't get much out of [the hotelier]," he wrote to Sander hopelessly, "he knows a black he says who has been all through the woods here, but he dont seem to bring him forward."[3] He found it especially hard to find or communicate with members of the local Black community, even as he knew he needed their expertise. On several occasions he did manage to work with Black guides and assistants in the mountains, and by and large he reported that their interactions went quite well. He believed the men treated him fairly; he was far more suspicious of the English people living locally. "[I]t is the English one has to watch out here or they do you," he wrote.[4]

I've found no explicit reference in the letters to the Black guide Linden mentioned, the friend of Silva who supposedly agreed to help Bungeroth find the cattleya Silva painted in 1888. But Ericsson did write often, in late 1891, of a Brazilian who lived locally and who worked with and for Bungeroth (and thus Linden). This is the man who liked to play cards, but lost regularly: "his agent hiere one Brazilian came up to Garanhuns where I was collecting, he did stay over the night, lost ten pound playing cards with some Brazilian next night morning he left for his home again, I will bye his plants if he can get any."[5]

Ericsson planned to undercut the agent, in other words, by buying plants while they were still *en route,* bribing locals to sell them to him instead of to "the Brazilian" with whom they'd contracted. The plant hunters often reported using this strategy, and it was also used against them, as we've seen; local people seem to have been swept up into conflicts between the sides, though battle lines were never clearly drawn and could easily be revised. Those transporting the orchids could always use possession of plants to their advantage; they had other tools at their disposal, too. Ericsson ended up buying orchids of lower quality on one occasion when told that otherwise the plants would be sold to "the Brazilian" and thence to Linden.[6]

The Brazilian agent seems to have had a very good handle on what the Sander men were up to, for all that he had a weakness for cards: "that

Gent he is still up the country playing cards with Lindens money," Ericsson wrote, crossly, in September 1891, "and when it is finished he will bye up all what I have not taken." Ericsson, watching the agent like a hawk, trying to interpret him, consoled himself with thoughts of his card-playing losses while worrying intensely about his strategies: "The Brazilian is still playing cards I have heard he lost 30 pounds the other night he is not bying any plants I think he has got letters not to collect any before the next spring and then Mr. Bung. may turn up." The agent was evidently an effective foe, and he seems to have been the man who either engineered or took advantage of the situation when Ericsson "got into trouble with a native." The Swede paid a hefty fine because otherwise "very likely he [the person who held his plants to ransom] had sold my plants to the Brazilian."[7]

Who was this local man, who lived somewhere else but rode over to Garanhuns? Though Ericsson does not name him in the letters of 1891, by 1892 *Ovidio Wanderley* is mentioned repeatedly in the letters of Forget, Oversluys, and Ericsson as the agent Bungeroth worked with closely when *C. warocqueana* was first "found." Oversluys wrote on January 9, 1892 that Bungeroth had employed a man named "Ovidio Wanderely" and that this man helped Bungeroth collect 5,000 plants. Ericsson repeated the same information on January 30: "the mans name helping Bung, and Class is Ovidio Wanderley." He wrote the name out very carefully, so Sander could read it. Forget also refers to "Linden's man Ovido" or "Oviedo" several times, repeating in June 1892 that "Orviedo was the guide of Bungeroth."[8]

Ovidio Bruno do Nascimento Wanderley, born around 1853, would have been in his late thirties at the time he worked with Bungeroth. He was a member of what his obituary would call a "distinguished family"—"*memro de distincta familia*"—from the region of Triompho (modern-day Triunfo) in Pernambuco; his father was a priest, and he was one of five children. Photographs of the parents, provided by a descendant, show a well-dressed, elegant, soulful couple. The father, with a beard, wears a pressed jacket and necktie, while Luiza Guilhermina do Nascimento, Ovidio's mother, with swept-back hair, is dressed in a stiff black dress with very large sleeves and a high neckline, adorned by a silver necklace.

Wanderley seems to have moved some distance from his family by the time the plant hunters arrived, settling in Canhotinho, east of Garanhuns.

Keep all granulosa I sent up to now
Dubosson: is a variety of it — herby 2
flowers —
January 6.th You say 31 cases bad
I don't understand that plants ought
all to be good now, all was that is lately
sent is only good stuff
Bung first time in Pernambuco in
end 89 beginning 90, the mans name
helping Bung, and Class is Ovidio
Wanderley
These time I send you here
49 cases which I hope will arrive safe
You ought to have enough of the Epid-
will the reddish leaves by this time
the Catasetums you will sell as soon
as you can that is only rubbish —
There is some cases of guttata ull
as well there are of them two varieties
one with greenish yellow sepals and petals
spotted with brown, and one other with
brownish sepal and petals spotted with dark
brown — Leopoldi ? !

Ericsson carefully writes out the name, in full, of the person who helped
Bungeroth: "the mans name helping Bung, and Class is Ovidio Wanderley."

The distance from Triunfo to what became the center of the hunt for the lost orchid was over 170 miles. Wanderley established a successful, respectable life in the region: he stood as godfather to friends' and relatives' children in Canhotinho in 1896 and 1910, according to baptismal registers preserved by the Catholic Church. A local newspaper, *Diario de Pernambuco*, records him several times as a subscriber to his local church; an honorific title, "Major," is used to reference him, which indicates he held a position in the National Guard. He appears several times in the early twentieth century in *A Provincia* and *Journal de Recife*, two other papers from the region, as a clerk serving in the registry office in Canhotinho.[9]

The letters of the European plant hunters typically include, as we've seen, a tangle of responses to Sander's demands, responses to his questions, and periodic blazes of anger or complaints about ill health; it can be difficult to distinguish quite who did what from their letters. But it is clear the Sander men viewed Wanderley as a competent collector and a significant adversary. Forget explicitly wrote to Sander that Wanderley "may be considered as a collector here," a judiciously applied descriptor indicative of respect. To be a "collector" meant to be something more than a hauler of plants; it meant to be involved in the business of finding, packing, and shipping out the orchids. Ericsson, when Forget and Oversluys left the region, explicitly asked Sander *not* to send a "collector" to replace them because he found it easier not to have to navigate with colleagues ("If you should take away Mr Oversluys and send someone to help me please don't send me a collector—").[10] When Forget notes that "Ovido has sent home 5000 plants," he means Sander to understand that Wanderley is a manager involved in all the details of the collection process and not just someone paid to follow orders by the day.[11]

While Ericsson seems to have hired someone working with Wanderley early in the process to help guide him to the spot where Bungeroth found *C. labiata*, Ovidio kept his own work for the Sander team to a minimum. There is evidence, in fact, that he felt deeply loyal to Linden. Wanderley continued collecting after Bungeroth left, and when Forget explicitly asked Ovidio, "the guide of Bungeroth," to start working for Sander going forward, Wanderley refused: "he told me if you were to give me five pound a day I should not do it for I am the agent of Mr. Linden." The rest of this particular letter from Forget is nearly indecipherable—the handwriting is so bad, the prose so incoherent and tangled, one suspects Forget was drunk or ill while writing it; still, it appears to report that Wanderley was working with a painter who was preparing plates of orchids and that the encounter between Wanderley and Forget (or possibly an encounter between Forget and one of Wanderley's associates) ended badly: "we pretty near got on fighting," Forget explained, adding: "he was going to do I don't know what but I told him to shut up."[12] For whatever reason, Wanderley—for all his need for money to cover gambling losses—seems to have been unwilling to switch sides.

I could find little on Wanderley beyond this. He died aged seventy, in penury—"*extrema pobresa*"—in Canhotinho. His obituary in the local newspaper noticed that, since his wife had predeceased him, his death left their two daughters, Maria and Laura, destitute.[13]

26

KING OF THE ORCHIDS

BY THE MIDDLE OF 1892, the hunt was essentially over. The lost orchid was found, and if the environment was damaged in the process, local communities impacted, these were points to note as interesting facts, at most, not to use as motivators for change or course correction. Another boundary had been pushed back, another great British discovery made. Sander's "win" was a "botanical sensation," a "triumph," the *St. James's Gazette* cheered. The lost orchid had helped to establish the orchid more firmly in Britain, argued Fredrick Boyle, leading Europeans to thousands of gorgeous blooms along the way: "In this manner the lost orchid has done immense service to botany and to mankind."[1]

Sander's attention turned to other orchids as soon his first import of *Cattleya labiata vera* hit the auction rooms, to other "introductions" and "discoveries" he and Godseff could trumpet to the world. In October 1891, barely a month after his supposed reintroduction of the lost orchid, Sander put up for sale specimens of the "Elephant Moth Dendrobe" (*Dendrobium phalaenopsis* var. *schroderianum*). This was pitched to consumers as "The Orchid Sensation of the Century," "the King of the Genus." The two orchids were put up for sale together at 205 Greenwich Street, New York, on November 6. (A full-page ad screamed of the sale in the *American Florist*.)

The dendrobe, latest sovereign of the orchid world, entered the auction rooms with yet another thrilling story of danger, conquest, and resilience accompanying it—"a tale of horror in several chapters," the *Gardeners' Chronicle* termed it, in an article that was another puff-piece for Sander's business. This time, the plant was plucked by collector Wilhelm Micholitz from the middle of tribal conflict in New Guinea: the collector was tested on the way home when his ship caught fire and all the plants were destroyed. "We received a cable gram with the words, 'Ship burned, everything lost, what do?'" Sander is quoted as explaining, adding proudly that he'd wired back "Return, try again." But it was the rainy season, Micholitz protested; Sander was undaunted. He wired peremptorily "Return, collect more," and "the intrepid collector obeyed our orders."

The article quoted Micholitz himself, in a letter to Sander; excerpts from this were proffered as evidence of the veracity of the tale. In it, the hunter explained finding more of the plants on his return to New Guinea twined among the bodies of dead tribespeople laid to rest on limestone along the sea. "At first the natives did not like the idea of collecting the plants off those rocks," Micholitz is quoted as observing, "they were afraid the souls of the departed whose bones were laying there bleaching in the sun would resent it; but when they saw the gorgeous handkerchiefs, beads, looking-glasses and my brass wire, I offered them for the plants, they did not trouble themselves any more about the souls of their ancestors, but boldly went and rooted out every plant to be found." Compensated with trumperies, the natives offered up a war dance as the orchids were packed up into baskets. Finally, they asked Micholitz to take a golden-eyed idol along with him to protect the plants. Micholitz's tone, in the reported letter, is of amused condescension toward the natives for their foolish spiritual practices; the scene of conflict apparently enacts and confirms his physical, intellectual, even spiritual strength.

In fact, Micholitz's letters, the Kew archive reveals, were no more likely to narrate the physical, intellectual, and economic dominance of plant hunters than those of any other collectors. Like the letters of Ericsson, Forget, Oversluys, and Digance, Micholitz's epistles were far from swaggering tales of confidence, romances "in several chapters" of a Briton's challenges inexorably overcome. Instead, they center the usual complaints about limited cash flow, struggles with infrastructure (finding boxes, packing specimens, working with shipping companies, waiting for the mail), fights with European spies and rivals, and circuitous musings on species and varieties. When Micholitz described negotiations with local people, it was without bombast: he wrote in November 1890 that he intended to pack "handkerchiefs etc" to help him negotiate with the people in Larat (Indonesia) because "Money is not known there." In February that year he reported on his experiences in New Guinea, and while he referenced the challenges of hunting in a period of tribal warfare, he spent as much time describing a fight with a local ship's agent over prices. Anger with Sander, meanwhile, threatened to disrupt his progress and his writing; when credit failed to materialize for months on end, Micholitz reached a breaking point, "I do not trust myself to write anymore."[2]

The letter "quoted" in the *Gardener's Chronicle* was presumably concocted, in other words, by Sander and Godseff. Some of its details were culled from Micholitz's actual letters—his reference to handkerchiefs, for instance, while he reports reading in a book in the Singapore Library that *Phalaenopsis* is "plentiful growing on the limestone rocks." But the amused, superior Micholitz character on the printed page is a Sanderian creation, and it startled the collector. Micholitz wrote on December 29, 1891, from Singapore that he had just received letters from Sander, together with press cuttings which, he noted, were "very interesting but good god what have you made of my letter? I am sure I wrote nothing about a war dance."[3]

There are signs of doubt and resistance to the Sanderian narrative from the *Gardeners' Chronicle,* too, for the article comments on the tale spun by the Sander office in ways that are both approving yet decidedly uneasy. The author evidently identifies the Sander tale as ripped from the pages of imperial romance: "Surely here are the materials for a thrilling romance indeed!" remarks the writer, pointing out the many familiar tropes referenced in the letter—"the lovely Orchids, the naked cannibals dancing amid the bones of their ancestors, the enterprising collector not oversure that he would not himself form part of their next meal." By terming it "a tale of horror in several chapters," the text signals to readers that Sander's supposed nonfiction is, in fact, a fictionalized creation.

Yet romances of the moment were not uncomplicated, either, not "mere" fiction—if by *fiction* we mean thoughtless entertainment. Fictional narratives, imperial romances, were far from univocal in their celebrations of imperialism. Even texts that cheered along British global expansion expressed concern about its impact in plots that push and pull and question. In H. G. Wells's "Aepyornis Island," for instance, an adventuring orchid hunter named "Butcher," shipwrecked in Madagascar, makes friends with a massive Elephant Bird, an extinct *Aepyornis* mysteriously brought to life in a sort of proto–Jurassic Park. When the bird begins to compete with him for resources, he butchers it to death—then, shattered, mourns it like a child. In Kipling's "The Mark of the Beast," butchery again concludes the tale when a British colonialist named Fleete is turned into a sort of werewolf after disrespecting the idol of a god named Hanuman in India. Fleete's two friends turn to unprintable torture in order to rescue their colleague from his freakish curse. Kipling, in typically slippery form, concludes by remarking "it is well known to every right-minded man that the gods of the heathen are stone and brass," even as his tale strongly suggests that the "gods of the heathen" are, in fact, intensely powerful, since they were capable of transforming a human being into a beast. Both texts press readers to question whether the Briton is as knowledgeable as he likes to think he is and point out the brutality that lurks beneath his supposedly civilized exterior and his practices—and this, after all, was the putative justification for empire in the first place: bringing civilization. What if that's *not* true, the texts tacitly inquire; what if there is no such justification after all?

The *Gardeners' Chronicle* points quietly to these more dialogic qualities of contemporary fiction as the author tries to make sense of Sander's Micholitz letter. There are clear shades of dismay at what the plant hunters have unleashed: the *Chronicle* author explains that the golden-eyed idol "was the very first article sold by the sacrilegious auctioneers!" for instance, as the god is converted into cash in materialist, capital-driven London. The sale is met, too, with "the laughter and jeers of a concourse of unsympathetic Orchidists," and briefly, in these moments, the narrator seems to part ways with the "Orchidists," to decry them (as loudly jeering, "unsympathetic").[4] Kipling's "Mark of the Beast" had been published just one year previously, in July 1890, and the disrespect showed to a golden-eyed idol particularly recalls that tale. Readers who recognized the figure and the plot might well be prompted to infer that bad consequences follow affronts to foreign gods: one of the careless auctioneers or unsympathetic orchidists would, in a Kipling narrative, likely encounter some strange, supernatural being or force, and the golden-eyed idol they had grossly disrespected would turn out to be the cause. Chaos, fear, and mortal danger would ensue; the auctioneer or orchidist would be forced to realize they had to placate the idol. In discovering this they would see, too, that their beliefs about life and death were flawed, that there are indeed "more things in Heaven and Earth" than materialistic Britons understand. Normalcy would be uneasily restored, though on the last page, the reader would be exhorted to understand that, going forward, the auctioneer/orchidist could never admit the truth, what *really* happened.

The author of the *Chronicle* piece comments on the Sander narrative in ways that point us firmly away from it as evidentiary document and toward, instead, contemporary fictions that qualify, question, and problematize the imperial project, its self-justifying narratives in particular. The fictionalized Micholitz letter is effectively treated in the article not as historical document so much as a trope of fiction—indeed, letters themselves were often used in nineteenth-century novels and short stories as framing devices (in, for instance, Sheridan Le Fanu's occultist collection *In a Glass Darkly*). Sander's effort to present his business as an apotheosis of the imperial project is accepted by the magazine, explicitly but awkwardly, the recognition of its fictitiousness directing readers to texts that were themselves probing Britain's economic and political exploitation of the globe.

Sander and Godseff, one suspects, saw little of this. In 1891, orchidelirium was at its height. The lost orchid was found on the heels of *Catasetum bungerothi*, a mere month before the Elephant Moth dendrobe. The Horticultural Society awarded Sander the Victorian Medal of Honor; Victoria appointed him Royal Orchid Grower. Boyle, visiting Sander's nursery, described the sight of imported orchids with slack-mouthed awe:

> Orchids everywhere! They hang in dense bunches from the roof. They lie a foot thick upon every board, and two feet thick below. They are suspended on the walls. Men pass incessantly along the gangways, carrying a load that would fill a barrow. And all the while fresh stores are accumulating under the hands of that little group in the middle, bent and busy at cases just arrived. They belong to a lot of eighty that came in from Burmah last night—and while we

look on, a boy brings a telegram announcing fifty more from Mexico, that will reach Waterloo at 2.30pm.[5]

Boyle gave the story of the lost orchid and its recovery a full, breathless chapter in his book-length work, seeing not waste, not vandalism, not exploitation but thrilling accumulation in the scenes of "orchids everywhere!"

Still, in its very height were the seeds of orchidomania's downfall. It would become harder and harder to package orchids—*any* orchid—as rare, exotic, and glamorous when they were becoming more readily available in suburban nurseries, when middle-class families could aspire to own and cultivate these once-rare plants. House plants were integral features of the middle-class suburban aesthetic: home designer Jane Ellen Panton advised her readers that "pictures, plants and ornaments" were positively "the hallmarks of a home . . . for these things judiciously chosen and arranged, and not overdone, are after all what turn the worst suburban villa that ever was designed into an artistic abode."[6] The suburbanite who wished to signal that they were "artistic" could realistically do so by adding orchids tolerant of lower temperatures to a room—and there was an ever-expanding market of orchid-related accoutrements to accompany them, too, such as patented orchid baskets, specialized orchid peats and composts, and teak rafts, boats, and cylinders to facilitate growth and display. Orchids were becoming just another commodity. A sale of orchids, cheerfully advertised in the *Sheffield Independent* on January 23, 1890, was slotted in between auctions of plumber's stock, boot stock, and tools. In the same year, readers of the *Graphic* were informed that orchids made out of velvet and satin were all the rage in the season's hats.

And there were fewer and fewer "new" orchids left to find. Forests were stripped, trees cleared, microclimates altered, ecosystems disrupted. Cattleya orchids are particularly vulnerable in the face of habitat loss, not only because, as epiphytes, they coexist with other plants, but also due to "high levels of endemism and their sporadic and wide [seed] dispersal"; consequently, "removal of one tree or a few plants can seriously cripple the reproductive potential and gene flow of an entire orchid population."[7] If Albert Millican is to be believed, a common late-century strategy was to cut down the trees on which *Cattleya* orchids grew, so as to make their harvesting easier: by the turn of the century, whole years passed without the arrival of a single new species or variety. Forget wrote of a dramatic scarcity of orchids in later 1890s Pernambuco, given the work he and his colleagues had already done; William Digance was right when he noted the area seemed "played out" on his ill-fated trip. Many firms gave up employing their own collectors and relied on freelance workers instead.

Orchid communities in Europe and North America endured, certainly—new periodicals would even be launched to cater to them: the *Orchid Review* subtitled itself (from 1897) "devoted to orchidology." But there was a clear recognition, in its editorials, that the days of breathless accumulation were over, that a new era had dawned: "the introduction of new species is falling off," the editors remarked in the first month of the new century. "The present year has served to emphasise an opinion which has been gradually gaining round for some time past . . . Most countries of the world have been visited by the collector in search of novelties, and the likelihood of the discovery of new species of Orchids of great horticultural merit is gradually becoming more remote."[8]

To some degree, the orchid's day was passing, too. The "wild garden" or "cottage garden" movements celebrated indigenous plants. As Arts and Crafts practitioners encouraged a new aesthetic, a return to older skills and workmanship, a move away from machine-produced textiles and furniture, so many gardeners looked to seasonal growth patterns and landscapes planned around naturalized plants. Greenhouse-reared, imported specimens were less and less consistent with the dominant tastes and preferences of the age; xenophobia also colored discussions of "exotics." A Fellow of the Horticultural Society, writing to the London *Times* to celebrate an exhibition of "British ferns" in 1892, noted distastefully that the "innumerable displays" of exotics like orchids to which he had been subjected all had one problem: the plants had "a considerable foreign element in their composition." Every specimen of fern on display, by contrast, had "either been found in all its beauty wild in our native lanes, woods, or other ferny habitats, or has been derived directly from such 'finds' by selective culture here." So much for exotics; to the Fellow, the "peculiarly indigenous feature" of the British ferns "entitles them to special study and consideration."[9]

In the face of this, the Sanderian approach to selling orchids became gradually and literally risible. The company's cleaving to the genre of popular romance started to seem positively hackneyed, and Sander began—orchids began—to fall out of fashion. It did not happen overnight, of course; in 1904, the sensationalist *Wide World Magazine* still told a wildly fictionalized version of Sander and Arnold's orchid-hunting in Caracas, complete with dramatic illustrations (a fierce-looking Arnold holds up his rival with a gun).[10]

But in 1909, the *Orchid Review* published a send-up of the language Sander & Co. used in advertisements. With arch amusement, apparently quoting from a source, the paper's author wrote of a supposed "New Orchid" recently found "after months of patient search":

> For many years Messrs. Sander's traveller has known of the species, but the real difficulty lay in obtaining the services of escorts of reliable natives.
>
> From time to time parcels of glass beads and watches, linen and bright handkerchiefs were sent from England to propitiate the natives. But they were suspicious and hard to pacify.
>
> The Orchid hunter bided his time, and with an escort of friendly natives seized an opportunity to secure a collection of the plants, and had them rapidly conveyed to the coast.
>
> The comforts of civilization were not for the intrepid hunter, whose bed was in the primeval forests, and whose breakfast—and whose life—depended on his gun. Fever lurked in the swamp and mountain and morass, and there were a hundred natural obstacles to be overcome.
>
> Messrs. Sander prefer not to state the place of origin of the new Orchid, of which they possess at present fifty plants.
>
> The agent who secured the new Orchid has now gone to New Guinea in quest of an Orchid that only grows among the bones of dead men.[11]

Here the narrative, discourse, and tropes of imperial romance, used by Sander and Godseff for a decade—the supposedly savage, yet easily tricked natives, the use of overblown language to describe the landscape ("primeval forests"), the mythologizing of the plant's habitat ("Messrs. Sander prefer not to state the place of origin"), the exoticizing of the spiritual practices of tribal people (the orchid grew only "among the bones of dead men")—are deployed as clichés from a moment utterly passé. The wizard's machinery is exposed, and with it the larger narrative of orchidomania, too, which evoked orchids as romantic, rarified exotics ordered and managed by the knowledge, resilience, and worthiness of Europeans. The familiar story now becomes pastiche. The 1909 writer does not fully reject the principles of imperialism, but he does obliquely point out to his readers the degree to which colonial adventuring was itself a story, told to an avid public in the service of commerce and economic exploitation.

The skewering of Sander, the jokes at the expense of orchid hunters, were not just topics for horticultural enthusiasts, either. By the Edwardian years, the very idea of a hunt for a lost orchid was so hackneyed it was fodder for music-hall hilarity. And so, in 1904, "The Orchid" opened in the West End at the Gaiety Theatre, a musical that turns on the side-splitting humor of a hunt to South America for a lost purple *Cattleya* orchid. "I've travelled far where panthers are," sings hunter Zaccary, in Act I:

> That jump on you and catch you!
> And snakes that twist about your wrist,
> And kill you if they scratch you!
> I've run for miles from crocodiles
> That came with jaws extended;
> But I have brought the flower I sought,
> The orchid rare and splendid!

Many of the costumes for this frothy production were dyed a shade of purple known as "cattleya": in a subplot, women gardeners at a horticultural college were dressed in school uniforms of mauve silk muslin in just this hue, "prettily made with many plissé and gathered frills and furbelows," noted the entranced fan magazine, the *Play-Pictorial*.[12] Associations of rarity and high value with the term *cattleya* were, by now, almost entirely lost, a point the musical's plot literally stages for its audiences: when Zaccary fails to find the rare cattleya he seeks, he buys one for £5 that turns out to be just the same. The musical—King Edward VII and Queen Alexandra attended opening night—subsequently moved to Broadway in 1907, where it ran until 1909 and then toured the world. It even visited Rio de Janeiro.

The cover of the fan magazine *Play-Pictorial* reproduces the lyrics of the 1904 musical *The Orchid*, with photographs, information about the actors, and more.

"THE ORCHID."

ZACCARY'S SONG.
(See p. 7 for illustration.)
"From Far Peru."

I've travelled far where panthers are
    That jump on you and catch you !
And snakes that twist about your wrist,
    And kill you if they scratch you !
I've run for miles from crocodiles
    That came with jaws extended ;
But I have brought the flower I sought,
    The orchid rare and splendid !

    Ah ! In the wilds of far Peru,
    It was there the orchid grew !
    Where the vampire bats flew
    Through the vapours of blue,
    In the woods of far Peru !

Gorilla hordes with poisoned swords
    By day and night attacked me !
At dawn and dark Peruvian bark
    I heard as bloodhounds tracked me !
I climbed for weeks the icy peaks
    And reached the top a victor ;
And, lastly, I was swallowed by
    A monstrous boa-constrictor.

    Ah ! In the wilds of far Peru,
    He had room inside for two,
    But my trowel I drew,
    And I dug my way through,
    To the light of far Peru.

FANCY DRESS SONG.
MISS CONNIE EDISS.

I've a passion for fancy dress more or less ! more or less,
I look sweet as a shepherdess that's made by a Dresden potter,
I have ribbons in bows and knots lots and lots ! lots and lots,
Like Elizabeth Queen of Scots, when Oliver Cromwell shot her !

Oh, only fancy, fancy dress, fancy me as good Queen Bess.
Only I never could get my breath with a waist like Queen Elizabeth.

I would dress like a girl of mark, Joan of Arc ! Joan of Arc,
Riding out in St. James's Park, and waving a flowing banner !
I'd have armour in lovely taste highly chased ! highly chased ;
If it pinched me about the waist, I'd loosen it with a spanner.

THE COUNTESS ANSTRUTHER—
MISS PHYLLIS BLAIR.

The names of the characters reading from left to right are, Miss Nancy Mansell, Miss Florence Dudley, Miss Madge Rossmore, Miss Ethel Christine, Miss Leone Roy.

Play-Pictorial, "Zaccary's Song": the dangers of orchid hunting are played, here, for laughs.

Play-Pictorial, "The Story of the Orchid in Four Chapters": a lost orchid is easily recovered—it costs just five pounds.

The day of the big Victorian orchid nurseries was coming to an end, too. When the *Orchid Review* mourned the absence of new "novelties" at the turn of the new century, it also mourned celebrated figures, including Georges Warocqué, after whom *Cattleya warocqueana* was named; William Protheroe, the celebrated orchid auctioneer; and four other members of the RHS Orchid Committee. James Herbert Veitch suffered a nervous breakdown and died in 1907, reportedly of syphilis. Veitch's brother John was unable to run the company, forcing their Uncle Harry to come briefly out of retirement to take up the reins. Even Sander suffered repeated bouts of ill health, and the royal attention he so loved seemed to be coming to an end. Sander was publicly snubbed by the king of Belgium, Leopold, in 1901—perhaps because of his relentless efforts to undermine Belgian business interests, though British sovereigns seemed to tire of him as well. Harry Veitch was knighted in 1912, but Sander was not. Tensions roiled both houses. The Sander sons believed Joseph Godseff to be dishonest, while Micholitz was increasingly convinced that someone in the Bruges office was leaking information to rival companies. Both he and Louis Forget, the only remaining Sander collectors, were repeatedly outplayed by rivals in the early years of the new century.

Harry Veitch, with no one to take over the family business, with gardeners around the country taking up arms as war approached, sold the business and put the entire Veitch holdings near London up for sale in 1913–14. Middlesex County Council bought some of it, to be used for building. Sander, in Bruges on August 4, the day hostilities were declared, predicted that the end had come. "We are ruined," he wrote, as his family fled Belgium.[13] He guessed that bloody, grinding years of battle would only make an obsession with orchids seem more ridiculous.

THE AGE OF INNOCENCE, Edith Wharton's lusciously nostalgic evocation of 1870s New York, was written from the postwar vantage point of 1920. In that novel, a rich Englishman named Julius Beaufort presents a mystery. Who is he, exactly? Is he even English? Heavyset and rather vulgar, Beaufort has something to do with finance, so far as the community can make out; he spends lavishly and is unfaithful to his wife. Little else is known about him, except that he is attracted to Ellen Olenska and is thus a thorn in Newland Archer's stiff side.

Wharton offers up signs of Beaufort's true character gradually, subtly, and partly through the flowers that are his personal *leitmotif*. Beaufort is an orchidophile, or at least he grows them and other exotics in his greenhouses. Early in the novel, we learn he has sent "wonderful orchids" to Ellen as a token of his affection; Newland, feeling at the time rather proud of the yellow roses he selected, realizes he's been outsmarted. "[N]o wonder mine were overshadowed by Beaufort's," Archer says, "irritably," and when later he visits Ellen, he is disturbed to see that behind her lies a "table banked with flowers," "orchids and azaleas which the young man recognised as tributes from the Beaufort hot-houses."[14]

In fact, almost every time Beaufort strolls into the narrative, he is accompanied by an orchid. When Newland sees Beaufort after a mysterious absence, which has set off "disquieting rumours" about what exactly he's been up to, Beaufort's complexion shows signs of dissipation, Newland muses. But still, in his buttonhole, just like always, rests "one of his own orchids." Wharton's narrator makes plain to us that Beaufort is not, in fact, just like always; he is hovering on the edge of financial ruin. The New Yorkers so carefully surveilling him, through New York and Newport, are onto something, for Beaufort is on the edge of a chasm. Trump-like, he chooses to spend lavishly, to convey the impression of great wealth in order to bolster his creditworthiness.

The orchid is a crucial prop in this little bit of personal theater. It is a way of performing that he still has money, that he—like Jay Gould, robber baron of the Gilded Age and known for his orchids—has so much he can spend it lavishly on decorative exotics.[15] Orchids and orchid cultivation are part of this carefully shaped theatrical display at many moments in the novel: "to every report of threatened insolvency," the narrator explains, "Beaufort replied by a fresh extravagance." This includes "the building of a new row of orchid-houses." In the world of Wharton's novel, where orchids are highly valued by the elite, Beaufort telegraphs status and financial stability through his orchids.[16] Needless to say, we never see him at the table, in the greenhouse, tending to his plants. We never see the collection or shipping of the orchids. The Beaufort orchids are a dilettante's playthings, imported by invisible collectors, tended to by paid experts, off stage. The history of their arrival in New York and the story of the labor that it took to source and cultivate them takes place in unseen, unknown ways, along with all the other dealings Beaufort uses to build up his careful performance of stupendous wealth.

Wharton picks up on the ongoing correlation of orchids and wealth, of course, giving her character a hobby readers would likely associate with Gilded Age excess. But Beaufort's attachment to orchids may also signal something else; his Sander-like posturing, his puffery and fake news. For if Beaufort is a Trump-like figure, he is surely Sanderian too: like Sander, he manipulates his public message carefully. He taps into what his culture values and works to navigate, just as carefully, what it fears. He grasps that orchids are weighted with cultural meanings of connoisseurship and control. But Wharton, Beaufort's creator, writing decades later, grasps that those meanings have substantially drained away, that the wizard's machinery has been exposed, the curtain pulled aside. The Beaufort orchids tell us, long before the 1870 characters grasp it, that their owner is a sham.

The golden age of orchidomania had passed by the time of Wharton's writing. In 1908, an article on "All the Cattleyas Worth Growing" in the New York–published *Garden and Home Builder* noted that *Cattleya labiata*, once rare, was now "common": what had been the preserve of the select few could be easily possessed. Even more could enjoy orchids in photographs, the multiple images accompanying articles like "All the Cattleyas Worth Growing" offering the pleasures of looking to a wide readership, far expanded from that specialist community once collected around the engraved, hand-finished botanical illustrations of (say) Lindley's *Collectanea Botanica* or Bateman's *The Orchidaceae of Mexico and Guatemala*. Sander's own monumental, multi-volume work on orchids, *Reichenbachia* (1890–1894)—each of the four volumes weighed in at forty-four pounds, with lavish, life-size illustrations, some hand-painted—may be seen as an effort to reconnect orchids with the grand days of Bateman at a time when their eliteness was, in fact, on the decline.[17] "[W]e may yet live to see orchids sold by the peck like potatoes," observed the *American Garden*.[18]

But orchids would continue to be admired and collected into the twentieth century and beyond. Through the Great War, the Royal Horticultural Society continued to meet and show orchids; and, after the war,

Leonard Barron's article in *The Garden* points out that the once-rare *Cattleya labiata* has become "common."

The true autumn flowering *C. labiata*, once rare, now common All the others shown are varieties of this. Season, October to December

many found signs of hope in fragile-seeming plants that had somehow, against all the odds, survived the onslaught.

First, though, it was necessary to take stock of what was lost. In the months immediately following the armistice, the *Orchid Review* published several pieces in a series titled "Orchids in Belgium" in which it published updates and excerpts of letters from collectors and horticultural houses on the continent that had, for over four years, been cut off from the rest of the world. Now that lines of communication were open again, collectors and nurserymen noted their gratitude to be reconnected with a community they loved. This was a very different kind of storytelling, the letters accounting not of gains and accumulation—"Orchids everywhere!"—but losses.

Collectors reported on years of horror—of burnings, bombardments, lootings, murders, and sacrifices undertaken to keep plants and people alive. But orchids were what made life worth living, perhaps. "Peace," one wrote, "has now relieved us of that awful nightmare, and I hope to be able to again read in the *Orchid Review* all about my beloved Orchids." Another—a former Sander collector named Pauwels—reported the night a bomb fell on his *Cattleya* house, a second two yards from the entrance to his house; somehow neither exploded. His relief was, unfortunately, short-lived. Pauwels's entire house and nursery were destroyed on November 8, 1918, in the final days of the war. The few plants he was able to save from the bombardment and 60,000 seedlings (including cattleya) saved from the blast were killed by frost in the early hours just before armistice, on November 11.

The magazine also reported on Sander's Bruges premises, again, a story of loss and destruction. The head clerk was killed the day before the re-entry of Belgian troops, while most of the greenhouses lost their panes. Many thousands of plants were lost—though the damage to the physical premises was surprisingly less than one might imagine, given that the Belgian front line passed through the Commune of St. André, outside Bruges. Still, Sander and his team were far from sanguine about the future, as a serious lack of fuel posed a problem to anyone struggling to cultivate fragile exotics. There is a different note in the voice of the postwar Frederick Sander. "[I]f the coming winter is not too severe, I shall manage to pull through till the spring," he wrote. A number of prominent horticulturalists were killed in battle. Sander's last hunter, Forget, died of a heart attack in 1915.[19]

But Sander's son, Charles Fearnley, who returned in 1919 to pick up the pieces, saw signs of hope amid the shatter of broken glass, the plants crushed by German cavalry, the tree sheds cleared out for horses, even the death, looting, and biting cold. The people of Bruges had apparently saved thousands of orchids by storing them in the cathedral and in churches during the war. The manager of the premises had shifted as best he could for four long years, growing vegetables and tobacco while taking out loans to cover the cost of a little fuel and labor to protect a portion of the stock. The business would return, and the Sander dynasty would continue, the family insisted—even as Frederick Sander himself sank: his wife suffered dementia, which distressed him greatly in the final years of his life. When he went into the hospital for an operation, he asked his son to leave out a letter he had written to be placed for her every morning on the sideboard. He was storytelling to the very end. "It will not matter I think if it is the same letter," he told his son: "she will not know."[20] Pneumonia set in after the operation, and he died two days before Christmas, 1920.

EPILOGUE
Fall—and Rise

TODAY, ORCHIDS ARE AMONG the best-selling potted plants in the world. Over 1.1 billion live plants were traded between 1996 and 2015. Most of these were orchids grown in commercial greenhouses, artificially propagated. Taiwan, Thailand, and the Netherlands are the largest exporters: Japan, the United Kingdom, and the United States are among the largest importers, with "new markets for orchids ... increasing annually." The *Phalaenopsis* genus alone "accounts for more than US$500 million in the global production and consumption markets." Orchids also make up about 10 percent of the international cut flower trade.[1]

To many consumers today, orchids are not a hushed, rare prize but a cheap and cheerful gift in plastic wrap from the local supermarket or home-improvement store.[2] What Sander did not fully predict was the orchid's next transformation, from precious exotic to ubiquitous office gift, a couple of steps up from scented candles and boxed soap. The thrill of the exotic endures, of course, and the market also runs the gamut, from ten-dollar grocery-store plants to rare specimens that cost thousands of dollars. "Rothschild's Orchid," *Paphiopedilum rothschildianum*, or the Gold of Kinabalu orchid from Malaysia, has sold for between five and six thousand dollars per plant.[3]

We are not the first generation to love shopping, the thrill of buying and bringing home something new. Industrial growth and mass production practices, new transportation networks, colonial and trade-route expansion, the growth of department stores and suburban high streets, print media in the form of "bills" and newspaper advertising, all brought shopping and its pleasures—and problems—into the reach of middle-class consumers in the nineteenth century. Orchids, according to Frederick Boyle, were the very greatest treat of all. He enumerated the technological developments that had given rise to orchidomania:

> Order and commerce in the first place; mechanical invention next, such as swift ships and easy communications; glass-houses, and a means of heating them which could be regulated with precision and maintained with no excessive care; knowledge both scientific and practical; the enthusiasm of wealthy men; the thoughtful and patient labour of skilled servants—all these were needed to secure for us the delights of orchid culture.

To own an orchid, to participate in orchidomania, was to feel oneself thrillingly a part of all this change and innovation, to benefit from modernity, to enjoy its rich "delights." Or was it? In a typically Victorian transition, Boyle suddenly suggests that perhaps orchids are not so much a reward from God as a "comfort" to those of us living in a difficult, "anxious age."[4]

Why did he and his colleagues feel such anxiety? It was an age that allowed for class mobility, that fundamentally reassessed merit and worthiness, that aspired to fling back the doors to ordinary people, the "million." But the dread alternative to social ascent was failure, as Mr. Tulliver learns all too painfully in George Eliot's tragic *The Mill on the Floss* (1861). As class status became increasingly unmoored from birth, as social mobility became not only possible but culturally valued, failure was attended with shame, humiliation. Sending one's son to a better school, marrying one's daughter to a wealthier family, owning better linens and china—as Eliot's poor, silly Mrs. Tulliver so longs to do—and, perhaps, purchasing a purple and red orchid were ways to gain traction in the terrible, great race of life.

The Great Exhibition of 1851 in London and the department store stand as two enduring symbols of, and stages for, the nineteenth-century love affair with things. This was so very much the era of the thing that, Thomas Richards explains, "a new class of words came into being to de-

Epilogue: Fall—and Rise

scribe things in general—words like gadget, dingus, thingamajig, jigger."[5] Yet Victorians worried, deeply, all along the political spectrum, about their obsession with things and accumulation, about the ways in which the acquisition of things was culturally and socially required. Marx and his followers wrote of the ways in which commodities became alienated from their human and geographical origins in commodity culture, abstracted into monetary value. Dickens's *Christmas Carol* exhorts readers to beware of putting money before people, family, the values of the home. Victorians thrilled to Tennyson's "ringing grooves of change" yet stood aghast at the pace of it all, fearful of the rise of capitalism and commodity culture, anxious about the ascendence of the machine, fearful of the loss of anchorage in a supposedly stable, rural past.

Many turned, in reaction, to horticulture. Orchids were a potential means of reconnecting with nature in new ways in an age that virtually defined itself as separating from rural and agrarian traditions. But as they arrived courtesy of rail and steamships, were piled high in nurseries, haggled over in auction rooms, placed in mass-produced greenhouses, advertised in bills of sale, newspaper columns, and suburban nurseries, they became too obviously commodities once more; products of the age, rather than its escape and solace. The romance of the lost orchid was a fantasy whose power inhered—as the wily Sander surely recognized—in its claim to resolve this problem. As orchids became too easily duplicated, too accessible, too evidently commodities, the "lost" orchid stood briefly, apparently, mysteriously, outside and beyond culture. A dream, its very inaccessibility was a sign that it was *not* a commodity; but this would be revealed as just a story, too.

Orchids, as "exotics," were "fictions of themselves," Elizabeth Chang explains in her analysis of plant life in nineteenth-century novels. Plants that were "imported, assimilated, hybridized, acculturated, and cultivated" were given names like "hybrid, exotic, invasive, alien, even 'weed,'" that assign the plant a "relational, situational, and contextual ... also always sociocultural" meaning.[6] As we've seen, this contingent identity rendered orchids available to play other, different roles too. Like the heroine of the era's most sensational fictions—a sort of horticultural Lady Audley, heroine of Mary Elizabeth Braddon's thrilling 1862 novel—Swainson's orchid acquired multiple different names on its arrival as a lustrous stranger. As *Cattleya labiata*, it was also the "Splendid-flowered

Catleya," *C. labiata vera, C. labiata autumnalis, C. warocqueana, C. labiata warocqueana, C. labiata* var. *warocqueana,* Gardner's labiata, Swainson's labiata, the "lost orchid," and more. (It had other names in Brazil, of course, including *Flor Natal / Flor de Guaresma,* rather as Lady Audley / Lucy Graham was Helen Maldon / Talboys in her previous life.) In each guise, the orchid appeared in slightly different character, in a different role, as part of a story shaped according to subtly different authorial designs. Early botanists offered it as a sign of intellectual and national triumph and, too, of personal achievement and worthiness for professional advance. In the emerging middle-class periodical, the lost orchid was framed as a signal of modernity at the midpoint of the century, an opportunity for the "million" to embrace the transformative delights of a fast-paced, globalizing modern world. Women's relationship to orchids, at least so far as we can infer through their collection practices, were different: orchids were decorative objects *they* controlled, whose spectacle they were in charge of, whose thriving was an opportunity to claim "mastery" on an equal platform with male peers. In US texts, *C. labiata* was represented as an opportunity for displaying expertise, a signal of cultural maturity, a means of connecting across states and nations. To the Edwardians, in the music hall at least, the whole *Cattleya* genus was a joke, a way of pointing to advance on the silly era that preceded them. To businessmen and commercial hunters, at least in letters, it was a baffling challenge, a constant reminder of colonial and infrastructural limitations, and at times, in the loneliness of relentless competition, a "friend." No one, at least in Europe, thought to ask communities in Brazil what the orchid meant to *them.*

We, every bit as much as the Victorians, struggle to see the costs of extraction still. Orchids today are "among the most threatened of all flowering plants," especially vulnerable because, as Forget and others observed, they are so intimately connected to other species in their environment.[7] They rely on local fungi for germination, growth, and nutrients, and many have very precise and specific pollination systems, requiring a healthy population of local insects. Epiphytes may be especially susceptible to habitat loss as they have evolved to thrive on particular trees, rocks, shrubs; if those are damaged or cleared, the orchid population dwindles. Increased access to remote spaces only makes orchids more accessible to collectors, even as the rarity of any "new" orchids that, by some astonishing miracle, remain to be "found" renders them, to some, objects of intense desire. Illegal collection practices and underground markets continue to flourish. Until very recently, it was considered acceptable for scientists and collectors to remove wild orchids from their habitat. Climate change is likely only to continue orchids' march toward extinction, both by increasing the risk of fires and droughts to orchid habitats and by further impacting orchids' symbiotic relationships with fungi and pollinators.[8]

The problems of clearing orchid populations were evident in the nineteenth century; many noted the terrible waste, though few advocated for real change. "[I]t can be safely estimated," confessed a contributor to the Minnesota State Horticultural Society in 1892, "that for every orchid that is to-day flourishing in the greenhouses in collections in America and Europe there have been at least thousands pulled from their native haunts."[9] Air pollution, too, was a problem for plants wrenched from their habitats and thrust into industrial cities: even the plants that made it across oceans might fail on arrival or during cultivation. Even before the catastrophic losses of the Great War, Boyle's gardener mourned the loss of two hundred specimens of *Cattleya* after a terrible fog, in October 1899: "In some cases plants looked as though they had been sprayed with paraffin and soot."[10] The soot-drenched orchid stands as a terrible betrayal of the Victorian dream of a plant supposedly beyond modernity and culture. It is hardly surprising that orchidomania, the dream of it, began to wane around this time, though the wasteful, exploitative practices themselves endured.

And had the lost orchid indeed, after all that vandalism, the trespassing, and the violence, the fevers and deaths, been found?

What does the question even mean? Again, as we've seen, *C. labiata* is a group, an extensive and diverse group at that. Orchids are predominantly allogamous, reproducing by outcrossing, and allogamous species experience particularly high levels of morphological variation (in contradistinction to autogamous plants, which self-fertilize). *C. labiata* is subject to high degrees of variation also because epiphytes have evolved a range of survival strategies based on their different environments. A Brazilian botanist in 2002 suggested that it's "possible to distinguish the flowers of C. labiata in accordance with its geographical occurrence. Thus the flowers of the Ceará State populations . . . have a darker coloration, are smaller" and are better shaped "when compared with the populations existing in the States of Paraíba, Pernambuco, and Alagoas."[11]

Moreover, wild plants tend to be morphologically distinct from plants grown in cultivation: scientists refer to this as "domestication syndrome." As Darwin argued just a few years after his orchid book, in *The Variation of Animals and Plants under Domestication* (1868), animals become less aggressive in captivity—and plants change too, becoming shorter, changing color, developing different flowering patterns. A recent study of multiple *C. labiata* plants, from a range of places in Brazil, found that "[c]ultivated plants were morphologically distinct from wild plants, mainly in characteristics such as flower number and shape, consistent with the domestication syndrome."[12] By the early 1890s, the domesticated *C. labiata* had been living in European greenhouses for fully seven decades.

Still, *C. warocqueana* and *C. labiata* specimens preserved in the Kew herbarium from 1890–1 do, the orchid biologist Cássio van den Berg confirmed to me, agree morphologically with one another and with a midcentury specimen from Hooker's herbarium. They are from the same population, the same area: "there is no question about the fact they are conspecific."[13] Rolfe's excited handwritten scrawl, "These eight flowers are the type specimens!" may still be read alongside dried, faintly purple flowers sent from L'Horticulture in May 1890 to Kew, which to this day rest on one of the Botanic Garden's remarkable herbarium specimen sheets, above a clipped-out, annotated column of Rolfe's article from the *Gardeners' Chronicle*. These strangely beautiful objects embody centuries of botanic engagement, each a sort of intergenerational assemblage of dried flowers, seeds, scraps of letters and articles, with stickers, labels, barcodes, and ruminative scribbles that evidence dialog, human questioning, and error even as they aspire to record scientific fact. (See Plates 12–13.) One states Sander's view of history: "From a plant bought at Mr. Sander's first sale of this species when re-discovered," while on another Rolfe notes, prosaically, "gluer has mounted this upside down."

Still, their preserved, ghostly flowers suggest that, indeed, the lost orchid was truly found. But when the hunters went in pursuit of the lost orchid, what *precisely* were they seeking? The specimens of *C. labiata* Victorians knew best, for most of the century, were likely descended— given the challenges of reproducing from seed—from the original: as the

Epilogue: Fall—and Rise

consequence of propagation by division, they would have been identical genetic copies. Of course hunters would not hope to find biological clones of the Swainson orchid growing in the wild, so that in one very particular sense, the lost orchid could not be found again. But also, time had passed; differences accumulate over the generations through the actions of evolution. Environmental changes effected by humans on plants, on habitats and climate, also affect—may accelerate—the processes of natural selection, so that the populations Ericsson and Bungeroth rifled would have changed in subtle ways.[14]

Lewis Castle wondered whether the "links in the chain of evidence" in the story of the lost orchid would one day be recovered. This was my opening inquiry, too. But the very question reflects a search not just for the recovery of facts, ordered by the sense-making operations of the story, but for closure, the powerful relief of the unknown's vanquishing at *The End*. Victorian novels worked toward a final moment of clarity, too, when the lost are recovered, the locked box stands opened, the mysterious stranger is revealed.

But Darwin's account of evolution presses us in a different direction; away from conclusion and resolution towards growth and process. Darwin explicitly rejected the traditional image of the hierarchical chain as a central organizing image in the *Origin of Species* in favor of a tree—vast, spreading, and changing, always:

> From the first growth of the tree, many a limb and branch has decayed and dropped off. . . . As buds give rise by growth to fresh buds, and these, if vigorous, branch out and overtop on all sides many a feebler branch, so by generation I believe it has been with the great Tree of Life, which fills with its dead and broken branches the crust of the earth, and covers the surface with its ever-branching and beautiful ramifications.[15]

The lost orchid could not be recovered any more than a final, completed story of its loss. Time passes, documents vanish and reappear, texts generate and will continue to generate new directions, different readings, fresh possibilities. Deviation, "ever-branching" transformation instead of stasis is nature's organizing principle. And that, as Darwin shows us, is perhaps the most beautiful story of them all.[16]

NOTES

Page vii epigraphs: Lewis Castle, "Orchids: Cattleya Labiata Vera," *Journal of Horticulture, Cottage Gardener, and Home Farmer* 23 (September 24, 1891): 262–263; 262. There is debate about how, precisely, to interpret the Heraclitus fragment; see Daniel W. Graham, "Does Nature Love to Hide? Heraclitus B123 DK," *Classical Philology* 98, no. 2 (2003): 175–179.

PROLOGUE

1. "The Queen's Jubilee Bouquet," *The Times*, no. 32105, June 22, 1887, 4.
2. Sander estimated spending £1,800–£3,000 per hunter per year, or about £250,000–£417,000 ($320,000–$535,000 in today's money; "Orchids and Orchid Hunters. An Interview with Mr. Frederick Sander, The Orchid King," *Tit Bits*, repr. *Otago Witness*, no. 2202, May 14, 1896, 49. Sander was given to exaggeration, but his correspondence suggest he spent in excess of £1,000 per hunter per annum, or over £140,000 per hunter today ($179,000). August Strunz, a merchant who advanced funds to hunters in Baranquilla, Colombia, referenced receiving checks for £75 monthly for a hunter named Kerbach; Strunz also regularly wrote to ask Sander for more money because the allowance had run out. Ultimately, as we'll see, the high cost of supporting plant hunters contributed to the near-disappearance of the profession. See Strunz to F. Sander, September 2, 1893, 629; February 3, 1894, 634, in F. Sander Correspondence, box 22, Royal Botanic Gardens, Kew (hereafter SC / RBGK).

Here, and throughout, I use www.measuringworth.com to establish the purchasing power of nineteenth-century monies. Calculations factor in the percentage increase of the RPI (retail price index, a measure of inflation) between then and 2023. The website helpfully contextualizes each answer, analyzing not just the "real price" of each commodity today, but also the labor value, income value, and economic share.

3. William Digance's father died in the workhouse in 1911. Digance's first charge from Sander was to source *Cattleya walkeriana*; he was still struggling to find it in the last weeks of his life.

4. W. Digance to F. Sander, October 5, 1890; November 1, 1890; and December 11, 1890, SC / RBGK, box 1, folder 3, 140, 141, 144–145. All documents, including letters, are quoted here with their original spelling, capitalization, and punctuation.

5. C. Ericsson to F. Sander, May 23 [1891]; Ericsson to Sander, January 4, 1891, SC / RBGK, box 1, folder 4, 237, 231.

6. Ericsson to Sander, March 25, 1892, SC / RBGK, box 1, folder 4, 234.

7. "The Showy Cattleya, Queen of the Orchid Beginners' Handbook—XV," American Orchid Society, https://www.aos.org/orchids/additional-resources/cattleya-queen-of-the-orchids.aspx.

8. James Veitch & Sons, *A Manual of Orchidaceous Plants* (London: James Veitch, 1887–1894), v. 1, "Cattleya and Laelia," 15. The volume to which this citation refers is a collection of multiple earlier, separately published parts, brought together in 1894. Information on "Cattleya and Laelia" was published originally in Part II of X.

9. This orchid of great "size and gorgeous beauty, was almost single-handedly responsible for starting the orchid craze of the nineteenth century," Daniel Karlin observes in *Proust's English* (Oxford: Oxford University Press, 2005), 86.

10. Jim Endersby reports that the most money ever spent on a single orchid was 650 guineas; see his *Orchid: A Cultural History* (Chicago: University of Chicago Press, 2016), 108. Prices dropped once larger numbers of the orchid entered the market. Endersby relates 650 guineas to £298,000 / $471,000 today, though measuringworth.com gives the 2023 "real prices" cited in my text.

11. Ericsson to Sander, April 9, 1891, SC / RBGK, box 1, folder 4, 235.

12. Ericsson to Sander, June 20, 1890, SC / RBGK, box 1, folder 4, 239.

13. Nineteenth-century migrations threw "a netting of transoceanic kinship connections" across the globe; see Jürgen Osterhammel and Niels P. Petersson, *Globalization: A Short History*, trans. Dona Geyer (Princeton: Princeton University Press, 2003), 78. The pressure on countries outside the British Empire to adapt to British values was a key part of the "collective self-transformation" of the era (72).

14. Key works from the last forty years on this self-perpetuating cycle (empire facilitates the extraction of plants, which in turn facilitates the expansion of empire) include Alfred W. Crosby, *Ecological Imperialism: The Biological Expansion of Europe, 900–1900* (Cambridge: Cambridge University Press, 1986);

Richard Drayton, *Nature's Government: Science, Imperial Britain and the 'Improvement' of the World* (New Haven: Yale University Press, 2000); and Richard Grove, *Green Imperialism: Colonial Expansion, Tropical Island Edens and the Origins of Environmentalism, 1600–1860* (Cambridge: Cambridge University Press, 1995) and Richard Grove, *Ecology, Climate and Empire: Colonialism and Global Environmental History, 1400–1940* (Cambridge: White Horse, 1997). Indeed, recognition of the ways in which "the emergence of a pan-European world-economy, stretching from the Baltic to the Americas, was at once cause and consequence of an epochal reorganization of world ecology," as Jason Moore puts it, shapes and motivates much of the field of environmental history, which emerged in something like its current form in the late twentieth century; see Jason W. Moore, "*The Modern World-System* as Environmental History? Ecology and the Rise of Capitalism," *Theory & Society* 32, no. 3 (2003): 307–377, 312. Clarence J. Glacken, *Traces on the Rhodian Shore: Nature and Culture in Western Thought from Ancient Times to the End of the Eighteenth Century* (Berkeley: University of California Press, 1967) was also a pioneering study of the conception of the environment in Western civilization. Studies of environmental history, its foundations, and new directions include Richard White, "American Environmental History: The Development of a New Historical Field," *Pacific Historical Review* 54, no. 3 (1985): 297–335; Donald Worster, *Nature's Economy: The Roots of Ecology* (1977), repr. as *Nature's Economy: A History of Ecological Ideas* (Cambridge: Cambridge University Press, 1985); and Alfred W. Crosby, "The Past and Present of Environmental History," *American Historical Review* 100, no. 4 (1995): 1177–89. Sverker Sörlin and Paul Warde examine and critique the movement in "The Problem of the Problem of Environmental History: A Re-Reading of the Field," *Environmental History* 12, no. 1 (2007): 107–130, arguing for more robust engagement with social and political theories.

15. James Beattie, Edward Melillo, and Emily O'Gorman, eds., *Eco-Cultural Networks and the British Empire: New Views on Environmental History* (London: Bloomsbury, 2016), 9.

On "informal empire" and the efforts to gain "indirect political hegemony" in South America, see J. Gallagher and R. Robinson's classic "The Imperialism of Free Trade," *Economic History Review* 6, no. 1 (1953): 1–15, 8; see also P. J. Cain and A. G. Hopkins, *British Imperialism: Innovation and Expansion, 1688–2015* (London: Routledge, 2016), esp. 265–299; and Jessica Reeder, *The Forms of Informal Empire: Britain, Latin America, and Nineteenth-Century Europe* (Baltimore: John Hopkins University Press, 2020), esp. 1–34. See too Mary Louise Pratt's important work on explorers and naturalists as the "capitalist vanguard," in *Imperial Eyes: Writing and Transculturation* (1992; repr. London: Routledge, 2008), esp. 141–168.

16. For more on the rapid growth of mechanical papermaking, the ways in which this both answered the needs and facilitated the rise of a complex, rapidly industrializing, imperial economy, and also on its impact on free trade after 1861, see Timo Särkkä, *Paper and the British Empire: The Quest for Imperial Raw Materials, 1861–1960* (New York: Routledge, 2021).

17. Many writers in our own time state that the orchid was first sent to collector William Cattley, that he was the one to bring it to flower. The claim is reprinted in a range of sources, including Merle A. Reinikka, *A History of the Orchid* (1972; repr. Portland: Timber Press, 1995), 23; Susan Orlean, *The Orchid Thief: A True Story of Beauty and Obsession* (1998; repr. New York: Ballantine, 2014), 69; Ambra Edwards, *The Plant Hunter's Atlas: A World Tour of Botanical Adventures, Chance Discoveries and Strange Specimens* (London: Greenfinch, 2021), 290; and Sue Shephard, *Seeds of Fortune: A Gardening Dynasty* (London: Bloomsbury, 2003), 186. The claim is, as we'll see, almost certainly incorrect.

18. For a few examples of texts that take on the tales circulating in Europe, see—from a French perspective—"Les Nouveaux Cattleya Labiata," *Orchidophile* 11 (1891): 310–313; from the British, see Frederick Boyle, *About Orchids: A Chat* (London: Chapman & Hall, 1893), 179–180.

19. Endersby, *Orchid*, 113.

20. See L. C. Menezes, *Cattleya Labiata Autumnalis* (Brasilia: Charbel Grafica, 2002), 70.

21. This story was told in the *Gardeners' Chronicle*, October 31, 1891, and in Boyle's *About Orchids*, 181; it is repeated in A. A. Chadwick and Arthur E. Chadwick, *The Classic Cattleyas* (Timber Press, 2006), 55 (the source of the quotation in my text); and Edwards, *The Plant Hunter's Atlas*, 292.

22. George Levine, ed., *One Culture: Essays in Science and Literature* (Madison: University of Wisconsin Press, 1987). The collection both begins with the phrase and quickly problematizes it. I don't mean to imply that all scientists and novelists were engaged in exactly the same enterprise, under the same circumstances, for the same reasons; I use the phrase "one culture" still because it quickly and effectively communicates that our prized disciplinary boundaries were experienced differently by our forebears. See Levine, *One Culture*, 3–6; and Gillian Beer, *Darwin's Plots: Evolutionary Narrative in Darwin, George Eliot and Nineteenth-Century Fiction* (1983; repr. Cambridge: Cambridge University Press, 2009), 5. Devin Griffiths finds a deeply collaborative process of sense-making at the heart of what he calls *collective authorship*: "As naturalists and literary authors turned toward each other in their efforts to shape a historical understanding suited to their different ends, collaboration became central to their work." Devin Griffiths, *The Age of Analogy: Science and Literature between the Darwins* (Bal-

timore: Johns Hopkins University Press, 2016), 9. Griffiths also discusses scholars who have pushed back against "one culture" (8–9).

23. Many studies have taken up the analysis of Darwin's impact on literature and vice versa. These include Beer, *Darwin's Plots*; and George Levine, *Darwin and the Novelists: Patterns of Science in Victorian Fiction* (Cambridge, MA: Harvard University Press, 1988). Levine importantly explores not Darwin's direct "influence" so much as the ways in which fiction took up, absorbed, and tested Darwinian principles thereafter (3). Other key scholarly studies of Darwin and literature include Sally Shuttleworth, *George Eliot and Nineteenth Century Science: The Make-Believe of a Beginning* (Cambridge: Cambridge University Press, 1984); and Gowan Dawson, *Darwin, Literature and Victorian Respectability* (Cambridge: Cambridge University Press, 2007). On evolution and storytelling see Brian Boyd, *On the Origin of Stories: Evolution, Cognition, and Fiction* (Cambridge, MA: Harvard University Press, 2009). For a contemporary examination of the scholarly field, its preoccupations, and key questions, see Devin Griffiths, "Darwin and Literature," in *Cambridge Companion to Literature and Science*, ed. Steven Meyer, 62–80 (Cambridge: Cambridge University Press, 2018).

24. Michael Hanne, *The Power of the Story: Fiction and Political Change* (Providence: Bergahn, 1994), 8. *Narrativization* is the term scholars use to refer to the arranging of facts and events in a sense-making enterprise shaped by the teller's understanding of what is (or what they hope or imagine to be) true. Narrative itself is a "linked sequence of events," as James Phelan and Peter J. Rabinowitz put it, with constitutive elements "shaped in the service of larger ends." Narrative is constructed "to affect readers in particular ways," "authorial designs . . . conveyed through the occasions, words, techniques, structures, forms, and dialogic relations of texts as well as the genres and conventions readers use to understand them." See Phelan and Rabinowitz, "Introduction: the Approaches," *Narrative Theory: Core Concepts and Critical Debates*, ed. David Herman et al. (Columbus: Ohio State University Press, 2012), 3–28; 3, 5.

25. Lynn Voskuil, "Victorian Orchids and the Forms of Ecological Society," in *Strange Science: Investigating the Limits of Knowledge in the Victorian Age*, ed. Lara Karpenko and Shalyn Claggett, 19–39 (Ann Arbor: University of Michigan Press, 2017), 23. Voskuil explores the ways in which orchids are represented in a range of texts, suggesting that humans' close relationships with orchids provoked ecological engagement. Lindsay Wells also examines how living with houseplants fostered broader engagement with the environment in "Vegetal Bedfellows: Houseplant Superstitions and Environmental Thought in Nineteenth-Century Periodicals," *Victorian Periodicals Review* 54, no. 1 (2021): 1–23.

Key book-length scholarly explorations of nineteenth-century plants in literature and in public, sociocultural discourse include Amy M. King, *Bloom: The Botanical Vernacular in the English Novel* (Oxford: Oxford University Press, 2004), and Elizabeth Chang, *Novel Cultivations: Plants in British Literature of the Global Nineteenth Century* (Charlotte: University of Virginia Press, 2019).

26. Cannon Schmitt in *Darwin and the Memory of the Human: Evolution, Savages, and South America* (2009; repr. Cambridge: Cambridge University Press, 2012) makes a similar point, though Schmitt's focus is on published works by scholars like Darwin. Still, he too notes the importance of scenes in naturalists' writing in which "the self is overwhelmed by what it comes into contact with" and that "among the most perdurable elements of encounter was its ability to place verities in question" (11).

27. Christopher Hager, *I Remain Yours: Common Lives in Civil War Letters* (Cambridge, MA: Harvard University Press, 2018), 10.

28. Hager, *I Remain Yours*, 5–6. Hager, in laying out ways of reading letters, stresses the value of excavating the lives of the writers, family members, and addressees, and of putting letters into dialogue with one another in order to trace the emotional tenor of the overarching narrative unfolding between the figures (8–9)—an approach I follow here.

29. Frederick Sander, quoted in Arthur Swinson, *Frederick Sander: The Orchid King* (London: Hodder & Stoughton, 1970), 46.

30. I also explore the slippage between public representations of plant hunters as brave heroes, and their self-presentation in their letters, in Sarah Bilston, "The Infrastructures of Plant Hunting," *Victorian Literature and Culture* 52, no. 3 (forthcoming, 2024).

31. Jacob Rees-Mogg, *The Victorians: Twelve Titans Who Forged Britain* (London: Penguin, 2019), xix.

32. The proper construction of narratives about the past is, of course, a longstanding topic of debate. Hayden White argues: "It is sometimes said that the aim of the historian is to explain the past by 'finding,' 'identifying' or 'uncovering' the 'stories' that lie buried in chronicles"; "This conception of the historian's task, however, obscures the extent to which 'invention' also plays a part in the historian's operations." Hayden White, *Metahistory: The Historical Imagination in Nineteenth-Century Europe* (1973; repr. Baltimore: Johns Hopkins University Press, 2014), 6. More recent scholars have argued that, for all the ways in which historical narratives draw on the kinds of rhetorical and literary tools White identifies, historical narrative does not *need* to be reduced to this, to "mere" fiction, pointing out the ways in which historical narrative also aims (for instance) to meet the demands of experts. See, for instance, Chris Lorenz, Stefan Berger, and Nicola Brauch, "Narrativity and Historical Writing: Introductory Remarks,"

in *Analysing Historical Narratives: on Academic, Popular, and Educational Framings of the Past*, ed. Stefan Berger, Nicola Brauch, and Chris Lorenz, 1–28 (New York: Bergahn, 2021).

33. "Narrative imitates life, life imitates narrative," as Jerome Bruner succinctly puts it in "Life as Narrative," *Social Research* 54, no. 1 (1987): 11–32, 13. On post-White approaches to narrative, including by Bruner, see Hanna Meretoja, *The Ethics of Storytelling: Narrative Hermeneutics, History, and the Possible* (Oxford: Oxford University Press, 2018), 8–11. Narrative structures, Meretoja argues, "can be oppressive, empowering, or both"; she encourages authors to be "sensitive to the ways in which narratives as practices of sense-making are embedded in social, cultural, and historical worlds" (2).

34. Marci Shore, "Can We See Ideas? On Evocation, Experience, and Empathy," in *Rethinking Modern European Intellectual History*, ed. Darrin M. McMahon and Samuel Moyn, 193–211 (Oxford: Oxford University Press, 2014), 195, 200. But also see Meretoja, who questions the idea of encouraging empathy to provoke change: "there is a significant difference between embracing the perspectives of others—or imagining what one might do in hypothetical scenarios—and actually carrying out concrete actions in the real world," she argues. Still, she suggests that encouraging understanding of the experiences of others "may be a *necessary* condition for moral agency, even if it is not a *sufficient* condition . . . It has transformative potential" (4).

In my own effort to convey the "historical heaviness" of the nineteenth-century mania for resource extraction, I seek to follow in the footsteps of authors like Alexander Nemerov, *The Forest: A Fable of America in the 1830s* (Princeton: Princeton University Press, 2023), ix, and Randall Fuller, *The Book That Changed America: How Darwin's Theory of Evolution Ignited a Nation* (New York: Viking, 2017). Both embrace the power of story, structured and underpinned by scholarly research, to make the stakes of nineteenth-century natural history and of lives intertwined with the natural world available to all. With Nemerov, "I hope that something real is revealed"; and with the editors of *Narrative in Culture* I suggest that "more often than not, a story reveals as much about life as a microscope." Astrid Erll and Roy Sommer, "A Tale of Two Concepts: Ansgar Nünning at Sixty," in *Narrative in Culture*, ed. Astrid Erll and Roy Sommer (Berlin: De Gruyter, 2019).

35. Lewis Castle, "Orchids: Cattleya Labiata Vera," *Journal of Horticulture, Cottage Gardener, and Home Farmer* 23 (September 24, 1891), 262–263, 262.

Part I epigraph: Márcia Wayna Kambeba, *Ay Kakyri Tama—Eu Moro Na Cidade* (Manaus: Gráfica e Editora, 2013); trans. Rosalien Bacchus, https://rosalienebacchus.blog/2019/11/17/silent-warrior-by-indigenous-brazilian-poet-marcia-wayna-kambeba/#more-4737.

1. ORIGINS

1. Jim Endersby discusses the treatment and appearance of orchids in literature over millennia; see Endersby, *Orchid: A Cultural History* (Chicago: University of Chicago Press, 2016), 4–5, 8, 11–23; on Theophrastus, see 14–16, 18. *Orkhis* means "testicle"; some orchids possess two tubers.

2. For more on the rock art of the region, see George Nash, "Serra da Capivara (North-East Brazil) and the Lamirí Basis (Chile): A Tale of Two Rock Art Landscapes," in *Archaeologies of Rock Art: South American Perspectives*, ed. Andrés Troncoso, Felipe Armstrong, and George Nash (London: Routledge, 2018), 150–176.

3. On the language of the convert versus the cannibal in European-authored literatures, see Hal Langfur, "Recovering Brazil's Indigenous Pasts," in *Native Brazil: Beyond the Convert and the Cannibal, 1500–1900*, ed. Hal Langfur (Albuquerque: University of New Mexico Press, 2014), 1–28. See also Tracy Devine Guzmán, *Native and National in Brazil: Indigeneity after Independence* (Chapel Hill: University of North Carolina Press: 2013), esp. 9–12.

4. Langfur explores the challenges of recovering Indigenous histories while stressing that those challenges have been exaggerated; see Langfur, "Recovering," esp. 10–12. For more on the complexity of Indigenous beliefs and practices and the myriad ways in which Indigenous peoples of Brazil endured and responded to European colonialism, see Langfur, *The Forbidden Lands: Colonial Identity, Frontier Violence, and the Persistence of Brazil's Eastern Indians, 1750–1830* (Stanford: Stanford University Press, 2006); on the enslavement of Indigenous peoples in Bahia, see Stuart B. Schwartz, *Sugar Plantations in the Formation of Brazilian Society: Bahia, 1550–1835* (1985; repr. Cambridge: Cambridge University Press, 1998), 28–50; on the relationship between the romanticization of "Indians" in Brazilian literature, culture, and politics and Indigenous lived experiences, see Guzmán, *Native and National*. The author examines Indigenous movements in Brazil today and modern Indigenous self-representation (159–206).

5. Pero Vaz de Caminha, "The Letter of Pero Vaz de Caminha," repr. in *Early Brazil: A Documentary Collection to 1700*, ed. Stuart B. Schwartz, 1–9 (Cambridge: Cambridge University Press, 2010), 2, 7, 6, 4. For more on the 1500–1501 encounters between Portuguese fleets and Indigenous tribes, see Langfur, "Recovering," 1–8.

6. Amerigo Vespucci, "Third Voyage of Amerigo Vespucci," *The Letters of Amerigo Vespucci and Other Documents Illustrative of His Career*, ed. and trans. Clements R. Markham, 34–41 (London: Ashgate, 2010), 38.

7. Amerigo Vespucci, "Letter on His Third Voyage from Amerigo Vespucci to Lorenzo Pietro Francesco Di Medici," in Markham, *Letters of Amerigo Vespucci*, 42–52, 47.

8. Langfur, "Recovering," 5.

9. Caminha, "Letter," 6.

10. Vespucci, "Letter to Di Medici," 45.

11. On the numbers of Indigenous peoples living in Brazil at the time of first contact with the Europeans, and the range of environments they inhabited, see Ann Marie B. Bahr, *Indigenous Religions* (New York: Chelsea House, 2005), 9.

12. Many scholars warn that contemporary environmentalist accounts of a near-pristine Brazilian landscape destroyed by colonialism risks reproducing an older stereotype (of the virgin forest) that was itself a colonialist invention; we must pay close attention to uncovering the many and diverse ways in which Indigenous peoples in Brazil interact and, in precolonial times, interacted with the environment. For more on the impact of colonialist agricultural practices on Indigenous peoples today, see Seth Garfield, "Indigenous Peoples and the Environment: Views from Brazil," in *Environment and Belief Systems*, ed. G. N. Devy, 167–189 (Oxford: Routledge, 2020); and William M. Denevan, "Pre-European Forest Cultivation in Amazonia," in *Time and Complexity in Historical Ecology*, ed. William Balée and Clark L. Erickson, 153–164 (New York: Columbia University Press, 2006).

The field of historical ecology, Balée and Erickson explain, offers a means of understanding how humans have left marks on the world around them: fundamentally interdisciplinary, it deploys anthropology as well as geography, archaeology, biology, and more to uncover and recognize those signs and further understand human activity that may not have left signs of its presence in text form. See William Balée and Clark L. Erickson, "Preface," in *Time and Complexity in Historical Ecology*, 1.

13. Guzmán, *Native and National*, 163. Guzmán also examines the reductive ways in which nonnative scholars have examined Indigenous communities and metaphysics (160–162), warning in particular against "translations" made by anthropologists who effectively claim to understand Native peoples better than they understand themselves (160).

14. For a brief overview of some classificatory systems used by Indigenous peoples in the Amazon, for instance, see Miguel Pinedo-Vasquez, Susanna Hecht, and Christine Padoch, "Amazonia," in *Traditional Forest-Related Knowledge: Sustaining Communities, Ecosystems and Biocultural Diversity*, ed. John A. Parrotta and Ronald L. Trosper, 119–156 (London: Springer, 2012), 120–122. Summarizing the work of earlier scholars, the authors note that "rural Amazonian

communities, whether indigenous or not, use complex classification systems that often include multiple criteria, including local geomorphology, physiognomy, soils, disturbance, and indicator plants (especially palms) and animal species to classify their environments; local communities often distinguish dozens of forest types over small scales" (12). (Note that Ka'apor is unrelated linguistically to the Kayapó.) The classificatory systems of the Ka'apor are also examined by Darrell A. Posey, *Kayapó Ethnoecology and Culture*, ed. Kristina Plenderleith (London: Routledge, 2002); on bees, see 107–108, on the forest and Cerrado types, see 169. William Balée's *Footprints of the Forest: Ka'apor Ethnobotany: The Historical Ecology of Plant Utilization by an Amazonian People* (New York: Columbia University Press, 1994) also focuses on the Ka'apor, examining how their knowledge of the natural world helped the tribe maintain control of their territory.

15. Guzmán, *Native and National*, 163. Guzmán explores the challenges Indigenous communities in Brazil have faced as they seek to play more central roles in politics and culture and promote their communities' interests; see esp. 180–181.

16. Caminha, "Letter," 6.

17. As Garfield explains, the Tupi-Guarani peoples used fire to clear quite small areas for cultivating crops for subsistence, with long fallow periods of twenty to forty years to allow the regeneration of the forests and the land. The deforestation practices of the Portuguese settlers, by contrast, involved massive destruction of the forests to prepare the way for monocrop agriculture and mining. Garfield, "Indigenous Peoples and the Environment."

18. Caminha, "Letter," 9.

19. On the oldest orchid fossil, a pollinia of *Mellorchis caribea* on the back of a bee, see David L. Roberts and Kingsley W. Dixon, "Orchids," *Current Biology* 18, no. 8 (2008): R325–R329, fig. 2.

20. Death was not the only fate of Brazil's Indigenous peoples in colonial times; they confronted the new arrivals in a range of ways—including adopting leadership positions in the new social order, renting out land, cutting timber, and generally adapting, as well as mounting active resistance, as Langfur explains in *Forbidden Lands*, presenting them as "full and deft participants in the contested process of territorial consolidation" (10).

21. See Posey, *Kayapó*, 158, on the use of orchids and other plants as contraception. The relocation of orchids by a member of the Kayapó tribe is referenced in Darrell Addison Posey, "Interpreting and Applying the 'Reality' of Indigenous Concepts: What Is Necessary to Learn from the Natives?" in *Conservation of Neotropical Forests: Working from Traditional Resource Use*, ed. Kent H. Redford and Christine Padoch, 21–34 (New York: Columbia University Press, 1992), 22.

22. For more on engagement with orchids in Brazil in the early nineteenth century, see Carlos Ossenbach, "Orchids in the Era of Grigory von Langsdorff: Two Golden Decades in the History of the Botanical Exploration of Brazil (1813–1830)," *Lankesteriana* 18, no. 2 (2018): 111–149; on Langsdorff more generally, see Roderick J. Barman, "The Forgotten Journey: Georg Heinrich Langsdorff and the Russian Imperial Scientific Expedition to Brazil, 1821–1829," *Terrae Incognitae* 3, no. 1 (1971): 67–96. Langsdorff, as we shall see presently, worked with Swainson in Rio.

23. Jorge Cañizares-Esguerra, "How Derivative Was Humboldt? Microcosmic Nature Narratives in Early Modern Spanish America and the (Other) Origins of Humboldt's Ecological Sensibilities," in *Colonial Botany: Science, Commerce, and Politics in the Early Modern World*, ed. Londa Schiebinger and Claudia Swain, 148–165 (Philadelphia: University of Pennsylvania Press, 2005), 152. Cañizares-Esguerra engages head on claims that Caldas was a mere disciple of Humboldt's by exploring the ways in which Humboldt's mapping of the Andes was indebted to traditions of local Spanish American intellectuals. For more on Humboldt's transformative impact on nineteenth-century natural science, see Andrea Wulf, *The Invention of Nature: Alexander von Humboldt's New World* (New York: Knopf: 2015).

24. For more on the habitats preferred by *Cattleya labiata* and its growth patterns, see L. C. Menezes, *Cattleya labiata autumnalis* (Brasilia: Charbel Grafica, 2002), 177, 179.

25. For more on the flowering seasons of *C. labiata*, see Menezes, *Cattleya labiata*, 185; I am indebted to Menezes's detailed accounts of the orchid's seasonal and growth patterns.

2. A NATURALIST ARRIVES

1. For more on the geography of the region and the dramatic events unfolding around the time of Swainson's visit, see Caitlin A. Fitz, "'A Stalwart Motor of Revolutions': An American Merchant in Pernambuco, 1817–1825," *The Americas* 65, no. 1 (2008): 35–62, 37.

Swainson's journal of the trip—transcribed in Geoffrey M. Swainson, ed., *William Swainson, Naturalist & Artist: Diaries 1808-1818* (Palmerston North, NZ: Swift Print, 1989), hereafter *Diaries 1808-1818*—begins on January 22, 1817 (95).

2. Theodore Gill, "Swainson and His Times," *Osprey* 4, no. 7 (March 1900): 104–108, 106, 107.

3. Swainson, March 28, 1814, *Diaries 1808-1818*, 72; W. Swainson to Sir James E. Smith, April 22, 1815, Linnean Society of London (hereafter Linn Soc), GB-110 / JES / ADD / 101.

4. "Epidendrum Nutans," *Botanical Register* 1 (1815): 17.

5. J. Shepherd to W. Swainson, n.d., Linn Soc, Correspondence of William Swainson R-S, MS 273, 816; Swainson to Smith, April 22, 1815, Linn Soc GB-110 / JES / ADD / 101.

Because Shepherd's first letter to Swainson is undated, I can't be certain it is from 1815. However, by the time of Swainson's next trip, the naturalist was so confident in his plant-packing skills he wrote a whole treatise on the subject on his return. It makes sense to assume this letter was written earlier. Shepherd also sounds in this letter as if he is addressing someone he does not know well, while his 1817 correspondence with Swainson has a familiar, confident tone.

6. For a more detailed investigation of the place of orchids in culture, their early arrivals in Europe and representations in print texts before Swainson, see Jim Endersby's excellent *Orchid: A Cultural History* (Chicago: University of Chicago Press, 2016), 11–63. Key scholarly books on the close relationships among the rise of botany as a science, a broadening environmental consciousness, and, through the early nineteenth century, imperialism and colonial profit-making include: Richard Drayton, *Nature's Government: Science, Imperial Britain and the 'Improvement' of the World* (New Haven: Yale University Press, 2000); Richard Grove, *Green Imperialism: Colonial Expansion, Tropical Island Edens and the Origins of Environmentalism, 1600–1860* (Cambridge: Cambridge University Press, 1995); Richard Grove, *Ecology, Climate and Empire: Colonialism and Global Environmental History, 1400–1940* (Cambridge: White Horse, 1997); Londa Schiebinger, *Plants and Empire: Colonial Bioprospecting in the Atlantic World* (Cambridge, MA: Harvard University Press, 2004); and Londa Schiebinger and Claudia Swain, eds., *Colonial Botany: Science, Commerce, and Politics in the Early Modern World* (Philadelphia: University of Pennsylvania Press, 2005).

7. "A species very recently introduced by Mr. William Swainson from the Botanic Garden at Palermo. Native of the coast of Barbary, and probably of Sicily." "Orchis Longicornu. Long-Spurred Orchis," *Botanical Register* 3 (1817): 202. Smith's reply to Swainson is dated January 15, 1816; Swainson's certificate of election as a fellow of the Linnean Society is dated December 17, 1816—he had become an associate on April 21, 1812. Today, by convention, plant binomials follow a standard for capitalization and are italicized, and common names of orchid species are usually lowercased. However, the convention was not consistently in force in the nineteenth century, so binomials in quoted texts appear as they are named, spelled, and formatted in the original; see also, for example, note 4, above.

8. The trips Henry Koster describes in *Travels in Brazil* (London: Longman, 1816) date to 1809–1815.

9. For more on the circumstances leading up to the moving of the Royal Family, see Lilia M. Schwarcz and Heloisa M. Starling, *Brazil: A Biography* (New York: Picador, 2015), 162–175.

10. For more on the relationship of Britain and Brazil in the period, see Leslie Bethell, *Brazil: Essays on History and Politics* (London: University of London Press, 2018), esp. 57–60. For more on Swainson's early years, the influence of Koster, and the context of the opening relationship between Britain and Brazil, see Paul Lawrence Farber, "Aspiring Naturalists and Their Frustrations: The Case of William Swainson (1789–1855)," *Archives of Natural History* 1985, no. 1 (1985): 51–59. On the role played by a diverse array of Latin American scientists in disseminating knowledge to arriving Europeans, see Jorge Cañizares-Esguerra, "How Derivative Was Humboldt? Microcosmic Nature Narratives in Early Modern Spanish America and the (Other) Origins of Humboldt's Ecological Sensibilities," in *Colonial Botany: Science, Commerce, and Politics in the Early Modern World*, ed. Londa Schiebinger and Claudia Swain, 148–165 (Philadelphia: University of Pennsylvania Press, 2005). On orchid explorations at the time specifically in Brazil, see Carlos Ossenbach, "Orchids in the Era of Gregory von Langsdorff: Two Golden Decades in the History of the Botanical Exploration of Brazil (1813–1830)," *Lankesteriana* 18, no. 2 (2018): 111–149.

11. William Swainson, "Sketch of a Journey Through Brazil in 1817 and 1818. By Mr. Swainson of Liverpool. In a Letter to Professor Jameson," *Edinburgh Philosophical Journal*, vol. 1, 1819, 369–373; 370 (hereafter "Sketch").

12. W. Swainson, "The Autobiography of William Swainson, F.R.S., F.L.S." (1840), repr. in Geoffrey M. Swainson, *William Swainson, Naturalist and Artist: Family Letters & Diaries 1809-1855*, 4–5 (Palmerston North, NZ: Swift Print, 1992), 4; hereafter *Family Letters & Diaries*.

13. W. Swainson, "Sketch," 370. He was, his Brazilian diaries make plain, living with Koster for the early months of his trip. See Swainson, *Diaries 1808-1818*, 95–96.

14. Quoted in David Elliston Allen, *The Naturalist in Britain: A Social History* (1976; repr. Princeton: Princeton University Press, 1994), 39. Allen dates the beginnings of natural history as a wider "social activity" to the seventeenth century (3).

15. Farber, *Aspiring Naturalists*, 51. He continues: "These changes, combined with the many social and cultural changes that altered the fabric of English life in the post-Napoleonic period, created new possibilities and new pitfalls for those who wanted to pursue natural history" (51–52).

16. For more on the increasing accessibility of information about gardening and horticulture in the eighteenth century, see Andrea Wulf, *The Brother Gardeners: Botany, Empire, and the Birth of an Obsession* (New York: Vintage, 2008), esp. 46–47.

17. Swainson, "Sketch," 370.

18. For more on the ways in which British communities in Brazil both tried to maintain their exclusivity but were also shaped by and around local Brazilian communities, see Bethell, *Brazil*, 82–83.

19. For discussion of Pernambuco's history, see Jeffrey Carl Mosher, *Political Struggle, Ideology, and State Building: Pernambuco and the Construction of Brazil, 1817–1850* (Lincoln: University of Nebraska Press, 2008).

20. Koster, *Travels*, 2.

21. Fitz, "Revolutions," 37. I owe my calculation of the racial make-up of Recife to Fitz; see too Mosher, *Political Struggle*, 22. Swainson discusses the sedan chairs in his journal; see Swainson, *Diaries 1808–1818*, 167–168.

22. Daniel Barros Domingues da Silva and Daniel Eltis, "The Slave Trade to Pernambuco, 1561–1851," in *Extending the Frontiers: Essays on the New Transatlantic Slave Trade Database*, ed. David Eltis and David Richardson, 95–129 (New Haven: Yale University Press, 2008), 104, 106. The authors estimate that over three centuries, 814,617 enslaved Africans arrived in Pernambuco (112); on the mortality rates on the journeys, see 113–114. Da Silva and Eltis's research draws on the resources made available at https://www.slavevoyages.org/.

23. On the slave trade and its supposed cessation in the 1810s, see Schwarcz and Starling, *Brazil*, 208–209. During the 1790s, "cotton surpassed sugar as Pernambuco's most valuable export," da Silva and Eltis explain ("Slave Trade," 111).

24. For discussion of the lodges and republican ideals, see Mosher, *Political Struggle*, 26, 25–26.

25. I'm particularly indebted in my account of the situation in Recife at the time of Swainson's visit to Fitz, "Revolutions"; Mosher, *Political Struggle*, 16–38; and Laurentino Gomes, *1808: The Flight of the Emperor: How a Weak Prince, a Mad Queen, and the British Navy Tricked Napoleon and Changed the New World* (Guilford, CT: Lyons, 2013).

26. W. Swainson, April 24, 1817, in Swainson, *Diaries 1808–1818*, 98. Swainson mentions the famine in several diary entries, for instance on February 14, 1817.

27. The diary is "an uncertain genre uneasily balanced between literary and historical writing, between the spontaneity of reportage and the reflectiveness of the crafted text, between selfhood and events, between subjectivity and objectivity, between the private and the public" and thus "constantly disturbs attempts to summarize its characteristics within formalized boundaries." Rachael Langford and Russell West, "Introduction: Diaries and Margins," *Marginal Voices, Marginal Forms: Diaries in European Literature and History*, ed. Rachael Langford and Russell West, 6–21 (Amsterdam: Rodopi, 1999), 8. Almost all diaries, that is, exhibit some of the confusions and fragmentations of Swainson's diaries; I call at-

tention to them here to point out the ways in which the naturalist recrafts his life experiences for his anticipated audience in the published work.

For more on the genre of the early nineteenth-century diary specifically and the ways in which it shaped, facilitated, and demanded forms of self-narrativization, see Rebecca Steinitz, *Time, Space, and Gender in the Nineteenth-Century British Diary* (New York: Palgrave Macmillan, 2011), 2.

28. W. Swainson, March 2, 1817, in Swainson, *Diaries 1808–1818*, 96.

29. For scholarly (English-language) accounts of the revolt and its wellsprings, see Fitz, "Revolutions," esp. 38–40; and Mosher, *Political Struggle*, esp. 16–38. Gomes's highly readable *The Flight of the Emperor* places the revolt in the context of events unfolding from 1808. Schwarcz and Starling emphasize the unpopularity of the governor and the influence of US ideals as articulated by the founding fathers (*Brazil*, 209–210). Mosher, *Political Struggle*, describes the key players in the catalyzing events of March 6 (26–27), the march on the fortress (27–28), and the spread of the movement (29–31).

30. W. Swainson, "Sketch," 370. See "Notes on the Revolution" in Swainson, *Diaries 1808–1818*, 109–117.

31. W. Swainson, diary entries on April 24, May 20, and May 14, 1817, in Swainson, *Diaries 1808–1818*, 98, 99, 99.

The diaries were, Geoffrey M. Swainson's preface indicates, updated by William after they were written, thus entries both reference contemporary events and also reflect back on them.

32. On the collapse of the rebellion, region by region, city by city, see Mosher, *Political Struggle*, 31–35.

33. W. Swainson, diary entries May 24 and May 29, 1817, in Swainson, *Diaries 1808–1818*, 100, 102. On reprisals against the rebels, including the public display of their body parts, see Mosher, *Political Struggle*, 34–35.

34. W. Swainson, diary entry June 24, 1817, in Swainson, *Diaries 1808–1818*, 105.

35. Swainson states he left in June; see W. Swainson, "Sketch," 370. His journals and diaries suggest a departure date of July, however, since he reported his departure on July 18–19, 1817, in Swainson, *Diaries 1808–1818*, 118.

36. W. Swainson, "Note: Customs House," in Swainson, *Diaries 1808–1818*, 171.

37. The *Betsey's* June 9 departure from Rio and arrival in Liverpool were reported in *Lloyd's List* on August 11, 1818.

38. W. Swainson, diary entry July 27, 1817, in Swainson, *Diaries 1808–1818*, 122.

39. W. Swainson, diary entry July 27, 1817, in Swainson, *Diaries 1808–1818*, 122.

40. W. Swainson, "Sketch," 371.

41. W. Swainson, diary entry January 20, 1818, in Swainson, *Diaries 1808-1818*, 157.

42. W. Swainson, "Slave Eating Earth," diary entry November 6, 1817, in Swainson, *Diaries 1808-1818*, 146.

43. W. Swainson, diary entry October 9, 1817, in Swainson, *Diaries 1808-1818*, 140.

Menezes lists known habitats of the orchid today, in Pernambuco and Ceará, Paraîba, and Alagoa (*Cattleya labiata autumnalis*, 187–189). Leading expert Cássio van den Berg confirmed to me, in private correspondence on March 28, 2023, that the Bahia region is not a known habitat of the orchid today; thus of the two places Swainson mentions, this seems the less likely collection site.

44. W. Swainson, "Sketch," 371.

45. W. Swainson, "Sketch," 372.

46. J. Paxton and J. Lindley, "Varieties of the Ruby-Lipped Cattleya (Cattleya Labiata)," *Paxton's Flower Garden*, vol. 1, 117–118 (London: Bradbury and Evans, 1853), 117.

47. Cássio van den Berg explained to me that "when a very large population of Cattleya [is] in flower, there is the odd plant in flower even out of season ... This is easy to notice in a large population in a rock or field, but not very likely to be spotted in a high canopy species such as [this]." Van den Berg, email to author, March 27, 2023.

48. Sir Joseph Banks was president of the Royal Society from 1778 to his death in 1820; he also served as botanist on the *Endeavour*. W. Swainson to Joseph Banks 181?, Linn Soc MS/270, 56. Swainson does not include the final digit of the date, but a reply from Banks is dated February 16, 1819 (Linn Soc MS/270, 57); Swainson's letter also explicitly notes that the "parasitic roots" were sent from Brazil.

3. TROUBLE BREWING

1. Joseph Dalton Hooker describes his father's "muscular" physique and broken nose in *A Sketch of the Life and Labours of Sir William Jackson Hooker, Late Director of the Royal Botanic Gardens of Kew* (1903; repr. Cambridge: Cambridge University Press, 2010), lxxxv.

2. J. D. Hooker, *Sketch*, xxvii.

3. Jim Endersby explains in *Imperial Nature: Joseph Hooker and the Practices of Victorian Science* (Chicago: University of Chicago Press, 2008) that even as late as the middle of the century "social and scientific acceptance and authority

were closely linked." The position of the scientist was changing, but still, to be taken as an intellectual equal, it helped to be seen as a social equal also (7).

4. Smith wrote grandly to a bishop that Hooker was worthy of support because he "devotes an independent fortune to literary and other commendable objects." J. E. Smith to the Bishop of Carlisle, April 3, 1808, in Lady [Plesaunce] Smith, ed., *Memoir and Correspondence of the Late Sir James Edward Smith*, v. 1 (London: Longman, 1832), 565.

5. Banks was also president of the Royal Society, as noted in Chapter 2. For more on the "gentleman scientist" see Endersby, *Imperial Nature*, esp. 2–3, 8–12. For more on the growth of the organizations that evolved to support young men of science, especially the Linnean Society, see David Elliston Allen, *The Naturalist in Britain: A Social History* (1976; repr. Princeton: Princeton University Press, 1994), 37–44.

6. J. Banks to W. Hooker, October 1, 1809; summarized in Warren R. Dawson, ed., *The Banks Letters: A Calendar of the Manuscript Correspondence of Sir Joseph Banks* (London: British Museum, 1958), 421. William Hooker describes the sinking of the ship, the *Margaret and Anne*, in thrilling terms in *Journal of a Tour in Iceland in the Summer of 1809* (Yarmouth: J. Keymer, 1811), 295–297. Quotations from Banks are from J. Banks to W. Hooker, June 19, 1813, in Dawson, ed., *Banks Letters*, 422–423.

7. J. D. Hooker, *Sketch*, xix.

8. W. Hooker to Dawson Turner, July 29, 1818, WJH / 2 / 1 Royal Botanic Gardens, Kew (hereafter RBGK), 272.

9. Hooker to D. Turner, July 23, 1817 (WJH / 2 / 1, RBGK, 239); Hooker, *Sketch*, xxvii, xxv. Hooker anxiously discusses the cost of barley in a letter of April 1, 1818 (WJH / 2 / 1, RBGK, 263–264) and is exercised about duty in a February 8, 1820, letter in which he lists his debt (WJH / 2 / 1, RBGK, 297–298). Hooker also notes his debts to his father-in-law, despairingly, in the July 23, 1817, letter, but the scale of the catastrophe only seems fully clear in February 1820— by which time he had received a job offer, so perhaps he felt able to admit the full extent of the problem. He continued preparing *Muscologia Britannia* in his leisure time between 1817 and 1819.

10. Andrea Wulf, *The Brother Gardeners: Botany, Empire, and the Birth of an Obsession* (New York: Vintage, 2008), 22. For more on the impact of industrialization on orchid cultivation in Britain in the nineteenth century and the speeding up of the import process, see Jim Endersby, *Orchid: A Cultural History* (Chicago: University of Chicago Press, 2016), 65–79, esp. 69.

11. James Veitch & Sons, *A Manual of Orchidaceous Plants* (London: James Veitch, 1887–1894), vol. 1, "Cattleya and Laelia," 110.

12. On greenhouse and stove construction in the early years of industrialization, see James Mean and John Abercrombie, *Abercrombie's Practical Gardener, or Improved System of Modern Horticulture*, 2nd ed. (London: T. Cadell & W. Davies, 1817). This edition is updated with information on recent scientific and horticultural discoveries and thus offers a picture of the state of the field at the time Swainson journeyed to Brazil.

13. William Swainson, "Sketch of a Journey Through Brazil in 1817 and 1818. By Mr. Swainson of Liverpool. In a Letter to Professor Jameson," *Edinburgh Philosophical Journal* 1, 1819, 369–373; 370 (hereafter "Sketch"), 373.

14. William Swainson, *The Naturalist's Guide for Collecting and Preserving Subjects of Natural History and Botany* (London: W. Wood, 1822), 61.

15. William Hooker, "Cattleya Labiata," *Exotic Flora*, vol. 3 (Edinburgh: Blackwood, 1825), Tab 157.

16. Lynn Voskuil explores in detail how the Victorian fascination with orchids stemmed from the ways in which these most remarkable plants challenge the boundaries between the plant and animal worlds: see Voskuil, "Victorian Orchids and the Forms of Ecological Society," in *Strange Science: Investigating the Limits of Knowledge in the Victorian Age*, ed. Lara Karpenko and Shalyn Claggett (Ann Arbor: University of Michigan Press, 2017), 19–39.

17. See George Graves, *The Naturalist's Pocket-Book, or Tourist's Companion* (London: Longman, 1818), 289.

18. "*Aerides paniculatum*: Sir Joseph Banks's *Aerides*," *Botanical Register*, vol. 3 (London: James Ridgway, 1817): 220.

19. "The method [Banks] pursues, is, to place the plants separately in light cylindrical wicker baskets or cages of suitable widths, of which the frame-work is of long slender twigs wattled together at the bottom and shallowly round the side, the upper portion being left open that the plant may extend its growth in any direction . . . and yet be kept steady in its station. . . . a thin layer of vegetable mould is strewed on the floor of the basket, on which the rootstock is placed, and then covered over lightly with a sufficiency of moss to shade it, and preserve a due degree of moisture." "*Aerides paniculatum*," 220.

20. Hooker wrote a number of times to Sir James Smith (for example on November 27, 1817, and again in October 1820) to ask if he had any spare exotics in his collection; he offered to come to Norwich to pick up the plants. He also, as we shall see, wrote repeatedly to Swainson to ask him for Brazilian specimens.

21. W. Hooker to D. Turner, July 23, 1817, WJH / 2 / 1, RBGK, 239–240; 240.

22. The elaborate construction of a "stove," the principles of hothouses and forcing gardens, and the issues of using dry versus moist heat are described in Abercrombie and Mean, *Abercrombie's Practical Gardener*, 586–593: "The Horticulturalist, when he steps into this department, aspires to the top and master-

ship of his art" (586). Hooker wrote to Swainson before Swainson's trip to Brazil, proudly mentioning his heated facilities when he asked his friend to send Brazilian specimens to Suffolk: "I am a most zealous collector in all departments." W. Hooker to W. Swainson, January 27, 1817, MS / 271, Linnean Society of London, 354.

23. W. Hooker, "Cattleya Labiata," Tab 157. Few of Hooker's letters from late 1818 seem to have survived; there is a gap between August 1818 and February 1819 in an otherwise very full collection of correspondence at RBGK.

24. John Lindley wrote that *Oncidium barbatum* and *Catasetum hookeri* were sent by Swainson to Hooker at the same time as what would become the "lost orchid"; see John Lindley, *Collectanea Botanica* (London: Richard & Arthur Taylor, 1821–1826), Tabs 27, 33, 40. Lindley also notes Swainson was responsible for "several other fine Orchideous plants" (Tab 27). The precise publication dates of the final two parts of the *Collectanea* are tricky to establish, as I'll presently discuss.

25. W. Hooker, "Cattleya Labiata," Tab 157.

4. SWAINSON VANISHES

1. Frederick Boyle, *About Orchids: A Chat* (London: Chapman & Gall, 1893), 178.

2. "The Great Orchid Sale," *Standard*, no. 20992, October 17, 1891, 3.

3. W. Swainson to Joseph Banks, 181?, Linnean Society of London (hereafter Linn Soc) MS / 270, 56.

4. J. Thelwall to W. Swainson, January 5, 1819, MS / 274, Linn Soc, 888. Thelwall assured Swainson he could cure any gentleman so long as "he will place himself under my care & follow up my instructions with due diligence and docility." In 1825 Swainson investigated another possible therapist, who charged the rather more manageable two and a half guineas a week (c. £262 / $336 in today's money). Swainson, experiencing ever greater financial pressures, seems not to have taken up either offer. See "Note by G.M.S.," W. Swainson, "The Autobiography of William Swainson, F.R.S., F.L.S." (1840), repr. in Geoffrey M. Swainson, *William Swainson, Naturalist and Artist: Family Letters & Diaries 1809–1855* (Palmerston North, NZ: Swift Print, 1992), 15; hereafter *Family Letters & Diaries*.

5. A. Macleay to W. Swainson, August 28, 1819, MS / 272, Linn Soc, 597.

6. W. Swainson to A. Macleay, February 18, 1822 and March 6, 1822, MS / 272, Linn Soc, 599–600. The conflict got tangled in a separate dispute with Mr. Curtis about color plates. Curtis seems to have mentioned the Dixon dispute as a way of shoring up his own (negative) claims about Swainson's character. By

1822, Swainson felt that he had given Macleay ample evidence of his integrity and was infuriated to hear that Curtis was telling everyone Macleay took his view of the matter.

7. Swainson, *Family Letters & Diaries*, 6.

8. W. Swainson to J. J. Audubon, May 1830, Swainson, *Family Letters & Diaries*, 30.

9. For discussion on the back-and-forth between Banks, Brown, Hooker, and the duke, see David Elliston Allen, *The Naturalist in Britain: A Social History* (1976; repr. Princeton: Princeton University Press, 1994), 75.

10. Joseph Dalton Hooker, *A Sketch of the Life and Labours of Sir William Jackson Hooker, Late Director of the Royal Botanic Gardens of Kew* (1903; repr. Cambridge: Cambridge University Press, 2010), xxxi. For more analysis of the move, and its possible *downward* social implications, see Jim Endersby, *Imperial Nature: Joseph Hooker and the Practices of Victorian Science* (Chicago: University of Chicago Press, 2008), 9–10. John Lindley, another towering botanist of the era, would soon give the lost orchid its name.

11. Emily O. Wittman, "'Death before the Fact': Posthumous Autobiography in Jean Rhys's *Good Morning Midnight* and *Smile Please*," in *Modernism and Autobiography*, ed. Maria DiBattista and Emily O. Wittman (Cambridge: Cambridge University Press, 2014), 185–196, 195.

12. W. Hooker to Turner, March 29, 1820, WJH / 2 / 1, Royal Botanic Gardens, Kew (hereafter RBGK), 299. Turner came to the rescue, writing much of the address himself.

13. W. Hooker to Robert Gostling White, March 18, 1822, HA244 / 1 / 215, Suffolk Archives, Ipswich. White was a solicitor in Halesworth and father to Robert Meadows White, a professor at Magdalen College, Oxford. The March 18 letter reveals Hooker was angst-ridden about his old house and its condition as he battled to adjust to life in Scotland.

14. W. Hooker to J. E. Smith, February 8, 1820, Linn Soc GB-110 / JES / COR / 23 / 4; see also Endersby, *Imperial Nature*, 11. Hooker wrote to his father-in-law to explain he was off to ask Smith "to explain to me something of the nature of a lecture." W. Hooker to Turner, February 8, 1820, WJH / 2 / 1, RBGK, 298.

15. W. Hooker to R. G. White, October 15, 1821, HA244 / 1 / 215, Suffolk Archives, Ipswich.

16. W. Hooker to W. Swainson, December 7, 1822, Linn Soc MS / 271, 356. By January 1825, the situation was vacant again; a letter from Hooker to his father-in-law indicates that Turner assessed the two positions for the Glasgow Professor and considered the move to the British Museum worse. He advised

Hooker against taking it. See W. Hooker to Dawson, January 29(?), 1825, WJH / 2 / 1, RBGK, 352–353.

17. W. Hooker to W. Swainson, May 21, 1824, Linn Soc MS / 271, 359.

18. W. Hooker to W. Swainson, December 15, 1825, Linn Soc MS / 272, 366a-b.

19. W. Hooker to W. Swainson, August 16, 1826, Linn Soc MS / 272, 367.

20. C. L. Swainson, letter to John Parkes, July 9, 1839, *Family Letters & Diaries*, 61.

Swainson lambasted his eldest child for displaying his very own faults; young William struggled with writing. "Your failing is, as it ever has been, want of application," Swainson snapped in 1838: "You write, in fact, not better but worse than you did two years ago, and that you may see for yourself I enclose you a letter written to me in October 1836." W. Swainson to W. J. Swainson, July 23, 1838, *Family Letters & Diaries*, 56.

21. Swainson's "Introductory Observations," sent by Swainson to Lindley, are still held at the Royal Horticultural Society in the Lindley Library, London. Swainson tried and failed to persuade Lindley to adopt the system.

22. William Swainson, "Introductory Observations on the Natural System," *Royal Horticultural Society Botanical Tracts*, RHS Lindley Library (hereafter RHS / LL), 22:12, xlil.

Swainson's father-in-law begged him to reconsider his departure: "as you say, you do not consider yourself infallible, you may reasonably doubt whether every friend you have in the world is in the wrong, and only you in the right." John Parkes to W. Swainson, February 21, 1840, *Family Letters & Diaries*, 64.

23. W. Hooker to F. von Mueller, April 9, 1854, Correspondence of Ferdinand von Mueller Collection, Royal Botanic Gardens Victoria, Australia. In Ferdinand von Mueller, *Regardfully Yours: Selected Correspondence of Ferdinand von Mueller*, 3 vols., ed. R. W. Home, A. M. Lucas, S. Maroske, D. M. Sinkora, J. H. Voigt, and M. Wells (Bern: Peter Lang, 1998, 2002, 2006), vol. 1, 176.

24. William Hooker, "Cattleya Labiata," *Exotic Flora*, vol. 3 (Edinburgh: Blackwood, 1825), Tab 157.

5. "OLD ANTIQUITY"

1. For more on the growth of the nursery business through the eighteenth century, see Sarah Easterby-Smith, *Cultivating Commerce: Cultures of Botany in Britain and France, 1760–1815* (Cambridge: Cambridge University Press, 2018).

2. "Advertisement," *Ipswich Journal*, no. 4331, April 28, 1821, British Library Newspapers Microfilm Reel # 2874.

3. See esp. Easterby-Smith, *Cultivating*, 26.

4. William T. Stearn, "The Life, Times and Achievements of John Lindley 1799–1865," *John Lindley, 1799–1865: Gardener, Botanist, and Pioneer Orchidologist*, ed. William T. Stearn (Woodbridge, UK: Antique Collector's Club with the Royal Horticultural Society, 1999), 15–72; 18. See also Lord [Nathaniel] Lindley, "Sketch of My Father's Life, Written for My Sons and Daughters and My Grandchildren. February 1911," in *John Lindley (1799–1865), "Father of Modern Orchidology": A Gathering of His Correspondence*, ed. Robert Hamilton, typescript MS in four parts, 920 LIN JOH 1–4, RHS / LL, 1995–1998 (hereafter *A Gathering*), (Part 2), viii–xvii; xvi.

5. Lord Lindley, "Sketch," viii; Stearn, "Life, Times," 17. For discussion on grammar schools as engines of social mobility, see Michael Sanderson, *Education, Economic Change and Society in England, 1780–1870* (1983; repr. Cambridge: Cambridge University Press, 1995), esp. 30–33.

6. Jacob Wrench, giving evidence in an 1823 Parliamentary debate concerning a Weights and Measures Bill, referenced business dealings in the Netherlands and Holland.

7. W. Hooker to Lindley, January 29, 1818, Lindley Letters A-K, Royal Botanic Gardens, Kew (hereafter RBGK), 422–423.

8. See for instance, W. Hooker to D. Turner, March 7, 1819, WJH / 2 / 1, RBGK, 280.

9. See William Hooker, "Cattleya Labiata," *Exotic Flora*, vol. 3 (Edinburgh: Blackwood, 1825), Tab 157. Hooker mentions Swainson in his March 7, 1819, letter: he indicates to Turner that he has received plants from Swainson from Greece and is trying to catalog them.

10. Andrea Wulf calls Soho Square "the centre of botanical enquiry" from when Banks leased the premises in 1777. Andrea Wulf, *The Brother Gardeners: Botany, Empire, and the Birth of an Obsession* (New York: Vintage, 2008), 200. For more on the day-to-day work of collaborative botany in Banks's herbaria, see Edwin D. Rose, "From the South Seas to Soho Square: Joseph Banks's Library, Collection and Kingdom of Natural History," *Notes and Records* 73, no. 4 (2019): 499–526.

11. An 1820 illustration by Francis Boott shows an imposing, high-ceilinged herbarium with a balcony.

12. W. Hooker to Lindley, January 21, [1819], Lindley Letters A-K, RBGK, 426–427.

13. W. Hooker to Lindley, January 21, [1819], Lindley Letters A-K, RBGK, 426–427. Though Hooker does not indicate the year at the head of the letter, he references Lindley's very recent arrival in lodgings in London.

14. Lindley to W. Hooker, April 22, 1819, in Hamilton, ed., *A Gathering* (Part 1), 1-2.

15. Lindley to Brown, June 27, 1819, in Hamilton, ed., *A Gathering* (Part 3), 2.

16. See Jim Endersby, *Imperial Nature: Joseph Hooker and the Practices of Victorian Science* (Chicago: University of Chicago Press, 2008), 172.

17. On the organization and presentation of information in Soho Square, on paper, in botanical slips (as a means of coping with near-overwhelming amounts of information), see Isabelle Charmantier and Staffan Müller-White, "Carl Linnaeus's Botanical Slips (1767-1773)," *Intellectual History Review* 24, no. 2 (2014): 215-238.

18. On the horrors of the "horticultural orgy" see Wulf, *Brother Gardeners*, 59, 53.

19. Amy M. King, *Bloom: The Botanical Vernacular in the English Novel* (Oxford: Oxford University Press, 2004), 12, 6. The Linnaean system and the "botanical vernacular" it spawned opened up new narrative possibilities for the nineteenth-century *bildungsroman* and courtship plot, King argues. See also Anne Secord, "Corresponding Interests: Artisans and Gentlemen in Nineteenth-Century Natural History," *British Journal for the History of Science* 27, no. 4 (1994): 383-408.

20. G. Stanley Hall's foundational study *Adolescence: Its Psychology and Its Relation to Physiology, Sociology, Sex, Crime, Religion and Education* would not be published until 1904. For more discussion on emerging concepts of adolescence in the nineteenth century, see Sarah Bilston, *The Awkward Age in Women's Popular Fiction: Girls and the Transition to Womanhood* (Oxford: Clarendon, 2004).

21. For discussion of the Natural and Artificial Systems, see Endersby, *Imperial Nature*, 172-180, and Theresa M. Kelley, *Clandestine Marriage: Botany and Romantic Culture* (Baltimore: Johns Hopkins University Press, 2012), esp. 20-21.

22. The impact of Linnaeus was felt more broadly, too, in a culture that by the early nineteenth century considered botanizing an appropriate activity for the many, the new middle classes licensed by the comparative ease of his system to try it out for themselves. See Endersby, *Imperial Nature*, 173-177, and *Orchid: A Cultural History* (Chicago: University of Chicago Press, 2016), 50-57; see too John Gascoigne, *Joseph Banks and the English Enlightenment: Useful Knowledge and Polite Culture* (1994; repr. Cambridge: Cambridge University Press, 2003), 98-107.

23. J. Lindley to R. Brown, October 10, 1819, Robert Brown Correspondence, National History Museum, London; quoted in Rose, "South Seas to Soho

Square," 523n123. On the ongoing force of the Linnaean system into the 1830s in amateur circles, see King, *Bloom,* 57–58. For more on the slow fade of Linnaean influences on Banks, his librarians, and the work of his collections in London, see Rose, "South Seas to Soho Square," 518–523.

24. John Lindley, *Collectanea Botanica* (London: Richard & Arthur Taylor, 1821–1826), Tab 2.

25. W. Hooker to Lindley, July 2, 1820, Lindley Letters A-K, RBGK, 439–440.

26. "The millstone thus early hung round his neck caused him trouble throughout life," his son later explained; Lord Lindley, "Sketch," in Hamilton, ed., *A Gathering* (Part 2), ix.

6. THE RUSSIA CONNECTION

1. W. Hooker to Lindley, December 13, [1819?], Lindley Letters A-K, Royal Botanic Gardens, Kew (hereafter RBGK), 443–444. Hooker does not explicitly give the year, but the place of writing (Halesworth) and events recounted make 1819 most likely.

2. See Geoffrey Jones's account of British-Russian trade in the period in *Merchants to Multinationals: British Trading Companies in the Nineteenth and Twentieth Centuries* (Oxford: Oxford University Press, 2002), esp. 24–25.

3. Robert Campbell, *Ruling Cases,* vol. 4 (London: Stevens, 1895), 376. For more on the Cattleys and Russia—some family members were in the Russia Trade while others were in insurance, indirectly in business with Russia—see also Ernest Heatherington, "The Life and Times of William Cattley," *American Orchid Society Bulletin* 61 (1992): 1228–1243, 1240.

4. Charles G. Harper, *The Great North Road: The Old Mail Road to Scotland* (London: Cecil Palmer, 1901), 76.

5. For a broader discussion, see Sarah Bilston, *The Promise of the Suburbs: A Victorian History in Literature and Culture* (New Haven: Yale University Press, 2019).

6. W. Hooker to J. Lindley, July 2, 1820, Lindley Letters A-K, RBGK, 439–440.

7. W. Hooker to D. Turner, March 29, 1820, WJH / 2 / 1, RBGK, 299–300. Cattley's salary amounted to over £40,000 / $51,000 today.

Jung or *Jungermannia* refer to a work Hooker published in 1816; the illegible word may be an abbreviation for another published work, probably *Musci Exotici,* publ. 1818–1820.

8. John Lindley, "Cattleya labiata," *Edwards's Botanical Register,* n.s. vol. 9, o.s. 22, 1836, Tab 1859.

9. John Lindley, *Collectanea Botanica* (London: Richard & Arthur Taylor, 1821–1826), Tab 33.

10. William Cattley to John Lindley, June 30, 1821, Lindley Letters A-K, RBGK, 171–172.

11. W. Hooker to Lindley, August 11, 1821, Lindley Letters A-K, RBGK, 447.

12. Lindley, *Collectanea* [Part 6], 182; Tab 31, "Lissochilus Speciosus."

13. Each of the first six parts was accompanied by a blue wrapper with a date / publication page and a page addressed to the reader and the binder. The last two parts do not seem to have included the date / publication page, or it was not preserved at the RHS Lindley Library, thus I've been unable to confirm their exact dates of publication. Scholars generally assume the dates used here.

It seems likely Part 7 was published by 1822 because Hooker's herbarium of that date, housed at Kew, includes a "Catleya," suggesting familiarity with the information included in part 7. Of course, it is always possible the plant name was added later.

14. John Lindley, *A Natural System of Botany*, 2nd ed. (Longman: London, 1836), vii.

15. "Epidendrum Violaceum," *Botanical Cabinet* 4 (1819), no. 337.

16. Lindley, *Collectanea*, Tab 37.

17. Quoted in King, *Bloom*, 163.

18. Lindley, *Collectanea*, Tab 33. The sexual implications of the name will be discussed in more detail presently.

19. Lindley, *Collectanea*, Tab 37.

20. Nineteenth-century novels are near-obsessed with names: many are explicitly named after characters whose identities structure the plot, such as Charles Dickens's *Oliver Twist, Martin Chuzzlewit,* and *David Copperfield*; Charlotte Brontë's *Jane Eyre*; Anne Brontë's *Agnes Grey*; Mary Elizabeth Braddon's *Aurora Floyd*; George Eliot's *Adam Bede, Daniel Deronda,* and *Silas Marner*. Others turn on the significance of a name (Thomas Hardy's *Tess of the D'Urbervilles*) and / or the loss of a name (Dickens's *Bleak House*, Wilkie Collins's *No Name*). Many gradually reveal a concealed name (Mary Elizabeth Braddon's *Lady Audley's Secret*, Collins's *The Woman in White*), and names often point to hidden truths; Jane Eyre's last name hints at her urge to be free, while David Copperfield's initials, flipped, become the author's own C[harles] D[ickens].

The midcentury sensation novel's particular obsession with names and with unstable identities has been much discussed by scholars: "Sensation fiction is filled with impostors who don false names, appearances and social positions," observes Tara Macdonald in "Sensation Fiction, Gender and Identity," *The Cambridge Companion to Sensation Fiction*, ed. Andrew Mangham (Cambridge: Cambridge University Press, 2013), 127–140, 129.

21. W. Botting Hemsley, "William Cattley," *Gardeners' Chronicle* 23, February 12, 1898, 93.

22. John Lindley, *Digitalium Monographia* (London: Richard & Arthur Taylor, 1821), dedication.

23. Lindley, *Collectanea*, Tab 33.

24. W. Hooker, "Herbarium Hookerianum, July 1822," WJH / 1 / 1, RBGK. The ledger comprises lists of plants in columns. The "t" of "Catley" is quite differently shaped from the "t" in the column to its left.

25. Lars Bernaerts, Dirk de Geest et al., *Stories and Minds: Cognitive Approaches to Literary Narrative* (Lincoln: University of Nebraska Press, 2013), 2.

26. Frederick Boyle, *About Orchids: A Chat* (London: Chapman & Hall, 1893), 174.

27. "Cattleya Labiata," *Orchid Review* 1 (1893): 329–331; 329.

28. Cattley showed *Cattleya labiata* on November 4, 1829, according to J. C. Loudon's *Gardener's Magazine*.

29. On the relationship of the events in 1817 Pernambuco and Brazil's eventual independence, see Jeffrey Carl Mosher, *Political Struggle, Ideology, and State Building: Pernambuco and the Construction of Brazil, 1817–1850* (Lincoln: University of Nebraska Press, 2008), 39–50.

30. Karl Marx explicitly notes that "The nature of such wants [for commodities], whether, for instance, they spring from the stomach or from fancy, makes no difference" in *Capital: A Critique of Political Economy* (1867; repr. Chicago: C. H. Kerr, 1906), 41–42.

7. FIRST SIGHT

1. The majority of the illustrations in the *Collectanea* are done by Lindley but, for whatever reason, the gap in publication led him to outsource to friends and colleagues; five of six illustrations after the hiatus were completed by other artists, the first two—including *C. labiata*—produced by Curtis.

2. "Cattleya Forbesii," *Botanical Register* 11 (1825): Tab 953.

3. Half a guinea, or c. 10 shillings 6d, is approximately £65 / $83 today. "Cives In Rare," "A List of the Most Beautiful Tropical Orchideae in Cultivation," *Floricultural Cabinet* 3 (1835): 78–79.

4. *Botanical Register* also positioned the words "exotic plants" in the center of the title page from 1815–1827.

5. Edward W. Said, *Orientalism* (1979; repr. New York: Random House, 1994), 118, 72, 1.

6. Piya Pal-Lapinksi distinguishes, in her study of exotic women, between Victorian responses to the "near exotic" (which tend to be places in Southern

Europe) and the "'remote' Exotic" (e.g., India) in *The Exotic Woman in Nineteenth-Century British Fiction and Culture: A Reconsideration* (Durham, NH: University of New Hampshire Press, 2005), 3.

7. Peter Mason, *Infelicities: Representations of the Exotic* (Baltimore: Johns Hopkins University Press, 1998), 1–2, quoted on 1.

8. For further discussion on the exoticism of the Eastern European vampire, see Pal-Lapinksi, *Exotic Woman*, 3.

9. Jim Endersby, *Orchid: A Cultural History* (Chicago: University of Chicago Press, 2016), 63. Endersby scrutinizes the Orchis myth and Liger's possible reasons for constructing it (59–63).

10. John Lindley, *Collectanea Botanica* (London: Richard & Arthur Taylor, 1821–1826), Tab 6.

11. Indeed, one of the most "inexplicable" phenomena Lindley identifies is the "manner in which foecundation takes place in *Orchideae*." This particular question would not be resolved, as we'll see, until Darwin (Lindley, *Collectanea Botanica*, Tab 6).

12. Jonathan Smith reminds us that "images can constitute knowledge rather than merely re-packaging textual statements" in *Charles Darwin and Victorian Visual Culture* (Cambridge: Cambridge University Press, 2006), 34. My discussion here is shaped by a longstanding scholarly engagement with seeing and the visual, encompassing perhaps most famously John Berger's *Ways of Seeing* (New York: Penguin, 1972). Images try to prompt us to produce certain meanings, Berger argues: oil paintings of nude women anticipate a male heterosexual observer, for instance, and position the woman as an object of desire. As viewers we may find ourselves the expected gazer or not the expected gazer; we may accept or resist the way we are positioned by the visual text.

13. On traditions of flower painting and botanical illustration, see Gill Saunders, *Picturing Plants: An Analytical History of Botanical Illustration* (Berkeley: University of California Press, 1995).

14. Saunders, *Picturing*, 15.

15. In each entry, Lindley identifies and classifies the plant at the head of the page, indicating on the left "Nat Ord," or Natural Order, on the right the Linnaean system. Lindley's "Natural Order" designation builds on the scholarship of post-Linnaean scholars like Brown: Lindley references Brown's 1810 *Prodromus Florae Novae Hollandiae Et Insulae Van-Diemen* when he designates *C. labiata* "Nat. Ord. Orchidae Sect 5 ['Orchideae'] *Br*" (330), for instance. This directs the well-versed reader to Brown's definition of *Monandrae*.

16. Lindley also built on the scholarship of predecessors such as de Candolle and Jussieu, who "relied on the totality of the plant's organs" and not just its sexual parts; Jonathan Smith, *Charles Darwin and Victorian Visual Culture*

(Cambridge: Cambridge University Press, 2006), 13. See also Saunders, *Picturing*, 97.

17. "[B]est described as a modern Romance language of special technical application," William T. Stearn notes, "derived from Renaissance Latin with much plundering of ancient Greek, [botanical Latin] has evolved, mainly since 1700 and primarily through the work of Carl Linnaeus . . . to serve as an international medium for the scientific naming of plants in all their vast numbers and manifold diversity." Stearn, *Botanical Latin: History, Grammar, Syntax, Terminology and Vocabulary* (1966; repr. London: David & Charles, 1983), 6–7.

18. See Stearn, *Botanical Latin*, 6–18.

19. On the engagement with botany far outside the elite, see Anne Secord, "Science in the Pub: Artistan Botanists in Early Nineteenth-Century Lancashire," *History of Science* 32, no. 3 (1994): 269–315.

20. Sarah Easterby-Smith, *Cultivating Commerce: Cultures of Botany in Britain and France, 1760–1815* (Cambridge: Cambridge University Press, 2018), 80.

21. On Victorian visual culture see Kate Flint, *The Victorians and the Visual Imagination* (Cambridge: Cambridge University Press, 2000); and Carol T. Christ and John O. Jordan, eds., *Victorian Literature and the Victorian Visual Imagination* (Berkeley: University of California Press, 1995). On visualizing Victorian science, see Smith, *Charles Darwin*. On color in the nineteenth-century periodical press, from hand-coloring through to chromolithography, see Patricia Mainardi, *Another World: Nineteenth-Century Illustrated Print Culture* (New Haven: Yale University Press, 2017), esp. 7, 39, 115, 201–202, 239; and Laura Anne Kalba, *Color in the Age of Impressionism: Commerce, Technology, and Art* (University Park: Pennsylvania State University Press, 2017).

22. Brent Elliott, "The Artwork of *Curtis's Botanical Magazine*," *Curtis's Botanical Magazine* 17, no. 1 (2000): 35–41, 36.

23. For more discussion see Susan Dackerman, *Painted Prints: The Revelation of Color in Northern Renaissance & Baroque Engravings, Etchings, and Woodcuts* (University Park: Pennsylvania State University Press, 2002), 3.

24. Quoted in Dackerman, *Painted Prints*, 12. See also John Gage, *Color and Culture: Practice and Meaning from Antiquity to Abstraction* (1993; repr. Berkeley: University of California Press, 1999) on deep-rooted "perceptual prejudices" against color (7).

25. Mainardi, *Another World*, 2.

26. Lindley, *Collectanea*, Tab 1.

27. Lindley, *Collectanea*, Tab 2.

28. *Floral Cabinet, and Magazine of Exotic Botany* cost half a crown (2s / 6d, or 2 shillings sixpence), or £14.35 / $18.30 today, and advertised in the solidly middle-class 8d (8 pence) weekly *Literary Gazette* and 4d *Athenaeum* (£3.70 /

$4.72 and £1.95 / $2.50 respectively today). G. B. Knowles and Frederic Westcott, "Preface," *Floral Cabinet* 1 (1837): v–vi; vi.

29. Knowles and Westcott, "Preface," vi.

30. Loddiges's *Botanical Cabinet* also foregrounded plates on the title page, explaining to readers in 1817: "This is assuredly no dry or abstruse study; it is a perpetual spring of the most genuine satisfaction." The experience of looking at the gorgeous productions of Nature is also presented as a means of getting closer to "our Divine Saviour," as it is in the *Floral Cabinet*. "Introduction," *Botanical Cabinet* 1 (1817), n.p.

31. Flint, *Visual Imagination*, 5. I owe to Kate Flint the point about newspaper illustrations and the *Strand* (4).

32. "Cattleya Labiata," *Floral Cabinet* 1 (1837): 53–56; 53–54.

33. John Keats, "On Seeing the Elgin Marbles," *The Poetical Works of John Keats: With a Memoir* (Boston: Little, Brown, 1866), 389; Knowles and Westcott, "Preface," vi. The editors here quote—then proceed to misquote—four lines by Mark Akenside's *The Pleasures of Imagination* (London: printed for R. Dodsley, 1744). The lines are quoted by all sorts of naturalists in the era, usually incorrectly; the words are also sometimes wrongly attributed to William Cowper.

34. Joseph Paxton, "Cattleya Labiata, *var*. Atropurpurea," *Paxton's Magazine of Botany* 7 (1841): 73–74; 73.

35. The *Botanical Cabinet* as a whole pays, Easterby-Smith remarks, "only cursory attention to botanical taxonomies." It is rather a "lavish picture [book]" that portrays "some of the flashiest exotic ornamental specimens" (Easterby-Smith, *Cultivating*, 111).

36. In *Exotic Flora*, Hooker uses Lindley's description of the type-species and plant in botanical Latin, then transitions to English. Hooker also references the Linnaean system and Natural Order.

Dissections did not decrease under Hooker's hand: indeed, he "increased the quantity of dissections and the use of the leaf outline in the background ... Dissections came to be standard features of plates in *Curtis's Botanical Magazine*, so that the absence of a dissection became more remarkable than its inclusion," Brent Elliott explains ("Artwork," 37).

37. William Hooker, "Cattleya Labiata," *Exotic Flora*, vol. 3 (Edinburgh: Blackwood, 1825), Tab 157.

38. John Lindley, "Cattleya labiata," *Edwards's Botanical Register*, n.s. vol. 9, o.s. 22, 1836, Tab 1859.

The claim that this was the most beautiful orchid of all was made often: "The splendour and high character of the colour of the flowers of this plant, surpass anything we ever witnessed in the whole family of orchideous plants," cheered Paxton, in "New and Rare Plants Figured in the Three Leading

Botanical Periodicals and Florists' Magazines for June," *Paxton's Magazine of Botany* 3 (1837): 164–168; 167.

39. W. Hooker, "Cattleya Labiata. Crimson-Lipped Cattleya," *Curtis's Botanical Magazine*, 2nd ser., vol. 16 [vol. 69 of entire work] (1843): Tab 3998.

40. "Cattleya Labiata," *Botanical Cabinet* 20 (1833), no. 1956; Lindley, *Collectanea*, Tab 33.

41. Lord Grey of Groby died at thirty-three; his *C. labiata* won a Class 1 award in 1834 for "The best flowered specimen either of a scarce plant or one of recent introduction into Europe" at the Birmingham Botanical and Horticultural Society. "Birmingham Botanical and Horticultural Society," *Royal Lady's Magazine* 3 (1834): 228–229; 228; James Veitch & Sons, *A Manual of Orchidaceous Plants* (London: James Veitch, 1887–1894), vol. 1, "Cattleya and Laelia," 16.

8. FALSE TRAIL?

1. On the importance of signaling credibility through "gentlemanliness," see Jim Endersby, *Imperial Nature: Joseph Hooker and the Practices of Victorian Science* (Chicago: University of Chicago Press, 2008), esp. 29.

2. See Endersby, *Imperial Nature*, 10.

3. See Hugh M. Donald, *Rambles Round Scotland* (Glasgow: Thomas Murray, 1856), 342–343.

4. David Elliston Allen, *The Naturalist in Britain: A Social History* (1976; repr. Princeton: Princeton University Press, 1994), 95, 96.

5. Quoted in Jessica Reeder, *The Forms of Informal Empire: Britain, Latin America, and Nineteenth-Century Literature* (Baltimore: Johns Hopkins University Press, 2020), 11. Brazil was "a virtual British protectorate," according to P. J. Cain and A. G. Hopkins, "Britain's most accommodating and most successful satellite in South America." Cain and Hopkins, *British Imperialism: Innovation and Expansion, 1688–2015* (London: Routledge, 2016), 280. On Latin America as both "free and English"—supposedly—see 267. Pedro I was the first ruler of the new empire.

6. George Gardner, *Travels in the Interior of Brazil; Principally Through the Northern Provinces, and the Gold and Diamond Districts, During the Years 1836–1841* (London: Reeve, 1846), 3, 4, 2. Much like Swainson, Gardner wrote a narrative after his trip to publicize his achievements. See William Swainson, "Sketch of a Journey Through Brazil in 1817 and 1818. By Mr. Swainson of Liverpool. In a Letter to Professor Jameson," *Edinburgh Philosophical Journal* 1 (1819): 369–373.

7. Mary Louise Pratt explores how natural historians and travelers of the period presented their interests in South America as disinterested, "simulta-

neously innocent and imperial, asserting a harmless hegemonic vision" in spite of the ways in which they used and firmed up European hegemony in the region as they journeyed. Pratt, *Imperial Eyes: Writing and Transculturation* 2nd ed. (London: Routledge, 2007), 33.

8. Other explorers of Brazil included Allan Cunningham and James Bowie, who had been sent out by Joseph Banks.

9. Charles Waterton, *Wanderings in South America, The North-West of the United States, and the Antilles, in the Years 1812, 1816, 1820, & 1824* (1825; repr. London: R. Fellowes, 1839), 218–219.

10. Gardner, *Travels*, 85.

11. For instance, a long account of journeying is interrupted when Gardner observes: "Thus we find, that the structure of the rocks in this locality, is very similar to that of the chalk formation in England" (Gardner, *Travels*, 206).

12. Gardner, *Travels*, 223. This approach is common in emerging science writing; "The contributions of local people became invisible through the conventions in particular genres of scientific writing," Jane R. Camerini explains, noting that the narrating "I" of science writing texts often reinforced colonialist norms, since it was a common practice of the genre to present the first person speaker as sole practitioner of fieldwork—"It appears to have been an unremarkable part of the practice." Camerini, "Wallace in the Field," *Osiris* 11 (1996): 44–65; 61.

13. Cannon Schmitt, *Darwin and the Memory of the Human: Evolution, Savages, and South America* (Cambridge: Cambridge University Press, 2009), 9. "The quest for natural-historical knowledge," Schmitt continues, "while in principle distinct from the imperial project, often enabled the exploration of new territory, advanced the possibility of exerting control over that territory, and extolled the desirability of doing so" (9).

14. Gardner to William Jackson Hooker, August 25, 1837, Directors' Correspondence (hereafter DC), 67 / 33, Royal Botanic Gardens, Kew (hereafter RBGK).

15. Gardner describes the traumatic brain injury in a chapter on Ceará; Gardner, *Travels*, 214–216; the horse stealing too is evoked (172). During the rainy season of 1839 he lost two thousand specimens when his horses fell repeatedly on a bad road. See Gardner to William Jackson Hooker, November 3, 1839, RBGK, DC, 68 / 33.

16. Gardner, *Travels*, 178.

17. Unable to see the man's reasons for anger, Gardner reported the whole business to the minister of justice. Measured reflection at least slightly reduced his ire, though it did not reduce his sense of entitlement. At the time of the attack, he had wished for his pistol, though later, he conceded to Hooker: "now I do

not regret it, as I certainly would have shot him." Gardner to William Jackson Hooker, August 25, 1837, RBGK, DC / 67 / 33.

18. Gardner, *Travels*, 97–98, 261–262, 274, 283. Gardner used his medical training to treat both white people and free people of color, and he expressed himself a critic of slavery. However, he had only the barest grasp of the horrors of chattel slavery, and his text is full of generalizations about race.

19. Gardner, *Travels*, 29, 33.

20. George Gardner, "Contributions to a History of the Relation Between Climate and Vegetation in Various Parts of the Globe," *Journal of the Horticultural Society* 1 (1846): 191–198; 196.

21. Gardner to William Hooker, December 18, 1836, RBGK, DC / 68 / 17.

22. Gardner, *Travels*, 65. Gardner tells the story of recovering the plants twice in the December 18 letter, first as a sort of summary, then later with more details.

23. *C. labiata* is mentioned again in chapter 14 of Gardner's *Travels:* "On the stems of the trees in the forest, I met with many fine orchideous plants, one of the most abundant, and certainly the most beautiful, being the *Cattleya labiata*" (538).

24. *Advice to a Young Gentleman on Entering Society* (Philadelphia: Lea & Blanchard, 1839), 291. The UK edition was published by A. H. Baily.

25. *Etiquette for Gentlemen: With Hints of the Art of Conversation* (London: Charles Tilt, 1856), 22; *Advice*, 287.

26. Letters of "compliment and friendship" were especially hard to construct—"the severest trial of taste and delicacy, and almost the only test of inferiority of good-breeding." Surely striking fear into the heart of any anxious reader seeking instruction, the author explained that the only way to pass the test, to communicate elevated status, was by thoroughly reading "the best models in the epistolary manner" from the era of Charles II onward and thereby learning to internalize their methods (*Advice*, 291).

27. Endersby, *Imperial Nature*, 29.

28. Quoted in Endersby, *Imperial Nature*, 90. Emphasis in original.

29. Gardner, *Travels*, 561.

30. Gardner, *Travels*, 28–29, 33–34.

31. Swainson and Gardner were in touch in the 1840s; in a letter from Gardner to Hooker, held at Kew, Gardner mentions receiving a letter from Swainson in New Zealand.

32. Gardner, *Travels*, 561.

33. John G. Reddie to William Hooker, March 31, 1849, RBGK, DC 54 / 387; "George Gardner: Obituary," *Ceylon Times*, 1849, press clipping, RBGK, DC / 54 / 330a.

9. ORCHID VARIETIES

1. Plants varied considerably in how many "spikes" of flowers they shot up, for instance. Lindley explained: "There is ... a great difference in the degree of success with which these plants are managed even by excellent cultivators; for if we see C. labiata and crispa with two or three flowers in a cluster, so also do we see them with a larger number." One specimen Lindley saw on display in the Horticultural Society possessed "four flowers in a cluster, and I have seen it with six." Lindley, "Note upon Cattleya Guttata," *Transactions of the Horticultural Society of London*, vol. 2 (London: W. Nicol, 1842), 177–179 (his talk, delivered April 18, 1837, was published as an article in 1842); John Lindley, "Cattleya Labiata," *Edwards's Botanical Register*, n.s. vol. 9 [o.s. vol. 22], 1836, Tab. 1859.

2. On disagreements about the principles of the Natural System, and the ongoing pull of the Artificial (in spite of Lindley's claims to the contrary), see Jim Endersby, "Classifying Sciences: Systematics and Status in Mid-Victorian Natural History," *The Organization of Knowledge in Victorian Britain*, ed. Martin Daunton (Oxford: British Academy, 2005), 61–84.

3. "[T]he present species may be known, especially from C. labiata, its nearest affinity, by its elongated, branching stem, bearing many deeply sulcated pseudo-bulbs, by the much broader sepals and petals, which latter are unguiculated at the base, and by the colour and markings and size of the lamina of the labellum." [W. Hooker], "Cattleya Mossiae: Mrs. Moss's Superb Cattleya," *Curtis's Botanical Magazine*, 2nd ser., vol. 12 [vol. 65 of entire work] (1839): Tab 3669.

Hooker described multiple plants in *Curtis's Botanical Magazine*, including orchids, that were collected by Gardner in 1838–1839 in the Organ Mountains, though none of them is a pink-colored cattleya.

4. W. Hooker, "Cattleya Mossiae," Tab 3669. The woman after whom *C. mossiae* is named was the grandmother of Sir J. Edwardes Moss, then still living. Frederick Boyle, *The Woodlands Orchids Described and Illustrated, with Stories of Orchid-Collecting* (London: Macmillan, 1901), 32. Hannah's husband, John, was a banker, railway pioneer, and enslaver operating in British Guiana and St. Vincent.

5. W. Hooker, "Cattleya Labiata. Crimson-Lipped Cattleya," *Curtis's Botanical Magazine*, 2nd ser., vol. 16 [vol. 69 of entire work] (1843): Tab 3998.

6. See "Cattleya Intermedia; *var.* Angustifolia," *Curtis's Botanical Magazine*, 2nd ser., vol. 13 [vol. 66 of entire work] (1840): Tab 3711. The Intermedia, or "Middle-Size-Flowered Cattleya," is described in *Curtis's Botanical Magazine*, 2nd ser., vol. 2 [vol. 55 of entire work] (1828): Tab 2851.

7. Joseph Paxton, "Cattleya Labiata, *var.* Atropurpurea," *Paxton's Magazine of Botany* 7 (1841): 73–74.

8. Lindley, "Cattleya: C. *labiata*," *Edwards's Botanical Register* 30 (1844): 5: section 1.5. Lindley distinguishes the *Laelia* genus from *Cattleya* by the number of pollen masses—four in *Cattleya*, eight in *Laelia*.

9. "The Lobed Cattleya," *Gardeners' Chronicle*, n.v., 1848, 403.

10. Joseph Paxton and John Lindley, "Varieties of the Ruby-Lipped Cattleya," *Paxton's Flower Garden*, vol. 1, 117–118 (London: Bradbury & Evans, 1853), 118. The white orchid was from the Duke of Northumberland's collection at Syon, the pinker bloom from the collection of a private enthusiast, J. J. Blandy of Reading.

11. Jim Endersby, *Imperial Nature: Joseph Hooker and the Practices of Victorian Science* (Chicago: University of Chicago Press, 2008), 158. Endersby also notes that local botanists were more influenced by small differences in the plants they found and tried to establish new species, while metropolitan botanists with access to very large numbers of plants discounted small differences in favor of seeing large patterns.

12. Endersby, *Imperial Nature*, 150.

13. This orchid, with its eight rather than four pollinia, was not yet identified at the time of Gardner's travels, though it had been distinguished from *Cattleya* by the late century. The Brazilian *Laelia* has since then, on the basis of DNA evidence, been folded back into the genus *Cattleya*. See Cássio van den Berg, W. E. Higgins, et al., "A Phylogenetic Analysis of Laeliinae (Orchidaceae) Based on Sequence Data," *Lindleyana* 15 (2000): 96–114; "Nomenclatural Notes on Laeliinae—I," *Lindleyana* 15 (2000): 115–119; Cássio van den Berg and M. W. Chase, "A Chronological View of Laeliinae Taxonomical History," *Orchid Digest* 68 (2004): 226–254; Cássio van den Berg, "New Combinations in the Genus *Cattleya* Lindl. (Orchidaceae)," *Neodiversity* 3 (2008): 3–12.

14. L. F. [Louis Forget], "Erreurs géographiques concernant les orchidées," *Le jardin* 1 (1897): 246–248; Robert Allen Rolfe, "Cattleya Labiata," *Orchid Review* 15 (November 1907): 335–336.

15. But is the purple orchid—var. *atropurpurea*—one of the "attractive candidates" unable to compete with the queen for notice, or is it a variety of the queen herself, and so "unapproachable" by the rest? It is left unclear whether the species and all its (many) varieties take the crown, or whether there is one variety that stands above the rest. See Paxton, "Cattleya Labiata, *var*. Atropurpurea," 73–74.

10. "ORCHIDS FOR THE MILLION"

1. For more on Bateman and the remarkable garden he constructed at Biddulph Grange in Staffordshire—the ways in which it staged different geo-

graphical locations (China, Egypt, Italy, the Himalayas) through built structures and plantings—see Peter Hayden, *Biddulph Grange: A Victorian Garden Rediscovered* (London: George Philip / National Trust, 1989).

2. James Bateman, *The Orchidaceae of Mexico & Guatemala* (London: Ridgway, 1843), 1–6.

3. These magazines included *The Botanical Register* (started 1815, succeeded by *Edwards's Botanical Register*), *The Botanical Cabinet* (1817–1833, published by nursery Loddiges & Sons), *Curtis's Botanical Magazine* (edited by William Hooker from 1826), and the wonderfully subtitled *Companion to the Botanical Magazine; Being a Journal Containing Such Interesting Botanical Information As Does Not Come Within the Prescribed Limits of the Magazine* (1835–1837); *The Gardener's Magazine and Register of Rural & Domestic Improvement* (1826–1843, edited by Bayswater resident John Claudius Loudon); *The Gardeners' Magazine of Botany, Horticulture, Floriculture And Natural Science* (1850–1851), succeeded by *Garden Companion and Florists' Guide*; *The Floricultural Cabinet and Florist's Magazine* (1833–1859), later *The Gardener's Weekly Magazine*, later *The Gardener's Magazine* (ed. Shirley Hibberd, from 1861); *Paxton's Magazine of Botany* (1834–1849, edited by Joseph Paxton); *The Gardener's Gazette* (1837–1847, edited by Loudon from 1840); the *Magazine of Natural History and Journal of Zoology, Botany, Mineralogy, Geology, and Meteorology* (edited by Loudon, 1828–1840), succeeded by *Magazine of Natural History; The Magazine of Zoology and Botany* (1837–8); the *Annals of Natural History* (1838–1841), merging with *Magazine of Natural History* and others to become *Annals and Magazine of Natural History*, succeeded by the *Journal of Natural History; The Gardeners' Chronicle and Agricultural Gazette* (1841, founded by Paxton, Lindley, and others—from 1874 *The Gardeners' Chronicle*); *The Florist* (1848, later *The Florist and Garden Miscellany*); *The Floral World and Garden Guide* (1858–1880, edited from 1866 by Shirley Hibberd); *The Gardener: A Magazine of Horticulture and Floriculture* (1867–1882), and *The Garden: An Illustrated Weekly Journal of Gardening in All Its Branches* (1871–1827). I owe many of these dates and details to Anne Wilkinson, "The Preternatural Gardener: The Life of James Shirley Hibberd (1825–90)," *Garden History* 26, no. 2 (1998): 153–175; 161–162.

For a list of worldwide publications, see *Gardeners' Chronicle* 9 (January 24, 1891): 121. For discussion on agricultural and horticultural periodicals, see Nicholas Goddard, "The Development and Influence of Agricultural Periodicals and Newspapers, 1780–1880," *Agricultural History Review* 30, no. 2 (1983): 116–131.

4. John Charles Lyons's *Remarks on the Management of Orchidaceous Plants* (1843) is one of, if not *the* first work to try to make orchids accessible to a wide audience, though it is not clear how many people actually read it; see

Brent Elliott, "The Royal Horticultural Society and Its Orchids: A Social History," *Occasional Papers from the RHS Lindley Library*, vol. 2, 3–53 (London: Lindley Library, 2010), 11. (Lindley certainly knew it—it is collected in his papers at the RHS.) The "Orchids for the Million" articles likely made a larger impact because of the magazine's greater reach; by 1851, the *Gardeners' Chronicle* weekly circulation was c. 6,500 (Goddard, "Development and Influence," 122).

5. "Dodman" [Charles Bellenden Ker], "Orchids for the Million: I," *Gardeners' Chronicle* 10 (May 18, 1850): 308–309, 308. £2.10s corresponds to roughly £337 / $430 today.

6. Ward published a letter describing his discovery in the *Companion to the Botanical Magazine* (1836) and followed it up with *On the Growth of Plants in Closely Glazed Cases* (1842). Hooker wrote to approve the invention in a letter Ward published in his second edition: "Wardian cases . . . have been the means, in the last fifteen years, of introducing more new and valuable plants to our gardens than were imported during the preceding century; and in the character of 'Domestic Green-houses,' if I may so speak; *i.e.* as a means of cultivating plants in our parlours, our halls, and our drawing-rooms, that have constituted a new era in horticulture." Ward, *Companion to the Botanical Magazine*, 2nd ed. (London: John Van Voorst, 1852), 131–132. For more on the Wardian case, see Christopher Thacker, *The History of Gardens* (Berkeley: University of California Press, 1979), 236–238; and Endersby, *Imperial Nature*, 60–63.

7. Edward Sprague Rand, *Orchids: A Description of the Species and Varieties Grow at Glen Ridge, Near Boston* (1876; repr. Boston: Houghton Mifflin, 1888), 110.

8. "Dodman" [Charles Bellenden Ker], "Orchids for the Million," *Gardeners' Chronicle* 9 (April 14, 1849): 229. This was an initial piece on the subject; a full series, starting with Part I, began a year later. Ker described himself as a "humble beginner," though his father was a botanist and he himself an associate of Cattley's ("Dodman" [Ker], "Orchids: I," 308).

9. "Dodman" [Ker], "Orchids: I," 308.

10. Benjamin Williams, *The Orchid-Grower's Manual*, 2nd ed. (London: Chapman & Hall, 1862), vii–viii; *The Orchid-Grower's Manual*, 6th ed. (London: Victoria & Paradise, 1885), xi.

11. The *Orchid-Grower's Manual* was reprinted in seven separate editions between 1852 and 1894: 1st edition, 1852; 2nd edition, 1862; 3rd edition, 1868; 4th edition, 1871; 5th edition, 1877; 6th edition, 1885; 7th edition, 1894 (posthumous).

12. I explore Loudon's work on the suburban garden in more detail in *The Promise of the Suburbs: A Victorian History in Literature and Culture* (New Haven: Yale University Press, 2019), 20–36.

13. Ker made the same point in "Orchids for the Million": "If the house is built low, so much the better, the grower will have a greater command of heat and moisture. I think most of the houses in which the large collections are grown are too large," "Dodman" [Ker], "Orchids: I," 308.

14. Benjamin Williams, *The Orchid-Grower's Manual*, 4th ed. (London: Victoria & Paradise, 1871), x; Advertisement, "Exhibition of Orchids. B.S. Williams," *Gardeners' Chronicle* 17 (June 17, 1882), 794.

15. John Claudius Loudon, *The Suburban Gardener, and Villa Companion* (London: Longman, 1838), 35.

16. For more discussion on salaries, see Gregory Anderson's classic *Victorian Clerks* (Manchester: Manchester University Press, 1976), 20–27. Anderson reports that a bank manager could hope to command, on first appointment, £150; a clerk might start on as little as £60 (23).

17. Endersby's chapter on collection is superbly detailed on tools and practices; see *Imperial Nature*, 54–83.

18. Michael W. Hancock, "Boffin's Books and Darwin's Finches: Victorian Cultures of Collecting" (PhD diss., University of Kansas, 1995), 7–9; https://digitalcommons.imsa.edu/eng_pr/5/.

19. A classic work on collecting is Peter Conrad, *The Victorian Treasure-House* (London: Collins 1973). See too Endersby, *Imperial Nature*, 54.

20. Quoted by Hancock, *Boffin's Books*, 2; I am indebted to Hancock's fascinating study of Victorian collecting, esp. 1–16.

21. Tim Cresswell argues that mobility fundamentally shapes "what it is to be modern" in *On the Move: Mobility in the Modern Western World* (New York: Routledge, 2006), 48.

22. By the end of the century, this aspiration seemed less and less plausible. As collection culture became more popular, and reached deeper into the middle and working classes, it became also widely disparaged, the collector dismissed as a self-involved narcissist "who disregarded the alterity of objects and transformed artifacts into markers of taste rather than historical relics." See Kristin Mary Mahoney, "Nationalism, Cosmopolitanism, and the Politics of Collecting in *The Connoisseur: An Illustrated Magazine for Collectors, 1901–1914*," *Victorian Periodicals Review* 45, no. 2 (2012): 175–199; 175.

23. As David Mackay puts it, the "first task of the Banksian collection" was "to transcend the chaos of the lands they visited and reduce the natural world of empire to order." Mackay, "Agents of Empire: The Banksian Collectors and Evaluation of New Lands," *Visions of Empire: Voyages, Botany, and Representations of Nature*, ed. David Philip Miller and Peter Hannas Reill, 38–57 (Cambridge: Cambridge University Press, 2011), 54.

24. On the ways in which classification and collection practices in London's herbaria energized yet more resource extraction, see Endersby, *Imperial Nature*, 138. On eighteenth-century collection practices, see Barbara M. Benedict, *Curiosity: A Cultural History of Early Modern Enquiry* (Chicago: Chicago University Press, 2001), esp. ch. 4.

25. Harold Perkin, *The Rise of Professional Society: England Since 1880* (London: Routledge, 1989), 9.

26. Martin Wiener, *English Culture and the Decline of the Industrial Spirit, 1850–1980* (1981; repr. Cambridge: Cambridge University Press, 2004), 12.

27. On today's struggles, see Daniel Markovits, *The Meritocracy Trap: How America's Foundational Myth Feeds Inequality, Dismantles the Middle Class, and Devours the Elite* (New York: Penguin, 2019).

28. Rev. Abner W. Brown, *Village Horticultural Societies: The History, Rules, and Details of One Established in 1837, at Pytchley, Northamptonshire* (London: Wertheim & Macintosh, 1849), 16–19.

11. BATTLES AT THE GATE

1. John Ruskin, "Of Queens' Gardens," *Sesame and Lilies: Two Lectures Delivered at Manchester in 1864* (London: Smith, Elder, 1865), 142.

On women and gardening, see Ann Shtier, *Cultivating Women, Cultivating Science: Flora's Daughters and Botany in England, 1760–1860* (Baltimore: Johns Hopkins University Press, 1996); Catherine Horwood, *Gardening Women: Their Stories from 1600 to the Present* (London: Virago, 2010); and Fiona Davison, *An Almost Impossible Thing: The Radical Lives of Britain's Pioneering Women Gardeners* (Beaminster: Little Toller Books, 2023). For more on women's relationships to the natural world more broadly, see Barbara T. Gates, *Kindred Nature: Victorian and Edwardian Women Embrace the Living World* (Chicago: University of Chicago Press, 1998). For a selection of scholarship on women's literature of the garden, see Judith W. Page and Elise L. Smith, *Women, Literature and the Domesticated Landscape: England's Disciples of Flora, 1780–1870* (Cambridge: Cambridge University Press, 2011); Rachel O'Connell, "Love Scenes and Garden Plots: Form and Femininity in Elizabeth von Arnim's *Elizabeth and Her German Garden*," *Women: A Cultural Review* 28, no. 1–2 (2017): 22–39; and Sarah Bilston, "'Queens of the Garden': Victorian Women Gardeners and the Rise of the Gardening Advice Text," *Victorian Literature and Culture* 36, no. 1 (2008): 1–19.

2. "Horticultural Society," *Athenaeum*, no. 686, November 19, 1840, 1013. Lawrence's success was also written up in "Horticultural Society, Nov. 3," *Polytechnic Journal* (July–Dec 1840): 424. The first time Lawrence showed *Cattleya labiata*, as far as I can tell, was November 17, 1838, at the Horticultural Society, as reported in the *Gardener's Magazine*.

3. John Claudius Loudon, *The Suburban Gardener, and Villa Companion* (London: Longman, 1838), 576, 577. As I've discussed elsewhere, the text of this book is significantly changed in the revised, updated, and posthumously published edition, J. C. Loudon, *The Villa Gardener: Comprising a Choice of a Suburban Villa Residence,* 2nd ed., ed. Mrs. Loudon (London: W. S. Orr, 1850). Among other shifts, the earlier text's support for Mrs. Lawrence and her aesthetic is significantly reduced. The later version is edited by Loudon's wife, Jane Loudon, who perhaps admired the beautiful Mrs. Lawrence a little less than her husband. While the 1838 text asserts that Louisa's home is a model in taste (587), the 1850 text accuses it of a "profusion of ornament" and claims "its principal fault" is "its total want of repose" (278). For more on the relationship between the 1838 and 1850 texts, see Sarah Bilston, *The Promise of the Suburbs: A Victorian History in Literature and Culture* (New Haven: Yale University Press, 2019), 20–36; and Bilston, "'They Congregate . . . in Towns and Suburbs': the Shape of Middle-Class Life in John Claudius Loudon's *The Suburban Gardener,*" *Victorian Review* 37, no. 1 (2011): 144–159.

The Horticultural Society of London, founded in 1804, powered interest in orchids from the 1830s, and its garden at Chiswick held an important, world-famous collection. For more, see Brent Elliott, "The Royal Horticultural Society and Its Orchids: A Social History," *Occasional Papers from the RHS Lindley Library,* vol. 2, 3–53 (London: Lindley Library, 2010), esp. 7.

4. Loudon, *Suburban Gardener,* 589.

5. W. Hooker, "To Mrs. Lawrence, of Ealing Park," *Curtis's Botanical Magazine,* 2nd ser., vol. 15 [vol. 68 of entire work] (1842): dedication.

6. Edith Jackson, *Annals of Ealing: From The Twelfth Century to the Present Time* (London: Phillimore, 1898), 222. Jackson gives a detailed description of the house and its rooms in Mrs. Lawrence's time.

7. Kathleen Davidson and Molly Duggins, "Commodifying the Ocean World in the Long Nineteenth Century," *Sea Currents in Nineteenth-Century Art, Science and Culture: Commodifying the Ocean World,* ed. Kathleen Davidson and Molly Duggins, 1–20 (London: Bloomsbury, 2023), 8–9; Talia Schaffer, *Novel Craft: Victorian Domestic Handicraft & Nineteenth-Century Fiction* (Oxford: Oxford University Press, 2011), 33, 10.

8. Five women were elected to fellowship of the London Horticultural Society in 1830; Lawrence began exhibiting in 1833.

Women over thirty gained the vote in 1921; women were finally given suffrage on the same terms as men—that is, at the age of twenty-one—in 1928.

9. The awards for May 13, 1837, are detailed in "The London Horticultural Society and Garden," *Gardener's Magazine* 3 (1837): 379–383; 379. The value of the awards is detailed in "The Encouragement of Horticulture Generally By the

Award of Medals and Prizes for Absolute or Comparative Horticultural Merit: Report on the Progress of the Horticultural Society," *Transactions of the Horticultural Society of London*, vol. 3, 443–453 (London: W. Nicol, 1842), 446.

10. "Signed Minutes of Meetings of the Horticultural Society Council," RHS Lindley Library, London, RHS/Gov/3/1/1/10–12 (1830–37, 1838–47, 1847–55). Louisa's son, Sir Trevor Lawrence, was himself a noted orchid collector. He became president of the Royal Horticultural Society in 1885. *Cattleya lawrencena*, imported by Frederick Sander in 1884–1885, was named after him.

11. Coventry Patmore, "The Angel in the House," *The Angel in the House*, 2nd ed. (London: John Parker, 1858), 105; Sarah Stickney Ellis, *The Women of England: Their Social Duties and Domestic Habits* (London: Fisher, 1839), 53.

12. Ruskin, "Of Queens' Gardens," 147.

13. Ruskin, "Of Queens' Gardens," 149–150. On the importing of the *Amherstia* see Kate Colquhoun, *'The Busiest Man in England': A Life of Joseph Paxton, Gardener, Architect, & Victorian Visionary* (Boston: David Godine, 2005), 63–73; on the Paxton / Lawrence relationship, see 103; on the flowering of the *Amherstia*, 152–153.

14. Horwood, *Gardening Women*, 46.

15. D. Beaton, "Seasonable Scraps," *Cottage Gardener* 8 (1854): 367–368; 368.

16. Geoffrey Taylor, "Louisa Lawrence and Her Garden," *Journal of the Royal Horticultural Society* 80 (1955): 423–429, 423.

17. "The London Horticultural Society and Garden," *Gardener's Magazine* 2 (1836): 379–381; 380. On the following page, the secretary of the Horticultural Society responded to reports "maliciously" circulating that judges knew the names of the entrants and that prizes were not issued fairly. This was energetically refuted, but clearly there was at least *some* suspicion of favoritism on the part of the judges. George Bentham, Secretary, Letter, *Gardener's Magazine* 2 (1836): 381.

18. On the rise of the professional woman gardener, see Bilston, "Queens of the Garden," 1–19.

19. The magazine explained: "Suspended baskets must not entirely be left to depend on syringing and the necessary moist atmosphere, but should be taken down at least once a week, and dipped in water of the same temperature as the house." "The Garden—June," *Ladies' Companion and Monthly Magazine* 1 (1852): 335–336; 335.

20. The catalogs are held today at the RHS Lindley Library in London.

21. A letter in *The Times*, however, from the carrier suggests that the box was twelve feet and the palm was thirty feet tall. Quoted in Peter Hayden, *Biddulph Grange: A Victorian Garden Revisited* (London: George Philip / National Trust, 1989), 33–34.

22. "Removal of a Gigantic Palm-Tree," *Illustrated London News* 25 (August 5, 1854): 113.

12. A NEW DYNASTY

1. Thomas Veitch, quoted in S. Heriz-Smith, "The Veitch Nurseries of Killerton and Exeter, c. 1780 to 1863," *Garden History* 16, no. 1 (1988): 41–57; 43. I'm indebted to Heriz-Smith for her immensely careful accounts of the nursery's history.

2. I explore the possibilities of suburban modernity in more detail in Sarah Bilston, *The Promise of the Suburbs: A Victorian History in Literature and Culture* (New Haven: Yale University Press, 2019), esp. 3–7. "London grew by sucking in provincial migrants because jobs were either better paid there or thought to be so. . . . [London] offered a more persuasive legend of opportunity than could be found anywhere in the country," argue Harold J. Dyos and David A. Reeder in the classic "Slums and Suburbs," in *The Victorian City: Images and Realities*, ed. H. J. Dyos and Michael Wolff, vol. 1, 359–386 (London: Routledge, 1973), 362.

3. Quoted in S. Heriz-Smith, "Veitch Nurseries," 46.

4. John Britton, *The Original Picture of London Enlarged and Improved: Being a Correct Guide to the Stranger, as Well as For the Inhabitant, to the Metropolis of the British Empire* (London: Longman, 1826), 429.

5. "Messrs. Rollisson's Nursery, Tooting," *Journal of Horticulture and Cottage Gardener* 32 (March 15, 1877): 190–191; 190. The nursery is also often spelled "Rollison."

6. "Horticultural Society of London," *Floricultural Cabinet and Florist's Magazine* 3 (1835): 187–188; 187.

7. James H. Veitch, *Hortus Veitchii: A History of the Rise and Progress of the Nurseries of the Messrs. James Veitch and Sons* (London: James Veitch, 1906), 13. See Sue Shephard, *Seeds of Fortune: A Gardening Dynasty* (London: Bloomsbury, 2003), 65, on James Junior's London experiences and character and the conflict between John and his son and grandson.

8. "Melancholy Loss of the Collectors Messrs. Wallace and Banks," *Paxton's Magazine of Botany* 6 (1839): 135–137.

9. On the trip and the duke's reaction, see Shephard, *Seeds*, 74; Veitch quoted in Shephard, *Seeds*, 74.

10. The government attempted to abolish the Royal Botanic Gardens after a period of decline following Banks's death; Lindley and Paxton were both involved in a desperate (and ultimately successful) attempt to save it.

11. See Shephard, *Seeds*, 74–75; Veitch quoted in Shephard, *Seeds*, 78.

12. Veitch, *Hortus Veitchii*, 37. Letters between Veitch and Hooker suggest Lobb left in early 1841, though *Hortus Veitchii* and the company's *A Manual of Orchidaceous Plants* indicate Lobb was already examining the Organ Mountains in 1840.

13. James Veitch to William Hooker, March 31, 1841, RBGK, DC/16/357.

14. James Veitch & Sons, *A Manual of Orchidaceous Plants* (London: James Veitch, 1887–1894), vol. 1, "Cattleya and Laelia," 15.

The ten parts of the *Manual* were supposed to be bound into two volumes, according to "Directions to the Binder," but the parts were to be arranged neither sequentially nor in the order in which they were issued. Unsurprisingly, different parts received different treatment by different binders, and this makes referencing the text difficult. It is also hard to know exactly when each part was issued, though I've found discussion of a note in Veitch's "Cattleya" section in Edward Sprague Rand, "Nomenclature of Orchid Varieties," *American Garden* 9 (August 1888): 297–298.

As we shall see, by the time the *Manual* was complete, many believed the plant had in fact been discovered; the bound version, dated 1894, makes no mention of it.

15. "London Horticultural Society's Meeting, November," "Miscellany of Notes and Correspondence," *Floricultural Cabinet and Florists' Magazine* 14 (1846): 310–311; 310.

16. Shephard, *Seeds*, 63.

17. Mary Shelley, *Frankenstein: Or, the Modern Prometheus* (1818; repr. London: Routledge, 1888), 64.

18. Shelley, *Frankenstein*, 65.

13. HATCHING A PLAN

1. See Joseph Arditti, "An History of Orchid Hybridization, Seed Germination and Tissue Culture," *Botanical Journal of the Linnean Society* 89, no. 4 (1984): 359–381; 367.

2. Quoted in Alison Rix, "William Herbert (1778–1847) Scientist and Polymath, and His Contributions to *Curtis's Botanical Magazine*," *Curtis's Botanical Magazine* 31, no. 3 (2014): 280–298; 282.

3. Quoted in Rix, "William Herbert," 287; quoted in Tim Wing Yam et al., "History—Seeds: Orchid Seeds and Their Gemination: An Historical Account," *Orchid Biology: Reviews and Perspectives, VIII*, ed. Tiuu Kull and Joseph Arditti, 387–489 (New York: Springer Science, 2002), 425; Darwin quoted in Rix, "William Herbert," 280.

4. Scripture needs to be understood as text that explains "the immense operations of ages before the creation of man by expressions conformable to our

petty space of life," Herbert explained: "We must learn to understand the true force of the words of Scripture, and not derogate from the greatness of God by reducing it to the compass of our narrow conceptions." William Herbert, "On Hybridization Amongst Vegetables" (London: William Clowes, 1846), part 1, 1 (RHS / LL Botanical Tracts 58:12), 3.

5. Herbert, "On Hybridization," part 2, 103–104.

6. Veitch, *Manual*, 141.

7. Quoted in Tim Wing Yam and Joseph Arditti, "History of Orchid Propagation: A Mirror of the History of Biotechnology," *Plant Biotechnology Reports* 3 (2009): 1–56; 3–4.

8. Yam and Arditti, "History," 3–4.

9. Leanne Melbourne, "Edmond Albius: The Boy Who Revolutionized the Vanilla Industry," Linnean Society of London, blogs and essays, October 16, 2019, https://www.linnean.org/news/2019/10/16/edmond-albius; see also Arditti, "An History of Orchid Hybridization," 361. I am indebted to Arditti's painstaking research in my account of the early history of hybridization.

10. See Yam and Arditti, "History," 4.

11. D. Moore, "On Growing Orchids from Seeds," *Gardeners' Chronicle*, n.v., September 1, 1849, 549.

12. Herbert, "On Hybridization," part 2, 104.

13. C. Darwin to J. D. Hooker, June 2, 1847, Darwin Correspondence Project, University of Cambridge (hereafter DCP), LETT-1094, https://www.darwinproject.ac.uk/letter/DCP-LETT-1094.xml. On Darwin's first reactions to Herbert, see Darwin to Herbert, c. April 1, 1839, DCP-LETT-502, https://www.darwinproject.ac.uk/letter/DCP-LETT-502.xml.

14. D. Moore, "On Growing," 549.

15. J. Cole, "Orchids from Seed," *Gardeners' Chronicle*, n.v., September 15, 1849, 581–582.

16. Shephard, *Seeds*, 123.

14. DOCTORING ORCHIDS

1. Quoted in S. Heriz-Smith, "The Veitch Nurseries of Killerton and Exeter, c. 1780–1863: Pt. II," *Garden History* 16, no. 12 (1988): 174–188; 178.

2. "It Will Have Been Seen," *Gardeners' Chronicle*, n.v., April 23, 1853, 260.

3. Robert Allen Rolfe and Charles Chamberlain Hurst, *The Orchid Stud-Book: An Enumeration of Hybrid Orchids of Artificial Origin* (Kew: Frank Leslie, 1909), vii.

4. "Current Topics: Orchid Hybridisation," *British Gardening* 9, January 19, 1893, 29–30; 29.

5. See David Gorrie, "Remarks on the Hereditary Properties of Cultivated Plants," *Journal of Agriculture*, n.v. (1855–1857): 314–319; 315.

6. They are listed in the directory as "Harris John and Son (Jno. Wm.), 47 Southernay Place." William White, *History, Gazetteer, and Directory of Devonshire, and the City and County of the City of Exeter* (Printed on Subscription, 1850), 174.

7. See Joseph Arditti, "John Harris—Surgeon and Orchidologist," *Orchid Review* 87 (1980): 200–204.

8. John William had a son of his own in 1850, the Census reveals, called John Delpratt Harris, who also became a surgeon and lived on Southernhay in 1911. "Delpratt" was his mother's maiden name; a fifteen-year-old William Delpratt was living with the family in 1841.

9. James H. Veitch, *Hortus Veitchii: A History of the Rise and Progress of the Nurseries of the Messrs. James Veitch and Sons* (London: James Veitch, 1906), 99.

10. Mary Shelley, *Frankenstein: Or, the Modern Prometheus* (1818; repr. London: Routledge, 1888), 71.

11. James Veitch & Sons, *A Manual of Orchidaceous Plants* (London: James Veitch, 1887–1894), 142.

12. [John Lindley,] "On the 28th October, 1856," *Gardeners' Chronicle*, no. 1, January 2, 1858, 4.

13. Veitch, *Manual*, 142.

14. [Lindley,] "On the 28th October," 4.

15. See, for instance, R. A. Rolfe, *Journal of the Linnean Society* 24 (1888), 166; "Dies Orchidianae," *Orchid Review* 9 (1901), 65–66. The line appears too in Sue Shephard, *Seeds of Fortune: A Gardening Dynasty* (London: Bloomsbury, 2003), 127; and Susan Orlean, *The Orchid Thief: A True Story of Beauty and Obsession* (1998; repr. New York: Ballantine, 2014), 75.

16. C. W. C, "Home Correspondence: Hybrid Orchids," *Gardeners' Chronicle*, n.v., January 16, 1858, 38.

17. "Societies: Horticultural: *Nov. 10*," *Gardeners' Chronicle*, n.v., November 19, 1859, 929; "Cattleya Dominiana," *Gardeners' Chronicle*, n.v., November 26, 1859, 948.

18. [H. G. Reichenbach], "Garden Memoranda: The Royal Exotic Nursery, King's Road Chelsea," *Gardeners' Chronicle*, n.v., October 15, 1859, 831. Shephard, *Seeds*, 128, identifies this passage as by Reichenbach; it was published, like most of the magazine's columns, anonymously.

19. Shephard, *Seeds*, 129.

20. Gregory Anderson, *Victorian Clerks* (Manchester: Manchester University Press, 1976), 20–27. £500 corresponds to £59,300 / $75,800 today, £2,000 to £237,000 / $303,000.

21. "Serapius," "Dies Orchidianae—No. IX," *Gardeners' Chronicle*, n.v., December 19, 1863, 1204.

22. "Imported Orchids," *Gardeners' Chronicle*, n.v., September 27, 1862, 923; "Established Orchids," *Gardeners' Chronicle*, n.v., September 27, 1862, 923. "Serapius," "Dies Orchidianae," describes the scene in one collection: "Next [meaning in the hierarchy] to this Vanda, as regards striking effect, came the old Cattleya labiata" (1204).

23. Wilkie Collins, *Basil* (1852; repr. Leipzig: Tauchnitz, 1862), 62.

24. George Augustus Sala, *Gaslight and Daylight: With Some London Scenes They Shine Upon* (London: Chapman & Hall, 1859), 218.

25. "Serapius," "Dies Orchidianae," 1204.

15. DARWIN'S CRISIS

1. "Harvest Prospects," *Liverpool Mercury*, July 24, 1858, 5.

2. The House of Lords met to discuss "The State of the Thames" on Friday, June 25, at 5 P.M. See "Parliamentary Intelligence," *Times*, June 26, 1858, 6.

3. The event was reported in "State and Church," *Examiner*, July 3, 1858, 423–424.

4. Henrietta Litchfield, ed., *Emma Darwin, A Century of Family Letters, 1792–1896*, vol. 2 (London: Murray, 1915), 164. Many scholars have pointed out how much the Darwin family read and how literary discourse impacted the naturalist. Gillian Beer explores at length how Darwin drew on and was shaped by literary conceits, tropes, and narrative patterns in Beer, *Darwin's Plots: Evolutionary Narrative in Darwin, George Eliot and Nineteenth-Century Fiction* (1983; repr. Cambridge: Cambridge University Press, 2009). See also Devin Griffiths's examination of Darwin's reading of historical fiction in Griffiths, *The Age of Analogy: Science and Literature Between the Darwins* (Baltimore: Johns Hopkins University Press, 2016), esp. 230–237. Janet Browne's two-part biography of Darwin also situates him in relation to the world of fiction—of his life in 1857–1858, Browne notes: "If Charles Darwin had spent the first half of his life in the world of Jane Austen, he now stepped forward into the pages of Anthony Trollope." E. J. Browne, *Charles Darwin*, vol. 2, *The Power of Place* (Princeton: Princeton University Press, 2002), 3.

5. Darwin spent several decades developing the theory of natural selection, penning a sketch of the theory in 1842 that he expanded into a 189-page essay in 1844 (which, after writing, he seems somewhat anxiously to have hidden in a cupboard). For more on these early works, and on Darwin's communications with J. D. Hooker on transmutation before *On the Origin of Species*, see Duncan Porter and Peter Graham, *Darwin's Sciences* (Oxford: Wiley, 2015), 68–69; for a

book-length exploration of the relationships between Darwin and Wallace, see James T. Costa, *Wallace, Darwin, and the Origin of Species* (Cambridge, MA: Harvard University Press, 2014). More useful details, photographs, and links to letters may be found at "Darwin in Letters, 1858–1859: Origin," Darwin Correspondence Project, University of Cambridge (hereafter DCP), https://www.darwinproject.ac.uk/letters/darwins-life-letters/darwin-letters-1858-1859-origin.

6. "'Through Nature to Nature's God' was not only a common epigraph emblazoned on title pages, it was a deep, shared assumption of most gentleman naturalists," notes Richard England in "Natural Selection Before the *Origin*: Public Reactions of Some Naturalists to the Darwin-Wallace Papers," *Journal of the History of Biology* 30, no. 2 (1997): 267–290; 269.

7. John S. Wilkins, *Species: A History of the Idea* (Berkeley: University of California Press, 2011), 73.

8. C. Darwin to J. D. Hooker, June [8], 1858, DCP-LETT-2282, https://www.darwinproject.ac.uk/letter/DCP-LETT-2282.xml.

9. C. Darwin to Charles Lyell, [June] 18, 1858, DCP-LETT-2285, https://www.darwinproject.ac.uk/letter/DCP-LETT-2285.xml.

10. The precise nature of the relationship between the drought and the epidemic remains a matter of debate; see Susan Scott Susan and Christopher Duncan, *Human Demography and Disease* (1998; repr.: Cambridge: Cambridge University Press, 2005), 318.

11. C. Darwin to J. D. Hooker, [June 23, 1858], DCP-LETT-2290, https://www.darwinproject.ac.uk/letter/DCP-LETT-2290.xml.

12. Scholars have speculated that he may have been born with Down's Syndrome, given the relatively advanced age of his parents—his mother Emma was forty-eight at the time of his birth.

13. Litchfield, *Emma Darwin*, 162; C. Darwin, "Reminiscence of Charles Waring Darwin" [1858], John van Wyhe, ed., *The Complete Work of Charles Darwin Online* (2002), http://darwin-online.org.uk/content/frameset?pageseq=2&itemID=CUL-DAR210.13.42&viewtype=text.

14. "Charles Darwin's Memorial of Anne Elizabeth Darwin," April 30, 1851, https://www.darwinproject.ac.uk/people/about-darwin/family-life/death-anne-elizabeth-darwin.

15. C. Darwin, "Reminiscence of Charles Waring Darwin."

16. C. Darwin to J. D. Hooker, [June 29, 1858], DCP-LETT-2297, https://www.darwinproject.ac.uk/letter/DCP-LETT-2297.xml.

17. C. Darwin to C. Lyell, [June 25, 1858], DCP-LETT-2294, https://www.darwinproject.ac.uk/letter/DCP-LETT-2294.xml.

18. C. Darwin to C. Lyell, [June 26, 1858], DCP-LETT-2295, https://www.darwinproject.ac.uk/letter/DCP-LETT-2295.xml.

19. C. Darwin to J. D. Hooker, [June 29, 1858], DCP-LETT-2298, https://www.darwinproject.ac.uk/letter/DCP-LETT-2298.xml.

20. As Costa explains, none of the material shared with the Linnean Society was actually "intended for publication as written"; Wallace, who was away, had simply sent his essay to Darwin to show to Lyell, while Darwin had nothing concise ready to go. Darwin's material (whose dates also had to be provable, to establish "his claim of priority") was cobbled together from chapter 2 of his 1844 essay with a draft of a summary prepared for Asa Gray in 1857. See Costa, *Wallace, Darwin*, 178nn3–4.

21. "The Thames and the Victoria Sewer," *Times*, June 30, 1858, 9.

22. J. D. Hooker and Charles Lyell to the Linnean Society, June 30, 1858, DCP-LETT-2299, https://www.darwinproject.ac.uk/letter/DCP-LETT-2299.xml.

23. *The Autobiography of Charles Darwin 1809–1882*, ed. Nora Barlow (New York: Norton, 1969), 122.

24. Quoted by Tim Burt and Des Thompson, *Curious About Nature: A Passion for Fieldwork* (Cambridge: Cambridge University Press, 2020), 367.

25. Etty later recalled, gratefully, that her parents allowed her to take her kitten with her, though it surely "must have added to the troubles of travelling with a sick child." Litchfield, *Emma Darwin*, 163. Darwin reported the toll in a letter of July 30.

26. Lyell and Hooker seem to have suggested the idea of the "Abstract" to Darwin; perhaps they hoped that Darwin would find it easier to write if freed from the weight of a larger tome. "I am most *heatily* obliged to you & Lyell for having set me on this," he wrote to Hooker from the Isle of Wight; "[i]t seems a queer plan to give an abstract of an unpublished work; nevertheless I repeat I am extremely glad I have begun in earnest on it." C. Darwin to J. D. Hooker, [July 30, 1858], DCP-LETT-2314, https://www.darwinproject.ac.uk/letter/DCP-LETT-2314.xml.

27. Charles Darwin, *On the Origin of Species by Means of Natural Selection, or, The Preservation of Favoured Races in the Struggle for Life* (London: John Murray, 1859), 2.

28. Griffiths cites a letter from Darwin to John Murray, in which the naturalist stressed the ways he'd abbreviated and simplified his ideas for the *Origin*: "It is the result of more than 20 years work; but as here given, is only a popular abstract of a larger work on the same subject, without references to authorities & without long catalogues of facts" (Griffiths, *Age of Analogy*, 216). Griffiths also presents Darwin's *Orchids* as a critical extension of the *Origin*, a work that

lays out the "solid empirical grounds" the *Origin* merely sketched; I'm indebted to Griffiths's careful research.

29. [Richard Owen], "Darwin on the Origin of Species," *Edinburgh Review* 111 (1860): 487–535, 516. Owen quotes at this point in his review not from the *Origin* but from the paper submitted to the Linnean Society in July 1858, which was published, the following month, as "On the Tendency of Species to Form Varieties: and on the Perpetuation of Varieties and Species by Natural Means of Selection. By Charles Darwin and Alfred Wallace." See *Journal of the Linnean Society* 3 (1858–1859): 45–62.

Adam Sedgwick was also incensed by Darwin's supposed deviation from the inductive method; see Sedgwick, "Objections to Mr. Darwin's Theory of The Origin of Species," *Spectator* 33, March 24, 1860, 285–286. For further discussion see David L. Hull, "Darwin's Science and Victorian Philosophy of Science," *Cambridge Companion to Darwin*, 2nd ed., ed. Jonathan Hodge and Gregory Radick (Cambridge: Cambridge University Press, 2009), 173–198.

30. C. Darwin to Asa Gray, November 29, [1857], DCP-LETT-2176, https://www.darwinproject.ac.uk/letter/DCP-LETT-2176.xml.

31. Darwin encouraged Hooker to contact the nurseries in July; see [July 6, 1861], DCP-LETT-3200, https://www.darwinproject.ac.uk/letter/DCP-LETT-3200.xml. Darwin wrote to ask for names of possible suppliers on June 19 that year, and Hooker annotated that letter with the words "Parker & Williams / Holloway." See C. Darwin to J. D. Hooker, DCP-LETT-3190, https://www.darwinproject.ac.uk/letter/DCP-LETT-3190.xml.

16. "I MUCH WANT A CATTLEYEA"

1. At that time the Darwin family was tucked into a Georgian row house in Torquay, on the south coast of Devon, in an effort to improve Etty's health.

2. Joseph Hooker was furious; he was in the area searching for the same thing at the same time. Indeed, Joseph was not much of a fan of the Veitch family, regarding them as rapacious: he would later write, in 1864, that theirs was "the only firm of nurserymen who we do not get on with." J. D. Hooker to C. Darwin, March 29, 1864, Darwin Correspondence Project, University of Cambridge (hereafter DCP), DCP-LETT-4439, https://www.darwinproject.ac.uk/letter/DCP-LETT-4439.xml. However, personal dislike did not prevent Joseph from recommending the company to Darwin as a useful resource.

3. I'm indebted for this summary to Sue Shephard, *Seeds of Fortune: A Gardening Dynasty* (London: Bloomsbury, 2003), 158; and to S. Heriz-Smith, "Veitch Nurseries of Killerton and Exeter, c. 1780–1863: Pt. II," *Garden History*

16, no. 12 (1988): 174–188, esp. 177–185. Shephard discusses the tension between J. Hooker and Veitch over the blue orchid (106).

Benson and Ellis do not have their own, full chapters in *Hortus Veitchii* but are referenced in the work as collectors for Veitch's, Kew, and other nurseries. James H. Veitch, *Hortus Veitchii: A History of the Rise and Progress of the Nurseries of the Messrs. James Veitch and Sons* (London: James Veitch, 1906).

4. "Notice," *Gardeners' Chronicle*, n.v., April 13, 1861, 334.

5. C. Darwin to J. Lindley, October 18, [1861], DCP-LETT-3289, https://www.darwinproject.ac.uk/letter/DCP-LETT-3289.xml.

6. C. Darwin to D. F. Nevill, November 27, [1861], DCP-LETT-3414F, https://www.darwinproject.ac.uk/letter/DCP-LETT-3414F.xml.

7. C. Darwin to J. D. Hooker [July 17, 1861], DCP-LETT-3210, https://www.darwinproject.ac.uk/letter/DCP-LETT-3210.xml. Darwin credits Parker in Darwin, *On the Various Contrivances by Which British and Foreign Orchids are Fertilised by Insects, and on the Good Effects of Intercrossing* (London: John Murray, 1862), for his "extremely valuable series of forms," though he does not indicate what these were (158n). For more on Darwin and the catasetum, see Jim Endersby, *Orchid: A Cultural History* (Chicago: University of Chicago Press, 2016), 94–95.

8. Emma Darwin to Henrietta Litchfield, June 1887, in Henrietta Litchfield, ed., *Emma Darwin, A Century of Family Letters, 1792–1896* (London: Murray, 1915), vol. 2, 279.

9. Charles Darwin, *On the Origin of Species by Means of Natural Selection, or, The Preservation of Favoured Races in the Struggle for Life* (London: John Murray, 1859), 62; Darwin, *Origin*, 3rd ed. quoted by Alison Rix, "William Herbert (1778–1847) Scientist and Polymath, and His Contributions to *Curtis's Botanical Magazine*," *Curtis's Botanical Magazine* 31, no. 3 (2014): 280–298; 280.

10. C. Darwin to J. D. Hooker, June 19, [1861], DCP-LETT-3190, https://www.darwinproject.ac.uk/letter/DCP-LETT-3190.xml.

11. C. Darwin, *Contrivances*, 203.

12. C. Darwin, *Contrivances*, 1–2.

13. Devin Griffiths, *The Age of Analogy: Science and Literature Between the Darwins* (Baltimore: Johns Hopkins University Press, 2016), 243.

14. Griffiths, *The Age of Analogy*, 241, 239. On the power of collaboration, the degree to which Darwin's work was enabled by engagement with others, see also Gregory Radick, "Is the Theory of Natural Selection Independent of Its History?" *Cambridge Companion to Darwin*, 2nd ed., ed. Jonathan Hodge and Gregory Radick (Cambridge: Cambridge University Press, 2009), 147–172.

15. C. Darwin to J. D. Hooker [August 11, 1861], DCP-LETT-3226, https://www.darwinproject.ac.uk/letter/DCP-LETT-3226.xml.

16. C. Darwin, *Contrivances*, 157.

17. C. Darwin to J. D. Hooker [August 11, 1861], DCP-LETT-3226, https://www.darwinproject.ac.uk/letter/DCP-LETT-3226.xml.

18. Griffiths, *Age of Analogy*, 244.

19. C. Darwin, *Contrivances*, 160.

20. C. Darwin, *Contrivances*, 34, 164, 163. He mentions the discovery in a letter dated June 19, 1861.

21. C. Darwin to J. D. Hooker, July 13, [1861], DCP-LETT-3207, https://www.darwinproject.ac.uk/letter/DCP-LETT-3207.xml.

22. Griffiths points out that Erasmus Darwin too used the term "contrivance" in a discussion of Orphyrs (Griffiths, *Age of Analogy*, 239, 245).

23. For further analysis of chance and its role in variation, and the radical implications of this in Darwin's work, see John Beatty, "Chance Variation: Darwin on Orchids," *Philosophy of Science* 73, no. 5 (2006): 629–641.

24. C. Darwin to J. D. Hooker, January 25 [and 26], [1862], DCP-LETT-3411, https://www.darwinproject.ac.uk/letter/DCP-LETT-3411.xml.

25. C. Darwin to J. D. Hooker, January 25 [and 26], [1862], DCP-LETT-3411, https://www.darwinproject.ac.uk/letter/DCP-LETT-3411.xml; for his fuller discussion, see C. Darwin, *Contrivances*, 202. Endersby explores the relationship of moth and insect in *Orchid*, 101–102. For analysis of the rostellum and its role in pollination see Tim Wing Yam, Joseph Arditti, and Kenneth Cameron, "'The Orchids Have Been a Splendid Sport': An Alternative Look at Charles Darwin's Contribution to Orchid Biology," *American Journal of Botany* 96, no. 12 (2009): 2128–54, esp. 2135–36.

26. Darwin, *Contrivances*, 348–349.

27. "If the *Origin* sketched out the extensive dimensions of Darwin's castle in the air, the *Orchids* showed what that castle might look like if built on solid empirical grounds," Griffiths summarizes (*Age of Analogy*, 217); Griffiths also discusses the ways in which Darwin's book rendered research methods available to the layman (241–243).

28. Darwin was frustrated by those who did not recognize the importance of variations produced by *chance* in his theory: "I have over & over again, ad nauseam, directly said & by order of precedence implied (what seems to me obvious) that selection can do nothing without previous variability." C. Darwin to W. H. Harvey [September 20–24, 1860], DCP-LETT-2922, https://www.darwinproject.ac.uk/letter/DCP-LETT-2922.xml.

29. C. Darwin, *Origin* (1859), 435.

30. C. Darwin, *Origin* (1859), 490.

31. C. Darwin, *Contrivances*, 306–307.

32. The encounter, as recalled by the Duke of Argyll, was told in Francis Darwin, ed., *The Life and Letters of Charles Darwin*, vol. 1 (1887; repr. London: John Murray, 1888), 316n. Darwin was somewhat careful in his statements of religious doubt; see Michael Ruse, "Charles Darwin: Great Briton," in *Debating Darwin*, ed. Robert J. Richards and Michael Ruse, 1–82 (Chicago: University of Chicago Press, 2016), 2.

33. Endersby, *Orchid*, 90.

34. Philip Appleman, "Darwin: On Changing the Mind," *Darwin: A Norton Critical Edition*, ed. Philip Appleman, 3rd ed., 3–20 (New York: Norton, 2001), 8.

35. C. Darwin to Asa Gray July 23 [–4], [1862], DCP-LETT-3662, https://www.darwinproject.ac.uk/letter/DCP-LETT-3662.xml.

36. A. Gray to C. Darwin, July 7, 1863, DCP-LETT-4234," https://www.darwinproject.ac.uk/letter/DCP-LETT-4234.xml.

37. Matthew Arnold, "Dover Beach," *Selected Poems of Matthew Arnold* (London: Macmillan, 1893), 164–165; 165.

17. "THE WHOLE WORLD'S GONE CATTLEYA CRAZY"

1. Alfred, Lord Tennyson, "On the Jubilee of Queen Victoria," *Demeter and Other Poems* (London: Macmillan, 1892), 6–11; 10.

2. "The Orchid-Craze," *New Remedie* 10 (July 1881): 203.

3. Quoted in Thomas J. Mickey, *America's Romance with the English Garden* (Athens: Ohio University Press, 2013), 20.

4. Edward Sprague Rand Jr., *Orchids: A Description of the Species and Varieties Grow at Glen Ridge, Near Boston* (1876; repr. Boston: Houghton Mifflin, 1888), vii.

5. E. Holly, "Notes on Orchids," *Gardener's Monthly* 20 (1878): 200; W. Falconer, "Orchids," *American Garden* 6 (December 1885): 298–299; 299.

6. W. A. Manda, "Orchids for Beginners," *American Garden* 8 (December 1887): 384. Manda would presently leave his curatorial position; in later life he ran well-known nurseries in New York City and New Jersey.

7. W. A. Manda, "Orchids and Orchidophiles," *American Garden* 8 (December 1887): 384; "Flora," "Floral Fashions," *American Garden* 7 (February 1886): 50. W. J. Bean, a British botanist, also stressed the fabulous sums that could be made through hybridization in "Some of the Best New Orchids," *American Garden* 12 (1891): 88–92.

8. "A Fortune from Orchids," *Illustrated American* 8 (September 5, 1891): 125.

9. Rand, *Orchids*, xv, 140; "Editorial Notes," *Gardener's Monthly and Horticulturalist* 19 (February 1877): 44.

10. E. E. Maling, "Room Plants," *Lady's Friend* 2 (1865): 767.

11. Nancy Mann Waddle, "A Flower of the Air: The Cultivation of the Exquisite Orchid Explained," *Ladies' Home Journal* 10 (July 1893): 3. "Martha McCulloch-Williams" [Susan Martha Ann Collins], "The Flower of Paradox: The Wonders of Orchid Culture," *Godey's Magazine* 134 (February 1897): 136–143; 143.

12. Anna Nichols Goodno, "Orchids," *Friends' Intelligencer and Journal* 47 (1890): 174.

13. F. Sander to F. Arnold, n.d. [received Baranquilla, March (?) 28, 1882], "Miscellaneous Letters to Fritz Arnold," F. Sander Correspondence, Royal Botanic Gardens, Kew, box 2, 160. (A number of documents in the box are numbered 160. "Alle wellt ist Cattleyen verrückt geht" is on the reverse side.)

14. "S.," "An Orchid Nursery," *Garden and Forest* 2 (May 8, 1889): 227.

15. Rand wrote that he hoped US collectors would soon start to hire their own collectors. His wish was fulfilled; a Mr. J. N. Newsham, former orchid grower to a St. Louis enthusiast, offered his services to readers as a collector of species of *Cattleya* in the pages of *American Garden* in 1888.

18. FAKES

1. Frederick Boyle, "An Orchid Farm," *Longman's* 12 (1888): 139–151; 146; James Veitch & Sons, *A Manual of Orchidaceous Plants* (London: James Veitch, 1887–1894), vol. 1, "Cattleya and Laelia," 15fn.

Boyle also reported that Sander sent off a hunter in pursuit of *C. labiata* to Mount Roraima (at the apex of Brazil, Venezuela, and Guyana) after seeing a drawing of a possibly similar plant in the British Museum. This is perhaps a fanciful retelling of the plant hunter Fritz Arnold's journey.

2. Robert Warner, *Select Orchidaceous Plants* (London: Lovell Reeve, 1862–1865), "Cattleya Labiata," plate 88; W. H. G. [William Hugh Gower], "Cattleya labiata vera," *The Garden* 35 (January 5, 1889): 8; "Notes," *Garden and Forest* 3 (July 30, 1890): 376.

Authors tended to be a bit hazy about how long the lost orchid had been missing; some said forty, some seventy years.

3. Edward Sprague Rand, *Orchids: A Description of the Species and Varieties Grow at Glen Ridge, Near Boston* (1876; repr. Boston: Houghton Mifflin, 1888), 423.

4. "Oudeis," "Cattleya autumnalis," *L'orchidophile: Journal des amateurs d'orchidées* 7 (1887): 122–123; W. A. Manda, "Collections d'orchidées en Amérique," *L'orchidophile: Journal des amateurs d'orchidées* 7 (1887): 121.

5. W. H. G., "Cattleya labiata vera," 8.

6. A. A. Chadwick and Arthur E. Chadwick, *The Classic Cattleyas* (Portland, OR: Timber Press: 2006), 56; Nikisha Patel, personal communication with the

author, April 11, 2024. For a detailed enumeration of *C. labiata* varieties, see L. C. Menezes, "Cattleya labiata Lindley and Its Color Varieties," *Orchid Digest* 51 (1987): 119–138, and *Cattleya labiata autumnalis* (Brasilia: Charbel Grafica, 2002), 17–18, 72, 101–102.

For a book-length engagement with the idea of (and problems in the idea of) species, see John S. Wilkins: *Species: A History of the Idea* (Berkeley: University of California Press, 2009).

7. Alfred Wallace, "On the Tendency of Varieties to Depart Indefinitely from the Original Type," in *Darwin: A Norton Critical Edition*, ed. Philip Appleman, 3rd ed., 61–64 (New York: Norton, 2001), 61–62; Darwin, *Species*, 98, 52, 278.

8. C. Darwin to J. D. Hooker, December 24, [1856], DCP-LETT-2022, https://www.darwinproject.ac.uk/letter/?docId=letters/DCP-LETT-2022.xml.

9. Jody Hey, "The Mind of the Species Problem," *Trends in Ecology & Evolution* 16, no. 7 (2001): 326–329; 329.

10. Christopher Beauchamp, "Getting Your Money's Worth: American Models for the Remaking of the Consumer Interest in Britain, 1930s–1960s," in *Critiques of Capital in Modern Britain and America: Transatlantic Exchanges 1800 to the Present Day*, eds. Frank Trentmann and Mark Bevir, 127–150 (New York: Macmillan, 2002). On trade protection societies that worked to defend the interests of members against nonmembers, and more generally the origins and development of English consumer culture, see Tammy C. Whitlock, *Crime, Gender and Consumer Culture in Nineteenth-Century England* (New York: Taylor & Francis, 2005), 76–78.

11. Arthur Swinson, Sander's enthusiastic biographer, suggests that the St. Albans businessmen sent many other hunters in pursuit of the lost orchid. He also names Clarke, Chesterton, Bartholmeus, Kerbach, the brothers Klaboch, Oversluys, Smith, and Bestwood. I have not been able to confirm, from the Sander archive at Kew or other sources, that *all* these hunters were specifically in pursuit of *C. labiata*—see Arthur Swinson, *Frederick Sander: The Orchid King* (London: Hodder & Stoughton, 1970), 114—though Smith and Oversluys certainly were.

12. Jim Endersby, "Classifying Sciences: Systematics and Status in Mid-Victorian Natural History," in *The Organization of Knowledge in Victorian Britain*, ed. Martin Daunton, 61–84, (Oxford: British Academy, 2005), 71–73.

13. Coloring is challenging to interpret: "a range of flowers whose colors could be treated as mere chromatic variations of the type (pseudotype) flower are given fanciful names—a treatment that varies from region to region in Brazil—making a uniform, understandable classification of flowers of the species impossible" (Menezes, *Cattleya Labiata Autumnalis*, 103).

14. F. Sander to F. Arnold, [December?] 14, 1880; November 16, 1880, "Miscellaneous Letters to Fritz Arnold 1800–1890" (hereafter "Fritz Arnold"), F. Sander Correspondence, Royal Botanic Gardens, Kew (hereafter SC/RBGK), box 2, 106–7.

Sander's scrawled, crossed, wildly underlined letters are often challenging to read, and many are in German. I use translations prepared for me in 2022–3 by Gerda Dinwiddie and typescripts produced for and annotated by a member of the Sander family, most likely David Fearnley Sander (see "Fritz Arnold," SC/RBGK, box 2, 106–109). These typescripts were used by Swinson in *Frederick Sander*; I myself used Swinson's quotations when I was unable to find a letter, when the handwriting posed particular difficulties, when I could not find the typescript translation (some letters have found their way into the wrong files, while the typescripts also seem to be incomplete). I used the Swinson quotations on the basis that, so far as I can tell from the annotated transcripts, these are versions of Sander's words that the Sander family at least considered plausibly accurate.

15. The *Moselle*, built in 1871, was operated by the Royal Mail Steam Packet Company, founded in 1839 by a British plantation manager in the West Indies. The company ferried the mail through a contract with the government, together with passengers and freight. Ship specifications are listed, together with date and place of building and more, in the *Lloyd's Register of Shipping*.

16. F. Sander to F. Arnold, October 29, 1880; November 16, 1880; December [?] 14, 1880; December 29, 1880, "Fritz Arnold," SC/RBGK, box 2, 106–109.

17. Swinson, *Frederick Sander*, 47, 49. Swinson names the collector William Arnold, but I'm reasonably confident this is incorrect: he signed himself always and clearly in his own letters "F. Arnold," and the extensive file of letters to him at Kew include many saluting "Fritz" or "Frederic."

18. Sander warned Arnold in October 1880 that letters took four weeks to arrive—and the annotation on the documents indicates receipt, thus we can tell that a letter that left St. Albans on October 29, 1880, arrived in Caracas on November 26, for instance. Sander also had to time his letters to the sailings of the Royal Mail, which further delayed them.

On the growth of the electrical telegraph network, and the laying of submarine telegraph cables between Europe and the Americas, see Anton A. Huurdeman, *The Worldwide History of Telecommunications* (Hoboken, NJ: John Wiley, 2003), esp. 135–139.

19. The story of Arnold's tricking and subsequent threatening of White with a revolver is narrated by Frederick Boyle in *About Orchids: A Chat* (London: Chapman & Hall, 1893), 28, and by Sander in an interview in *Tit-Bits*, reprinted in the 1896 New Zealand paper *Otago Witness;* it was also repeated, with small

changes, by Swinson (*Frederick Sander*, 45). Arnold's letters are particularly difficult to read, and I have not been able to confirm all of these details, though Sander's letters to Arnold certainly indicate intense conflict with White and with Low's. See too the (highly fictionalized) account of the conflict in F. Sander, "Checkmate! The Romance of an Orchid," *Wide World Magazine: An Illustrated Monthly of True Narrative* 13 (1904): 295-298.

20. Arnold quoted in G. Wilson, "F. Arnold, Orchid Collector 1880-86," *Orchid World* 6 (November-December 1915): 41-43; 41; 62-64. A. Ernst to F. Sander, November 16, 1880, SC/RBGK, box 1, folder 4, 329.

21. Sander to Arnold, December 29, 1880, "Fritz Arnold," SC/RBGK, box 2, 108.

22. H. G. Reichenbach, "Cattleya (Labiata) Perciviliana, *n*. Var," *Gardeners' Chronicle* 17 (June 17, 1882): 796. See Gustavo Romero, *Venezuela: Paraíso de Orquideas* (Caracas: Armitano Editore: 1997), 46, on the short life of Arnold, the tussle with Hugh Low, and the naming of *C. percivaliana*.

Sander suspected Arnold's flower was *C. speciosissima*, later determined to be *C. lueddemanniana* by Reichenbach—see Sander to Arnold, January 3, 1881, "Fritz Arnold," SC/RBGK, box 2, 109. He asked Arnold to send more of these in early 1881. Many of his letters to Arnold that year muse on the relationships among species of *Cattleya* and their potential habitats. Ernst was certainly aware that the two orchids were the same, as evidenced in a letter to Sander (August 4, 1880, SC/RBGK, box 1, folder 4, 324).

23. Sander to Arnold, January 13, 1882, "Fritz Arnold," SC/RBGK, box 2, 160. I use Swinson's translation, quoted in *Frederick Sander*, 56.

24. Sander to Arnold, December 29, 1880, "Fritz Arnold," SC/RBGK, box 2, 118.

25. Sander to Arnold, n.d. [received Baranquilla, March (?) 28, 1882], "Fritz Arnold," SC/RBGK, box 2, 160. (A number of documents in the box are numbered 160.)

26. In November-December 1881, Alfred Booth & Co, shipping agents in Liverpool, had to track down another four cases of orchids intended for Sander that got mixed up with those belonging to a different importer on the journey. See Booth & Co, SC/RBGK, box 20, folder 3.

27. Sander to Arnold, December 29, 1880, "Fritz Arnold," SC/RBGK, box 2, 118.

28. Revolver, suitcase, money, and "Schein" (bill?) are crossed out on the memorandum: it's hard to know if this was because Palmer had accounted for them, and so was using the list to check things off, or because he had been *unable* to account for them, and so was indicating that they would not be shipped to England. C. Palmer to F. Sander, May 20, [1886,] SC/RBGK, box 15, 241.

Palmer wrote a long letter to the family on August 18, 1886; see "Fritz Arnold," SC / RBGK, folder 1, 93–96. Fond letters from Arnold's brother, illustrated and adorned with dried flowers, are also in the "Fritz Arnold" archive, attesting to the close ties of this particular family.

29. F. Sander to F. Arnold, December 29, 1880, "Fritz Arnold," SC / RBGK, folder 2, 108–109; W. Kirton to F. Sander, May 28, 1886, SC / RBGK, box 8, folder 2, 281; Palmer to Sander, 241. Though Kirton reported that Palmer would give Sander a fuller account of Arnold's death, unfortunately that letter seems not to have survived. A receipt signed by Charles Palmer for money forwarded by Kirton is in the archives at Kew, among the Palmer letters (box 15, 243).

30. F. Sander, "Checkmate! The Romance of an Orchid," *Wide World Magazine: An Illustrated Monthly of True Narrative* 13 (1904): 295–298; 297.

Contemporary writers repeat tales of dramatic death: Sander's biographer Swinson claimed, for instance:

> Klaboch, Roezl's nephew, was killed in Mexico; Arnold disappeared on a voyage up the Orinoco; Digance was shot by the natives in Brazil; Falkenerg died in Panama; Wallis was murdered in Ecuador, Hendres, in the Rio Hacha, and Brown in Madagascar. And Osmers, Sander's first traveller was killed in the Far East. (Swinson, *Frederick Sander*, 123)

Arnold did not disappear, Digance was not shot by natives, and Wallis died after an illness, according to the death notice issued by his family. These more shocking tales supported Sander's narrative (as we shall see in more detail presently) that plant hunting was a *Boy's Own,* Kipling-esque adventure narrative (SC / RBGK, box 18, folder 4).

31. Advertisement, "Hugh Low & Co," *Gardener's Monthly and Horticultural Advertiser* 28 (November 1886): 4.

32. Boyle, *About Orchids,* 188; Advertisement [Brackenridge], "Orchids: Hardy and Exotic, Stock Immense," *American Garden* 9 (1888): 115; Advertisement [Siebrecht & Wadley], "Orchids and Palms," *American Garden* 9 (1888): 161; Advertisement [John Cowan], "Orchids! Orchids!" *Orchid Review* 4 (1896) end pages; Advertisement, M. Godefroy Lebeuf, "Orchids!" *Naturalists' Universal Directory* (Boston: S. E. Cassino, 1883), end pages.

33. Swinson, *Frederick Sander,* 42.

19. DAMAGE

1. José Augusto Pádua, "The Dilemma of the 'Splendid Cradle': Nature and Territory in the Construction of Brazil," 91–114 in *A Living Past: Environmental Histories of Modern Latin America,* ed. Claudia Leal, John Soluri, and José Augusto Pádua (New York: Berghahn, 2018), 97.

2. John Soluri, Claudia Leal, and José Augusto Pádua, "Introduction. Finding the 'Latin American' in Latin American Environmental History," 1–22 in *A Living Past,* ed. Leal, Soluri, and Pádua,; 10–11.

3. "Dodman" [Charles Bellenden Ker], "Orchids for the Million: I," *Gardeners' Chronicle* 10 (May 18, 1850): 308–309; 309.

4. Quoted in Arthur Swinson, *Frederick Sander: The Orchid King* (London: Hodder & Stoughton, 1970), 42.

5. "Orchids: Sophronitis," *Gardening Illustrated* 11 (October 5, 1889): 411; "Trade Notes," *American Garden* 8 (April 9, 1887): 142.

6. The objectives of the society were reported in an American journal in "The Protection of Rare Species of Plants," *Popular Science* 30 (March 1887): 717–718. *Nature Notes* was, from 1890, published by pioneering conservationists the Selborne Society, an organization founded in 1885.

7. See Daniel Pick, *Faces of Degeneration: A European Disorder, c. 1848–1918* (Cambridge: Cambridge University Press, 1989). On American responses to degeneration and its embrace of nature as a response to modernism and urbanization, see Robin G. Schulze, *The Degenerate Muse: American Nature, Modernist Poetry, and the Problem of Cultural Hygiene* (Oxford: Oxford University Press, 2013), esp. 1–38.

8. Quoted in Patrick Brantlinger, *Rule of Darkness: British Literature and Imperialism, 1830–1914* (Ithaca: Cornell University Press, 2013), 240. I am indebted in my analysis to Brantlinger's scholarship on late Victorian adventure romances, also to Mary Louise Pratt's examination of the ways in which European travel literature "created the imperial order to Europeans 'at home' and gave them their place in it"—see *Imperial Eyes: Writing and Transculturation,* 2nd ed. (London: Routledge, 2007), 3.

9. For more on the ways in which South America prompted Europeans to see, or to claim to see, the past miraculously preserved, see Cannon Schmitt in *Darwin and the Memory of the Human: Evolution, Savages, and South America* (Cambridge: Cambridge University Press, 2009), 2–3.

10. Susan Stewart, *On Longing: Narratives of the Miniature, the Gigantic, the Souvenir, the Collection* (Durham, NC: Duke University Press, 1992), 133.

Georg Novalis introduced the symbol of the blue flower in *Heinrich von Ofterdingen* (1800): a hero yearns to see a flower he has heard of, then dreams of a quest. I'm grateful to Stefanie Markovits for pointing out parallels between the lost orchid and the blue flower. Plant hunters literally reported dreaming of finding a cattleya: one wrote to Sander, "Last night I dreamt I brought a new Cattleya home ... I hope it will come true." F[rederick] Dressel to F. Sander, August 10, 1887, F. Sander Correspondence, Royal Botanic Gardens, Kew, box 1, folder 3, 149.

20. ENTERING THE LISTS

1. Cecília Resende Santos, "The Brazilian Pavilion at the 1889 Paris Universal Exposition," https://www.mouse-magazine.com/the-brazilian-pavilion-at-the-1889-paris-universal-exposition/. I owe the details about the structure of the pavilion, and the kiosk, to Lilia M. Schwarcz and Heloisa M. Starling, *Brazil: A Biography* (New York: Picador, 2015), 348.

2. "La grotte et les jardins autour du pavillon et de ses annexes, son couverts de palmiers et d'orchidées du Brésil." S. Favière, "Brésil," *Livre d'or de l'exposition*, ed. Lucien Huard, vol. 2 (L. Boulanger: Paris, 1889), 654–655; 655; the catalogue itself is available at https://gallica.bnf.fr/ark:/12148/bpt6k9736976f/f1.double.

3. "La Brésil a l'Exposition Universelle de Paris en 1889," *Lindenia* 4 (1888): 30. Translations from the French are my own.

4. "Les Orchidées a l'Exposition Universelle de Paris en 1889," *Lindenia* 5 (1889): 10.

5. The Lindens' postcard to Sander was dated December 13, 1886, F. Sander Correspondence, Royal Botanic Gardens, Kew (hereafter SC / RBGK), box 10, folder 2, 117). It was presumably written by an employee at the Lindens' request; correspondence from Lucien was typically stamped "Le Directeur."

6. The whole scandal was written up in some detail in the *Gardeners' Chronicle* on May 17, 1890 ("Catasetum Bungerothi," *Gardeners' Chronicle* 7 (1890): 618–619.) I've been unable to locate Reichenbach's letter specifically, though Reichenbach does reference "Catasetum pileatum (Bungerothi)" in an 1887 card to Sander (SC / RBGK, box 16, folder 4, 343.) Month unclear, though probably November; Reichenbach's handwriting is exceptionally hard to read.

7. Frederick Sander, "Catasetum Pileatum," *Reichenbachia*, 1st ser., vol. 2 (St. Albans: F. Sander, 1890), 91–94; 92.

8. Sander, "Catasetum Pileatum," 91. The color plate continues to bear the name *Bungerothi*, but Sander claims this is because he didn't have time to correct the error before the book went to press.

9. "Catasetum Bungerothi," 618.

10. "Chronique orchidéenne mensuelle: L'Horticulture Internationale," *Journal des orchidées* 1 (June 1, 1890): 85–87; 87.

11. "Cattleya Warocqueana" [Advertisement], *Journal des orchidées* 1 (June 15, 1890): between 116–117.

12. Robert Allen Rolfe, "Cattleya Labiata, *Lindl.*, Var. Warocqueana, *n. var.* (?)," *Gardeners' Chronicle* 7 (June 14, 1890): 735.

13. "It is said to be from a part of South America previously unexplored, though what part it may be I have not the slightest suspicion." Rolfe, "Cattleya Labiata," 735. The translation appeared as R. A. Rolfe, "Les grandes introduc-

tions nouvelles: 1. - Cattleya Labiata Lindl. var. Warocqueana n. var." *Journal des Orchidées* 1 (July 15, 1890), 139-140.

14. [Lucien Linden?], Note to R. A. Rolfe, "Les grandes introductions nouvelles: 1. - Cattleya Labiata Lindl. var. Warocqueana n. var." *Journal des orchidées* 1 (July 15, 1890), 140.

15. The condition of *"la grande introduction"* is described in "Le Cattleya Warcocqueana," *Journal des orchidées* 1 (October 1, 1890): 219.

16. "Causerie sur les orchidées: VI: Les Cattleya Warocqueana et la presse horticole anglaise," *Journal des orchidées* 1 (November 1, 1890): 247-250.

17. "Cattleya Labiata Warocqueana," *Gardeners' Chronicle* 8 (December 6, 1890): 661.

18. "Cattleya Labiata Var. Warocqueana," *Gardeners' Chronicle* 8 (November 29, 1890): 624.

19. "Les Cattleya Warocqueana," *L'illlustration horticole* 37 (1890): 120. Translations from the French are my own.

20. William Hugh Gower, "Orchids: Cattleya Labiata," *The Garden* 38 (November 15, 1890): 470.

21. Linden, "Causerie sur les orchidées: VIII. Cattleya Labiata Autumnalis et Cattleya Warocqueana," *Journal des orchidées* 1 (December 1, 1890): 280-283.

21. ERICSSON AND BUNGEROTH

1. C. Ericsson to F. Sander, January 8, 1886, F. Sander Correspondence / Royal Botanic Gardens, Kew (hereafter SC / RBGK), box 1, folder 4, 183.

2. Ericsson enthusiastically expressed himself willing to set off by dinner tomorrow in his reply to the job offer from Sander (SC / RBGK, box 1, folder 4, 184).

3. Ericsson's address in Stockholm was Bryggaregaten 5.

4. Matthew Rubery, "Journalism," *The Cambridge Companion to Victorian Culture*, ed. Frances O'Gorman (Cambridge: Cambridge University Press, 2010), 177-194; 177.

5. For more on Victorian print advertising in the streets, see Kelley Graham, *Gone to the Shops: Shopping in Victorian England* (London: Praeger, 2008), 36-38. On the numbers of pages given over to advertising in Victorian periodicals, see Graham Law, "Distribution," 42-59 in *The Routledge Handbook to Nineteenth-Century British Periodicals and Newspapers*, ed. Andrew King, Alexis Easley, and John Morton (London: Routledge, 2016), 55.

6. On product placement in Victorian magazines see Graham Law, "Periodicalism," 537-554 in *The Victorian World*, ed. Martin Hewitt (London: Routledge, 2012), 548.

7. Vincent F. Hendricks and Mads Vestergaard, *Reality Lost: Markets of Attention, Misinformation and Manipulation* (Copenhagen: Springer, 2018), 6.

8. In 1883, historian J. R. Seeley remarked: "We seem, as it were, to have conquered and people half the world in a fit of absence of mind." His claim, in other words, was that the Empire had been built almost accidentally, and he sought to encourage a "great awakening" of imperial consciousness in the later nineteenth century. J. R. Seeley, *The Expansion of England* (1883; repr. London: Macmillan, 1890), 8.

Note that European countries were "scrambling" for influence in South America at this moment also—see P. J. Cain and A. G. Hopkins, *British Imperialism: Innovation and Expansion, 1688–2015* (London: Routledge, 2016), 272, 288.

9. Ericsson to Sander, January 4, 1891, SC / RBGK, box 1, folder 4, 231.

10. C. Palmer to F. Sander, November 10, 1890, SC / RBGK, box 15, 266.

11. Digance to Sander, December 11, 1890, SC / RBGK, box 1, folder 3, 144–145; Digance to Sander, February 6, 1891, SC / RBGK, box 1, folder 3, 148. Digance was also looking for a different *Cattleya* orchid, *Cattleya walkeriana*, though he usually referred to it in his letters as "walkeriana" (not "Cattleya"). Given the timing, and the fact that the Sander family later reported that Digance was hunting the "lost orchid" (according to Swinson), I think it likely the "Cattleya business" Digance mentions is indeed the hunt for *Cattleya warocqueana / labiata*.

12. "Why did you keep me so long in St. Albans, many times I wanted to go away but you did not let me," Ericsson complained to Sander. April 9, 1891, SC / RBGK, box 1, folder 4, 235.

13. "William Digance," Brasil, Rio de Janeiro, Registro Civil, 1829–2020, database with images, FamilySearch https://www.familysearch.org/ark:/61903/1:1:79Z9-HK3Z, trans. Bradley Hayes, December 12, 2022.

14. Ericsson to Sander, March 25, 1891, SC / RBGK, box 1, folder 4, 234.

15. Ericsson to Sander, April 9, 1891, SC / RBGK, box 1, folder 4, 235.

16. Ericsson to Sander, May 22, 1890, SC / RBGK, box 1, folder 4, 237. In the letter from Barbados, Ericsson reports on his time in Venezuela.

17. Ericsson to Sander, April 9, 1891, SC / RBGK, box 1, folder 4, 235; Ericsson to Sander, May 23, [1891], SC/RBGK, box 1, folder 4, 237.

18. Ericsson to Sander, May 23, [1891], SC/RBGK, box 1, folder 4, 237; Comte de Moran, "Une interview avec le 'Père des Orchidées,'" (July 15, 1891), *Journal des orchidées*, v. 2, 135–140; 139.

19. Ericsson to Sander, May 23, 1891, SC / RBGK, box 1, folder 4, 237.

20. "Vanda London" was the telegram code for Sander's St. Albans business.

21. Ericsson to Sander, June 20, 1891, SC / RBGK, box 1, folder 4, 239.

22. Ericsson to Sander, July 18, 1891, SC / RBGK, box 1, folder 4, 242.

23. The offices at Knowles & Co. were at 31, Rua dos Capellistas in Lisbon. Vessels left Southampton on the 9th and 24th of each month and took seventy-

five hours to reach Lisbon. They departed the city on the 11th and 27th, according to [John Murray], *A Handbook for Travellers in Portugal, with a Short Account of Madeira, the Azores, and the Canary Island* (London: John Murray, 1887), 11. Ericsson specifies that he wrote to Knowles in a letter to Sander, August 1, 1891, SC / RBGK, box 1, folder 4, 232–233.

The *Cintra*, commissioned two years previously, was operated by the South American S.S. Co. based in Hamburg. "S.S." stands for screw steamer, meaning that the ship used a propeller or screw.

24. Ericsson to Sander, August 1, 1891, SC / RBGK, box 1, Folder 4, 232–233. This letter is out of date order in the archive.

25. Ericsson to Sander, September 4, 1891, SC / RBGK, box 1, folder 4, 244–245.

26. Ericsson to J. Godseff, September 15, 1891, SC / RBGK, box 1, folder 4, 248.

27. Ericsson to Sander, October 9, 1891, SC / RBGK, box 1, folder 4, 251–252.

28. "Exportação" 3; Ericsson to Sander, November 7, 1891, SC / RBGK, box 1, folder 4, 253.

29. Ericsson to Sander, October 9, 1891, SC / RBGK, box 1, folder 4, 251.

22. THE GREAT DISCOVERY

1. Advertisement, "Friday Next: The True Old Autumn-Flowering Cattleya Labiata," *Gardeners' Chronicle* 10 (September 12, 1891): 295; Advertisement, "Rediscovery," *Gardeners' Chronicle* 10 (September 12, 1891): 298.

2. The Van Houtte nursery, and the life of Louis Van Houtte, is described in an obituary, "Louis Van Houtte," *Journal of Horticulture and Cottage Gardener* 30 (May 18, 1876): 388–390; the Verschaffelt premises are described in "A Belgian Nursery-Man," *Horticulturalist and Journal of Rural Taste* 27 (1872): 66–68.

3. "A Botanical Sensation," *St. James's Gazette* 23 (September 15, 1891): 5.

4. [Frederick Boyle], "The Lost Orchid," *Saturday Review* 72 (September 19, 1891): 327–328; 328; Frederick Boyle, *About Orchids: A Chat* (London: Chapman & Hall, 1893), 180. The chapter in the book-length work draws substantially from the anonymously authored piece in the *Saturday Review*.

5. Swinson tends to blame Joseph Godseff for the "taradiddles" in the press—meaning, the "fake news" stories of noble orchid hunters and remarkable discoveries (Arthur Swinson, *Frederick Sander: The Orchid King* (London: Hodder & Stoughton, 1970), 186). It is impossible to know who issued which statements to the press at this point. In my analysis, I tend to identify Sander as the prime mover since the company was his and since his letters reveal he was, at the very least, actively involved in crafting a narrative of the discovery of

the "*autumnalis vera.*" That said, the annotations on the letters, signed "J. G.," indicate Godseff was also very actively involved.

6. "The Coming Great Orchid Sale," *Florist's Exchange* 3 (October 31, 1891): 606. The motto of the *Florist's Exchange* was "we are a straight shoot, and aim to grow into a vigorous plant."

7. These photographs were first mentioned in *Journal of Horticulture and Cottage Gardener* 23 [3rd series] (September 17, 1891): 250.

8. Ellen Handy, "Dust Piles and Damp Pavements: Excrement, Repression, and the Victorian City in Photography and Literature," 111–132 in *Victorian Literature and the Victorian Visual Imagination*, eds. Carol T. Christ and John O. Jordan (Berkeley: University of California Press, 1995), 116.

9. Advertisement, "Friday Next. Great Sale of Orchids [F. Sander]," *Gardeners' Chronicle* 10 (December 12, 1891): 690. Here is it pitched as "The OLD LABIATA, in sheath: GARDNER'S LABIATA." This time, he also claimed the plants were "collected by the late W. Digance fifteen days before his death from yellow fever."

10. Advertisement, "Sander's Great Cattleya Sale," *Gardeners' Chronicle* 10 (October 3, 1891): 391.

11. Advertisement, "2000 Cattleya Labiata True," *Gardeners' Chronicle* 10 (November 14, 1891): 575.

12. "Cattleya Labiata, *Lindl.*," *Gardeners' Chronicle* 10 (October 31, 1891): 523.

13. Advertisement, "Spathoclottis Ericssoni," *Gardeners' Chronicle* 10 (November 14, 1891): 575.

14. L[ewis] Castle, "Orchids: Cattleya Labiata Vera," *Journal of Horticulture, Cottage Gardener, and Home Farmer* 23 (September 24, 1891): 262–263, and October 1, 1891, 283–284.

15. Patrick Brantlinger, *Rule of Darkness: British Literature and Imperialism, 1830–1914* (Ithaca: Cornell University Press, 2013), 11.

16. Ashmore Russan and Frederick Boyle, *The Orchid Seekers: A Story of Adventure in Borneo* (London: Frederick Warne, 1897), 19. Swinson confirmed that the novel draws heavily on Sander and Roezl (Swinson, *Frederick Sander*, 22–3).

17. "Petite Correspondence: A divers abonnés qui nous posent la même question," *Le journal des orchidées* 2 (January 1, 1892): after 324.

23. THE SECOND GREAT DISCOVERY

1. Adolpho Lietze, born in Germany, lived in Rio de Janeiro and was a renowned horticulturalist and hybridizer.

2. "Les Nouveaux Cattleya Labiata," *Orchidophile* 11 (1891): 310–313; 314.

3. Bungeroth published a series of articles on orchid collection January–March 1891 in *Journal des orchidées*. They were evidently edited; a few of

Bungeroth's letters to Sander have survived at Kew, and these are far less polished than those published under "E. Bungeroth, Collecteur de L'Horticulture Internationale."

4. Lucien Linden, "Histoire de la seconde découverte du vieux Cattleya Labiata," *Journal des orchidées* 2 (1891): 382–385.

5. An unsigned letter at Kew, probably by Forget, amplifies the point. Written on Christmas Eve 1891, it reports ongoing efforts to find the "famous Catt." and reports that "the Belgian" (at that point described as "a clever an very distinguished young man, and very devoted to his company") explained the "Brazilian flor" was recovered during "the Exhibition of 89": "it is there that he [Claes] got to know of the Catt." This, the letter continues, is why and how Linden knew where to search. Unsigned, December 24, 1891, F. Sander Correspondence, Royal Botanic Gardens, Kew, box 19, folder 2 [loose letter].

6. Edimilson de Almeida Pereira, and Steven F. White, "The Painter Estêvão Silva," *Callaloo* 24 no. 4 (2001): 1149.

7. Linden, "Histoire de la seconde découverte," 382–385.

8. The *Boy's Own Paper* ran a sequence of articles on orchid collection and cultivation in 1893, "The Boy's Own Orchid-House," by W. Watson of Kew.

9. Rolfe, "Cattleya Labiata: Lindl," *Gardeners' Chronicle* 10 (September 26, 1891): 366–368.

10. [M. G.], "La réintroduction du Cattleya labiata autumnalis," *L'illustration horticole* 38 (1891): 112–113.

11. Their arrival is listed on November 22, 1891, in the *Jornal do Recife*. It is perhaps another sign of Bungeroth's expertise that while Ericsson, Oversluys, and Forget all make appearances in the pages of the newspaper, as passengers or as shippers, Bungeroth does not.

Galahad is aided by Sirs Percival and Bors, who also arrive by ship. The three men are the only questers (of the one hundred and fifty knights) who succeed.

24. ERRORS

1. Digance to Sander, December 11, 1890, SC / RBGK, box 1, folder 3, 144.

2. The term *oecologie* was first used and defined by Ernst Haeckel in 1866; plant and animal ecology emerged in something like their current form as sciences in the 1890s, according to Frank N. Egerton in a study of the growth of ecology as a discipline and focus of inquiry; see *Roots of Ecology: Antiquity to Haeckel* (Berkeley: University of California Press, 2012), xi.

3. Nigel D. Swarts and Kingsley W. Dixon, "Terrestrial Orchid Conservation in the Age of Extinction," *Annals of Botany* 104 (August 2009), https://doi.org/10.1093/aob/mcp025.

4. L. F. [Louis Forget], "Erreurs geographiques concernant les orchidées," *Le jardin* 1 (1897): 246–248.

5. Lynn Voskuil argues that orchids throughout the century helped galvanize early conversations about ecology in "Victorian Orchids and the Forms of Ecological Society," 19–39 in *Strange Science: Investigating the Limits of Knowledge in the Victorian Age*, eds. Lara Karpenko and Shalyn Claggett (Ann Arbor: University of Michigan Press, 2017).

6. Ericsson to Sander, November 9, 1891, and November 23, 1891, SC/RBGK, box 1, folder 4, 254–255.

7. Forget to Sander, January 31, 1892, SC/RBGK, box 4, folder 1, 1.

8. "I wonder if F. Claes is Bung," Ericsson mused to Sander, adding "I know some people here in the town they will let me know so soon any cases are shipped for Linden"; November 23, 1891, SC/RBGK, box 1, folder 4, 255.

9. Ericsson to Sander, June 1, 1892, SC/RBGK, box 1, folder 4, 259; Oversluys to Sander, January 9, 1892, box 15, 110; Forget to Sander, January 31, 1892, and June 4, 1892, SC/RBGK, box 4, folder 1, 1 and 14–15.

10. Claes seems to have told Oversluys that he knew to expect the latter's arrival: "how the devil they should know this, and other things, I suppose there is some one at your establishment being in connection with Linden I am not sure but have a vague idea of this." Ericsson to Sander, January 9, 1892, SC/RBGK, box 1, folder 4, 111.

11. Ericsson to Sander, December 4, 1891, SC/RBGK, box 1, folder 4, 256.

12. Ericsson to Sander, November 7, 1891, SC/RBGK, box 1, folder 4, 253; Oversluys to Sander, January 9, 1892, SC/RBGK, box 15, 110; Ericsson to Sander, January 6, 1892, SC/RBGK, box 1, folder 4, 259. Digance also used this approach.

13. Sander to Strunz, Cable, May 19, 1892, SC/RBGK, box 22, 616. Strunz to Sander, June 22, 1892, SC/RBGK, box 22, 617.

14. Ericsson to Godseff, February 14, 1892, SC/RBGK, box 1, folder 4, 260.

15. Ericsson to Sander, January 30, 1892, SC/RBGK, box 1, folder 4, 258.

16. Ericsson to Godseff, February 14, 1892, SC/RBGK, box 1, folder 4, 260.

17. Oversluys to Sander, March 8, 1892, SC/RBGK, box 15, 117.

Ericsson reported to Sander and Godseff in two panicked letters, on a single day in February 1892, that Claes had vanished and that he was attempting to lure their collectors away; see Ericsson to Godseff, February 14, 1892, SC/RBGK, box 1, folder 14, 260–261. In a letter later that month, Ericsson reported that Kromer's presence was doubling the cost of plants in the region (Ericsson to Sander, February 27, 1892, SC/RBGK, box 1, folder 4, 262). Ericsson reported Kromer's money in a letter to Sander on March 13, 1892, SC/RBGK, box 1, folder 4, 263.

The archive suggests Kromer worked for Sander before he jumped ship to Linden; Kromer later ran a nursery of his own, the Roraima Nursery in West Croydon, specializing in orchids.

18. Ericsson reports Sander's response to his suggestion on January 6, 1892, SC/RBGK, box 1, folder 4, 259.

19. Ericsson to Sander, March 13, 1892, and March 16, 1892, SC/RBGK, box 1, folder 4, 263–264.

20. Oversluys to Sander, March 15–16, 1892, SC/RBGK, box 15, 118–119: "I get notice from Paquevira that people is waiting for me with a lot of plants"; "I am close to the spots where Bungeroth has collected such fine varietys."

The next month, he wrote that he had found an orchid "like the 'Warocqueana' from Lindens picture, the latter one I got many in Paquevira and this place I stood is not fare from the place where Lindens man Bungeroth has collected his." Oversluys to Sander, April 14, 1892, SC/RBGK, box 15, 125–126.

21. Oversluys to Sander, April 14, 1892, SC/RBGK, box 15, 125–126; Ericsson to Godseff, February 14, 1892, SC/RBGK, box 1, folder 4, 260.

22. E. Kromer's exports were listed on April 13 and May 13, 1892, in the *Diario de Pernambuco*. Kromer's legal troubles were reported, if rather less colorfully, in *Jornal de Recife* in 1892–3.

23. Ericsson to Sander, April 3, 1892, SC/RBGK, box 1, folder 4, 265. Kromer was still in the region in May but unable to pose much of a threat to the hunters because, having appealed his case to the high court, he was stuck battling it. He left in June. Ericsson was wrapping up in Pernambuco in July and was in Stockholm by Christmas.

24. Ericsson to Sander, April 10, 1892, SC/RBGK, box 1, folder 4, 266.

25. See L. Perthius to F. Sander, March 12 and April 22, 1893, SC/RBGK, box 16, 78–79. Perthius, who was certainly hunting *C. labiata*, recorded ongoing fights with Claes in the area. He also warned Sander that a Portuguese orchid hunter in the region was intending to use one of Ericsson's men, rather as Ericsson had earlier used Bungeroth's agent to lead him to *C. labiata*.

26. Oversluys to Sander, April 11, 1893, SC/RBGK, box 15, 163. I have been unable to locate information about Oversluys's death, though his birth date was probably 1862 or 1863. His father was likely Arnold Lodewijk Overluijs. A. L. Oversluijs wrote to Sander, expressing anxiety about his son's whereabouts on May 27, 1890.

27. Swinson notes that Sander family lore was that Ericsson "went native," a claim based on their belief that he had native wives and drank too much. Struggles with alcohol seem possible, based on the archive; the presence of multiple wives I am unable to confirm, though his references to getting in trouble with local

people could indicate that he was somehow abusing or exploiting local women. See Swinson, *Orchid King*, 124–125.

28. Léon Perthius to F. Sander, August 27, 1897, SC/RBGK, box 16, 115–116.

25. "SOME BRAZILIAN"

1. Albert Millican, *Travels and Adventures of an Orchid Hunter* (London: Cassell, 1891), 91.
2. Millican goes on an alligator hunt; see *Travels and Adventures*, 87–88.
3. Digance to Sander, September 17, 1890, F. Sander Correspondence, Royal Botanic Gardens, Kew (hereafter SC/RBGK), box 1, folder 3, 138.
4. Digance to Sander, October 5, 1890, SC/RBGK, box 1, folder 3, 140.
5. Ericsson to Sander, September 4, 1891, SC/RBGK, box 1, folder 4, 244–245.
6. In the same letter (September 4, 1891) Ericsson told his boss that "a shipment will leave via Lisbon this plants are not so good but as the people had brought them I had to take them, els they would have sold them to the Brazilian."
7. Ericsson to Sander, September 13, 1891, September 21, 1891, November 7, 1891, SC/RBGK, box 1, folder 4, 248, 249, 253.
8. Oversluys to Sander, January 9, 1892, SC/RBGK, box 15, 110–111; Ericsson to Sander, January 30, 1892, SC/RBGK, box 1, folder 4, 258; Forget to Sander, June 17, 1892, SC/RBGK, box 4, folder 1, 17–19.
9. I owe this detail to Ovidio's enormously helpful descendant, Samuel Sales Pinheiro Wanderley.
10. Ericsson to Sander, February 14, 1892, SC/RBGK, box 1, folder 4, 261.
11. Forget to Sander, January 31, 1892, SC/RBGK, box 4, folder 1, 1–2.
12. Forget to Sander, June 17, 1892, SC/RBGK, box 4, folder 1, 17–19. According to Perthius, Bungeroth gave up collecting after the discovery of *C. labiata* and went to Germany "to build a farm with his brother." He returned to orchid hunting later however—and worked for Sander himself. Perthius to Sander, July 20, 1893, SC/RBGK, box 16, 85–86.
13. "Diario Social: Fallecimentos [Deaths]," *Diario de Pernambuco (PE)*, April 21, 1923, issue 00091 (1920–1929), 2, http://memoria.bn.br/docreader/DocReader.aspx?bib=029033_10&pagfis=8856.

26. KING OF THE ORCHIDS

1. "A Botanical Sensation," *St. James's Gazette* 23 (September 15, 1891): 5; Frederick Boyle, *About Orchids: A Chat* (London: Chapman & Hall, 1893), 178.

2. W. Micholitz to F. Sander, November 12, 1890, F. Sander Correspondence, Royal Botanic Gardens, Kew (hereafter SC / RBGK), box 3, folder 8, 112; September 22-23, 1891, 142. "I wish to beg you will kindly let me have a small rise on my wages," Micholitz wrote plaintively from Manila, "I am now 7 and a half years in your service and had to commence with £50, and util now observing the strictest honesty, I have not been able to save much. Travelling is not a thing, which can be done without injuring one's health." Micholitz to Sander, July 21, 1889, SC / RBGK, box 3, folder 8, 90-91. Micholitz letters translated by Maren Talbot; see https://www.sandersorchids.com/sanders-orchids/the-orchid-hunters/letters-from-micholitz/.

3. "Dendrobium Phalaenopsis Var. Schroederianum," *Gardeners' Chronicle* 10 (November 29, 1891), 641-642; Micholitz to Sander, December 29, 1891, SC / RBGK, box 3, folder 8, 145.

4. Rudyard Kipling, "The Mark of the Beast," *Life's Handicap, Being Stories of Mine Own People* (London: Macmillan, 1892), 290-307; 306; "Dendrobium Phalaenopsis," 641-642.

5. Boyle, *About Orchids*, 188.

6. J[ane] E[llen] Panton, *Suburban Residences and How to Circumvent Them* (London: Ward & Downey, 1896), 142.

7. L. R. Pinheiro, A. R. C. Rabbani et al., "Genetic Diversity and Population Structure in the Brazilian *Cattleya labiata* (Orchidaceae) Using RAPD and ISSR Markers," *Plant Systematics and Evolution* 298 (2012): 1815-1825; doi: 10.1007/s00606-012-0682-9.

8. "Events of the Past Year," *Orchid Review* 8, no. 85 (1900): 1-4, 2.

9. [Charles] Druery, "Letter to the Editor. An Exhibition of British Ferns, *Times*, issue 33722, August 20, 1892, 12.

10. F. Sander, "Checkmate! The Romance of an Orchid," *Wide World Magazine: An Illustrated Monthly of True Narrative* 13 (1904): 295-298. The story ends with Arnold's tragic and mysterious death "in an open boat": "the cause of his death was never definitely known," explains the text—though, as we've seen, Sander knew perfectly well that Arnold died of a gastrointestinal complaint (298).

11. "Argus," "Dies Orchidiani," *Orchid Review* 17 (1909): 356-358; 356-357. The paper claims the tale was first published in another paper, though I have not been able to find the source. Swinson also reads it as a send-up, one of many "pinpricks in the press" at the time; Arthur Swinson, *Frederick Sander: The Orchid King* (London: Hodder & Stoughton, 1970), 186.

12. "The Orchid," *Play-Pictorial. Containing Six Plays* 4, no. 19 (1904): 21, xvi.

13. Quoted in Swinson, *Frederick Sander*, 213. Swinson fills in this brief picture of Sander & Co.'s decline, esp. 178-239; on the fading of the Veitch business, see Shephard, *Seeds*, 253-277.

14. Edith Wharton, *The Age of Innocence* (New York: Grosset & Dunlap, 1820), 80, 103.

15. Gould, it was said, overspent in just one area: orchids. According to a flower merchant, reported in the *Morning Tribune* in 1896: "Shrewd as Jay Gould was in every branch of finance ... he had no idea whatever of the value of orchids—his great hobby ... Careful, conservative buying by an orchid expert would have gathered [the collection] together at almost a fraction of the sum actually paid out." "Jay Gould's Orchids," *Morning Tribune* 14 (August 9, 1896): 1.

16. The top-of-the-trees van der Luydens grow orchids, too, in *The Age of Innocence*, and spread them around "conspicuously" when they host New York's finest. Wharton, *Innocence*, 210–211, 335.

17. Sander stressed in the introduction that what he called his "plant portraits" would be colored by lithography and / or hand-painted. Though *Reichenbachia* rejects the photograph as a medium for the expression of plant beauty, others embraced the new technology; see Rolf Sachsee, "Flowers Placed in the Medium," 23–28 in *The Art of the Flower: The Floral Still Life from the 17th to the 20th Century*, ed. Hans-Micahel Herzog (Kunsthalle Bielefeld: Edition Stemmle, 1996).

18. "Under the Hammer," *American Garden* 9 (November 1888): 398.

19. "Orchids in Belgium," *Orchid Review* 27 (1919): 22–24, 22; "Sander & Fils, St. Andre, Bruges," *Orchid Review* 26 (1918): 228.

20. Quoted in Swinson, *Orchid King*, 233. (Swinson quotes Fearnley's unpublished manuscript, 229.)

EPILOGUE: FALL—AND RISE

1. Amy Hinsley, Hugo J. de Boer et al., "A Review of the Trade in Orchids and Its Implications for Conservation," *Botanical Journal of the Linnean Society* 186, no. 4 (2018): 435–455; 436 ("Orchids are consistently ranked among the best sellers in the global plotted plant trade," 437); Lakshman Chandra De et al., *Commercial Orchids* (Berlin: De Gruyter, 2015), 14 (on importers and exporters, see 15); Fure-Chyi Chen and Shih-Wen Chin, eds., *The Orchid Genome* (Cham: Springer, 2021), xi; and Shi-Chang Yuan et al., "The Global Orchid Market," in *The Orchid Genome*, 1.

2. "As with flat-panel TVs and laptops, the once-rare orchid has become a mass-market commodity"; Eva Dou, "How the Precious Orchid Got So Cheap," *Wall Street Journal*, October 7, 2013, https://www.wsj.com/articles/SB10001424 052702304330904579137460770908586.

3. *Paphiopedilum rothschildianum* also disappeared from view for decades after Sander and Linden first sourced it, perhaps partly because Sander claimed

it grew in New Guinea. See H. G. Reichenbach, "*Cypripedium rothschildianum*," *Gardeners' Chronicle* 3 (1888): 457; R. A. Rolfe, "*Cypripedium kimballianum*," *Orchid Review* 3 (1895): 238; for contemporary discussion, see Antony van der Ent, Rogier van Vugt, and Simon Wellingo, "Ecology of *Paphiopedilum rothschildianum* at the Type Locality in Kinabalu Park," *Biodiversity and Conservation* 24 no. 7 (2015), doi:10.1007/s10531-015-0881-0.

4. Frederick Boyle, *The Woodlands Orchids Described and Illustrated, with Stories of Orchid-Collecting* (London: Macmillan, 1901), 3-4. These remarks were first published in 1891; Boyle quotes himself in the text.

5. Thomas Richards, *The Commodity Culture of Victorian England: Advertising and Spectacle, 1851-1914* (Stanford: Stanford University Press, 1990), 2.

6. Elizabeth Chang, *Novel Cultivations: Plants in British Literature of the Global Nineteenth Century* (University of Virginia Press, 2019), 12.

7. David L. Roberts and Kingsley W. Dixon, "Orchids," *Current Biology* 18, no. 8 (2008), https://doi.org/10.1016/j.cub.2008.02.026; see L. R. Pinheiro, A. R. C. Rabbani et al., "Genetic Diversity and Population Structure in the Brazilian *Cattleya labiata* (Orchidaceae) Using RAPD and ISSR Markers," *Plant Systematics and Evolution* 298 (2012): 1815-1825.

8. Francesca Carington, "Beauty Breeds Obsession: The Fight to Save Orchids from a Lethal Black Market," *The Guardian*, February 24, 2023, https://www.theguardian.com/lifeandstyle/2023/feb/24/orchid-show-new-york-botanical-garden-black-market; on the impacts of climate change, land clearing, and collection practices on orchids, with a particular focus on how conservationists can respond, see Jenna Wraith and Catherine Pickering, "Quantifying Anthropogenic Threats to Orchids Using the UUCN Red List," *Ambio* 47 (2018): 307-317.

9. W. A. Manda, "Orchids," *Annals of the Minnesota State Horticultural Society* 21 (1892): 369-373; 370.

10. "Notes," *Orchid Review* 7 (1899): 321-322; 322.

11. L. C. Menezes, *Cattleya labiata autumnalis* (Brasilia: Charbel Grafica, 2002), 183.

12. Pinheiro et al., "Genetic Diversity," 1815-1825.

13. Cássio van den Berg, private correspondence with the author, March 5, 2023.

14. Liza Gross, "Climate Change Could Change Rates of Evolution," *PLOS Biology* 9 (2011), https://journals.plos.org/plosbiology/article/figure?id=10.1371/journal.pbio.1001015.g001. As mentioned earlier in the text, L. C. Menezes, Brazilian forestry engineer and orchidologist, notes that to this day, Lindley's coloring from the type plant remains "practically unknown to Brazilian collectors of the species" (Menezes, *Cattleya labiata autumnalis*, 103).

15. Charles Darwin, *On the Origin of Species by Means of Natural Selection, or, The Preservation of Favoured Races in the Struggle for Life* (London: John Murray, 1859), 129–130.

16. "[D]eviation, not truth to type, is the creative principle," summarizes Gillian Beer in *Darwin's Plots: Evolutionary Narrative in Darwin, George Eliot and Nineteenth-Century Fiction* (1983; repr. Cambridge: Cambridge University Press, 2009), 59.

"Darwin evades any suggestion that the world is now accomplished and has reached its final and highest condition," Beer argues (*Darwin's Plots*, 58). On the "emphasis on growth and process rather than on conclusion and confirmation" in Darwin's own intellectual life, see 102, where Beer discusses the naturalist's indebtedness to Wordsworth and Coleridge. Beer also discusses Darwin's rebuttal of the Lamarckian idea of the chain (and the *scala naturae*, the chain or ladder of being) in favor of the tree: "The idea of the great chain places forms of life in fixed positions which are permanent and immobile . . . Darwin needed a metaphor in which degree gives way to change and potential, and in which form changes through time" (*Darwin's Plots*, 33).

ACKNOWLEDGMENTS

I began writing *The Lost Orchid* shortly before the pandemic shut down our lives in March 2020. That I was able to continue with the project at all is due to the kindness of librarians and fellow researchers all over the world. So many went the extra mile to help a stranger in the midst of global tragedy and chaos: I particularly remember that Isabelle Charmantier at the Linnean Society of London snapped photographs of Swainson's letters on her iPhone one evening and sent them to me when the library and almost everything else was locked up and dark. I'll never forget the thrill of Swainson's handwriting appearing on my screen.

Charmantier and her colleague Mark Spencer at the Linnean Society also gave me invaluable help transcribing several early nineteenth-century letters. Sandra Kurzban at the American Orchid Society answered a number of my naive questions and helped me locate useful articles on William Cattley. Jeff Liszka at the Trinity College Library in Hartford has offered unstinting help and support, as always, in finding digitized sources, while Fiona Davison, Nicky Monroe, and many other librarians at the Lindley Library, Royal Horticultural Society, located endless volumes for me and pointed me to resources I would never otherwise have found. Kat Harrington, Alice Nelson, and many others at the Royal Botanic Gardens, Kew, patiently chased down Sander, Hooker, Lindley, and Cattley letters for me over three years. Rosemary Stewart and Gemma Pardue helped me find William Hooker's letters about the sale of the Halesworth house in the Suffolk Archives in Ipswich. Chloé Besombes and the librarians at the Muséum National d'Histoire Naturelle treated me with great kindness and patience on several burning hot days in Paris, while Denis Diagre at the Jardin botanique de Meise hunted enthusiastically with me for Erich Bungeroth—though I was not surprised when that man of mystery eluded us once more.

A number of scholars have helped with translating letters, death notices, baptismal records, and more, especially Gerda Dinwiddie, Bradley Hayes, and Aldair Rodrigues. Aldair's help in navigating resources in Brazil was unstinting and transformative; he guided me to Ovidio Wanderley's family. I am so grateful to Samuel Sales Pinheiro Wanderley for generously sharing photographs and other family documents.

A huge thanks to the Sander family, without whom this book could not possibly have been written. The Sander & Co. materials at Kew form a treasure trove of documents that will yield insights into the mechanisms of global plant hunting for years to come. In particular, I'd like to thank Michael Sander and Peter Sander for kindly responding to my emails and sharing with me pertinent documents and memories. I'd also like to recognize the translations produced by Maren Talbot for Heritage Orchids—available at https://www.sandersorchids.com/sanders-orchids/the-orchid-hunters/—and those prepared in the 1960s by a Mrs. Pike, whose name Peter Sander provided. Pike's annotated typewritten pages rest in the files at Kew.

The death of the humanities is much lamented, but with a wealth of primary documents becoming ever more available at our fingertips, it's also an exciting time to be a scholar in the field. I'd like to pay tribute to all those responsible for digitizing primary resources, making new kinds of scholarship possible, offering access to those resting in George Eliot's "unvisited tombs." A few sites and projects I used repeatedly in this project are: JSTOR's Global Plants (www.plants.jstor.org), the Biodiversity Heritage Library (www.biodiversitylibrary.org), the Internet Archive (www.archive.org), the Darwin Project (www.darwinproject.org), Slave Voyages (https://www.slavevoyages.org), and the Brazilian Digital Library (BN digital, at www.bndigital.bn.gov.br).

Deepest thanks to Cássio van den Berg at the Universidade Estadual de Feira de Santana for answering so many questions about *Cattleya labiata* and for puzzling out Swainson's words with me, to Oscar Alejandro Pérez Escobar at the Royal Botanic Gardens, Kew, for answering my strangled messages, and to Gretel Kiefer at the Chicago Botanic Garden and Nikisha Patel at Trinity College for helping fine-tune my use of botanic terms. In this area, as in all others, mistakes and misunderstandings are entirely my own.

Many other scholars, friends, and colleagues have helped me work through the issues of the book over the last five years. Some asked helpful, probing questions; others read full drafts of the manuscript. All contributed incalculably to the development of my ideas, thus I also recognize, with gratitude: Zarena Aslami, Emily Bazelon, Katherine Bergren, Carolyn Betensky, Alison Booth, Daniel Botsman, Karen Bourrier, Alicia Carroll, Amy Chua, Dennis Denisoff,

Christopher Ferguson, Crystal Feimster, Kate Flint, Beverly Gage, Christopher Hager, Aeron Hunt, Andrea Kaston Tange, Amy M. King, Stefanie Kuduk Weiner, George Levine, Meira Levinson, Marc Lipsitch, Daniel Markovits, Stefanie Markovits, Samuel Moyn, Claire Priest, Jennifer Regan-Lefebvre, Jed Rubenfeld, Ananda Rutherford, Paul Sabin, Talia Schaffer, Marci Shore, Timothy Snyder, Jennifer Tucker, LeeAnne Richardson, Lynn Voskuil, Michael Warner, Tim Watson, Lindsay Wells, Hilary Wyss, and all the many who engaged with my "lost orchid" papers at conferences of the Northeast Victorian Studies Association, North American Victorian Studies Association, Interdisciplinary Nineteenth-Century Studies, the Victorians Institute, and the Modern Languages Association in 2021–2025. Thanks to the NAVSA V-Cologies Caucus for convening activities, panels, and more to foster understanding of environmental issues in the nineteenth century. Thanks also to the two anonymous reviewers of this manuscript at Harvard University Press, who pushed me to explain the stakes of the argument and encouraged me to lay out more clearly my narrative and scholarly approach.

Deep thanks to the National Endowment for the Humanities for funding a Summer Stipend in 2022, which allowed me to do research in London and Paris, and a Fellowship in the spring of 2023, which allowed me to finish writing the manuscript. In addition, I offer thanks to Sonia Cardenas and the Dean's Office at Trinity College in Hartford, Andrew Concatelli and Amy Myers in the Grants Office, and every one (literally) of my wonderful, generous, brilliant, supportive colleagues in the English Department at 115 Vernon Street. Many thanks to Emily Silk, Jillian Quigley, Kate Brick, and all the team at Harvard University Press for their good humor and commitment to the project, to Ian Malcolm for engaging so warmly with my ideas, and to Elizabeth Sheinkman, my agent at Peters Fraser and Dunlop, who first thought there might be a book idea in a tangled lunchtime story about orchids.

Finally, I want to thank my family—Barbara Bilston, who drove with me all around Suffolk in pursuit of William Hooker; Piers Bilston and family; Margaret and Katie Higginbottom; my Markovits in-laws, and Daniel Markovits and our three children, Maisie, Rosa, and Karl. All have been remarkably willing to go on the hunt with me these past five years. This book, and my deepest love, is, as always, for them.

ILLUSTRATION CREDITS

PAGE

27   George Graves, *The Naturalist's Pocket-Book, or Tourist's Companion* (London: Longman, 1817). Smithsonian Libraries and Archives

28   Henry Koster, *Travels in Brazil* (London: Longman, 1816), after 42. Darlington Digital Library, University of Pittsburgh Library System

108 Benjamin Williams, advertisement in the *Gardeners' Chronicle* 17 (1882): 794. University Libraries, University of Massachusetts, Amherst

178 *Godey's Magazine* 134 (1894). Pennsylvania State University Libraries

179 *American Florist* 1 (1885–6): 431. Division of Rare and Manuscript Collections, Cornell University Library

180 *American Florist* 11 (1895–6): 234. University Libraries, University of Massachusetts, Amherst

189 *Wide World Magazine* 13 (1904): 297. Bodleian Libraries, University of Oxford

193 *The Naturalists' Universal Directory* (Boston: S. E. Cassino, 1883), end pages. University of Virginia Libraries

205	"Exposição Universal de Paris: Exposição Brasiliera—Interior da Estufa" (catalog). National Archives of Brazil, BR RJANRIO 02.0.FOT.495/20
209	*Journal des orchidées* 1 (1890–1): after p. 100. LuEsther T. Mertz Library of the New York Botanical Garden
217	Sander Correspondence, Royal Botanic Gardens, Kew, box 1, folder 4, 231. Board of Trustees of the Royal Botanic Gardens, Kew
227	*Gardeners' Chronicle* 10 (1891): 298. University Libraries, University of Massachusetts, Amherst
230	*Gardeners' Chronicle* 10 (1891): 330. University Libraries, University of Massachusetts, Amherst
232	*American Florist* 7 (1891): 365. University Libraries, University of Massachusetts, Amherst
237	*The Orchid Seekers: A Story of Adventure in Borneo* (London: Frederick Warne, 1897). University Libraries, University of North Carolina, Chapel Hill
241	Sander Correspondence, Royal Botanic Gardens, Kew, box 1, folder 4, 258. Board of Trustees of the Royal Botanic Gardens, Kew
259	Percy Ainslie, *The Priceless Orchid: A Story of Adventures in the Forests of Yucutan* (1890). Rare Books and Manuscripts, Library, University Libraries, The Ohio State University
261	Albert Millican, *Travels and Adventures of an Orchid Hunter* (London: Cassell, 1891). John Hay Library, Brown University
265	Sander Correspondence, Royal Botanic Gardens, Kew, box 1, folder 4, 258. Board of Trustees of the Royal Botanic Gardens, Kew
277	*Play-Pictorial* 4, no. 19 (1904), cover. Author's collection
278	*Play-Pictorial* 4, no. 19 (1904): 7. Author's collection
279	*Play-Pictorial* 4, no. 19 (1904): 8. Author's collection
283	*The Garden* 6 (1908): 282. University of California Libraries

INDEX

abolition of enslavement in Brazil, 32, 243, 308n23
About Orchids (Boyle), 189–190, 228
Acton Green (London), Lindley in, 144
adolescence, emerging concept of, 58, 317n20
advertisements: for *Cattleya labiata*, 110, 193, 226–229; for *Cattleya warocqueana*, 208, 209f; in *Gardeners' Chronicle*, 108f, 109, 226–229; of Lebeuf, 193f, 194; of Linden, 229, 230f, 245; in newspapers and handbills, 215, 288; in *The Priceless Orchid*, 235; of Sander, 226–229, 227f, 233, 258, 268, 275; of Siebrecht & Wadley, 179f; in Victorian period, 185, 215, 353n5; of Williams, 108f, 109
Advice To a Young Gentlemen, 93
"Aepyornis Island" (Wells), 271
Aerides, 44, 312nn18–19
African people, enslavement of, 5–6, 23, 31–32, 34, 37, 88, 308nn22–23
Age of Innocence, The (Wharton), 280–282
Agoa Fria (Recife), 35
agriculture: colonialist practices in Brazil, 23, 303n12, 304n17; environmental damage from, 23, 196; labor of enslaved people in, 32, 37; plantations in (*see* plantations); of precolonial tribal communities in Brazil, 21, 22
Ainslie, Percy, 235, 258, 261f
air plants: *Aerides* as, 44; *Coelogyne* as, 67; Veitch offering for sale, 128
air pollution, 105, 129, 291

Alagoas (Brazil), 17, 292; orchid habitat in, 249, 310n43; rebellion and provisional government in (1817), 33, 34
Albert, Prince, 121
Albius, Edmond, 138, 337n9
Alexandra, Queen, 276
Allen, David Elliston, 87, 307n14
alligators, 260
allogamy, 162, 163, 291
"All the Cattleyas Worth Growing," 282
amaryllis, 136
Amazon, Indigenous people of: classification systems on natural world of, 21–22, 303–304n14; pre-colonial land uses of, 21, 303n12
American Florist, 179f, 180f, 229, 268
American Garden, The, 175, 176, 197, 282
American Gardener, 175
Amherstia nobilis, 120, 121
Anderson, Benedict, 111
Andes, 24, 305n23
"Angel in the House" (Patmore), 120
Angraecum sequipedale, 167
Annals de la Sociétié Entomologique de France, 233
Annals of Natural History, 329n3
anthropology, 21, 303n12
anthurium, 249
Antrobus, Edmund, 119
Apothecaries Act (1815), 143
Appleman, Philip, 170
archaeology, 17, 21
Arethusa, 162

Index

Argyll, Duke of, 169, 345n32
aristocracy, 49; art collection in, 243; decline of, 113; hobby of collecting in, 110; horticultural competitions in, 121; newness as concern in, 148; orchid cultivation in, 105, 107, 175; print media in, 215
Arnold, Fritz, 186–193; in Colombia, 186, 188, 262; death of, 192–193, 220, 349–350nn28–30, 361n10; and Sander correspondence, 348n14, 348n18; Swinson on, 348n14, 348n17; in Venezuela, 186, 188, 262, 275; and White, 188–190, 348–349n19; Young & Elliott receiving orchids from, 191, 192
Arnold, Matthew, 170
Artificial system, and Natural system, 59, 66, 317n21, 327n2
Arts and Crafts movement, 274
Aspiring Naturalists (Farber), 307n15
Association for the Protection of Plants, 197
Athenaeum, 115, 322–323n28
auction prices for orchids, 4–5, 122, 206
Audubon, John James, 48
Australia, 51–52
autogamy, 291
awards and medals in horticultural competitions, 115, 118, 119, 129, 133

Bahia (Brazil), 17; Claes in, 253; enslavement of Indigenous people in, 302n4; orchid habitat in, 310n43; rebellion and provisional government in (1817), 33, 34; Swinson in, 37, 47
Balée, William, 303n12
Banks, Joseph: Brown as librarian for, 48, 55–56, 60; cultivation of epiphytes, 44, 312n19; death of, 60, 63, 64, 68, 335n10; on *Endeavour* (ship), 59, 310n48; gatherings at Soho Square, 73; as gentleman scientist, 41, 311n5; as Hooker supporter, 41, 48, 49, 63; in intellectual community, 24; library and herbarium at Soho Square, 55–57; Lindley as assistant for, 55–57, 60, 63–64, 66; on organization of Kew specimens, 59; and Swinson correspondence, 39, 310n48
Banks, Peter, 130
Barbados, 1, 5, 219
Barker, George, 84
Barnet, Cattley at, 63; Lindley position with, 60–61, 62–66; and orchid flowers, 52, 61
Barron, Leonard, 283f
Basil (Collins), 147
Bateman, James: and Darwin, 167; on elite nature of orchids, 103–104, 105; garden of, 328–329n1; on hybridization of orchids, 147; *Orchidaceae of Mexico and Guatemala*, 103, 207, 282; price of orchid sold by, 109; Ure-Skinner employed by, 160
Battersea, Ericsson in, 214
Bauer, Ferdinand, botanical illustrations of, 55, 60
Bedford, Duke of (Lord John Russell), 87, 90, 92, 103
Beer, Gillian, 9, 298n22, 299n23, 339n4
bees, in pollination of orchids, 166–167
begonias, 149, 249
Belgium: Hooker in, 54; Sander business in, 2, 256, 280, 284–285; Sander on failures of, 216; Sander snubbed by King of, 280
Bell, Thomas, 157
Bellier-Beaumont, Ferreol, 138
Bennet, J.J., 156
Benson (Colonel in Burma), 160
Bentham, George, 95, 334n17
Berg, Cássio van den, 292, 310n43, 310n47, 328n13, 363n13
Berger, John, 321n12
Betsey (brig), Swainson travel on, 36, 309n37
Bezerros (Brazil), 36, 39
Bible, and hybridization concerns, 135, 137, 336–337n4
bifoliate plants, 67
birds, 20, 38, 138
Birmingham Botanical and Horticultural Society, 324n41
Black communities in Brazil: artists in, 242, 243, 244; enslaved, 5–6, 23, 31–32, 34, 37, 88, 308nn22–23; guides and assistants to European plant hunters in, 244, 263
Blandy, J. J., 328n10
Bletia, 139
Bligh, William, 42
Bloom (King), 58, 317n19
blue flowers: in Silva painting, 240, 242; symbol of, 200, 351n10; of *Vanda caerulea*, 160
Bolivia, Bungeroth in, 5
Bom Conselho, 252
Booth, James Godfrey, 92
Botanical Cabinet, 67, 83, 84, 323n30
botanical illustrations, 74–85; of *Cattleya labiata*, 73, 74, 77, 81–82, 83–84, 85, 185, 320n1; color in, 78–80, 81–82; of Curtis, 73, 83, 320n1; of Lindley, 63, 65, 185, 282; roots included in, 76; Saunders study of, 77, 321n13; value-laden choices in, 76, 321n12
Botanical Register: on air plants, 44; on *Cattleya* orchids, 73; on hopelessness of

orchids in England, 26; on "Orchis Longicornu" of Swainson, 27, 306n7
botany: Latin language in, 77–78, 83, 89, 322n17; in medical education, 86–87, 143; naming and classification in, 58, 100, 317n19, 328n11
Boyle, Frederick: *About Orchids*, 189–190, 228; on Arnold and White, 189–190; on comfort from orchids, 288; on lost orchid, 69, 182, 268, 273; on orchidomania, 288; *Orchid Seekers, The*, 235–236, 237f; as Sander supporter, 190, 193, 228, 242, 244, 272–273
Boy's Own Paper, 12, 235, 245, 350n30, 357n8
Brackenridge & Co / Rosebank Nurseries (Baltimore MD), 179, 193
Braddon, Mary Elizabeth, 242, 289
Brantlinger, Patrick, 234
Brazil: agriculture in, 21, 22, 23, 196, 303n12, 304n17; and Britain, 29, 31, 32, 70, 87–88, 307n10, 308n18, 324n5; Bungeroth in (*see* Bungeroth, Erich); deforestation in (*see* deforestation); Digance in (*see* Digance, William); diversity of flora and fauna in, 23, 195; drought in, 32, 240, 255; enslaved people in, 31–32, 34, 37, 88 (*see also* enslavement); Ericsson in (*see* Ericsson, Claes); Forget in (*see* Forget, Louis); Gardner in (*see* Gardner, George); geographic variations of orchids in, 291–292; Indigenous peoples in (*see* Indigenous peoples); Koster in (*see* Koster, Henry); Lietze in, 239; Lobb in, 131–133, 182, 336n12; local people working with European plant hunters in, 258–267; lost orchid of (*see* lost orchid); natural resources of, 14, 20, 195–200 (*see also* natural resources); Osmers in, 186, 187; Paris World's Fair (1889) exhibition of, 203–206, 239–240, 352n1; Peace and Friendship Treaty (1810) with Britain, 32; Perthius in, 256–257; and Portugal, 17–20, 29, 32, 33, 70, 87; rebellion and violence in (1817), 31–34; republican idealism in, 32, 33, 308n24; sources of information on life in, 12; Swainson in (*see* Swainson, William); viewed as empty and undiscovered by plant hunters, 88, 89, 90; Waterton in, 88–89, 90
Brazilian duplicates of Swainson, 50–51
Brejo, 250
brewery business of Hooker, 39, 40, 41, 42, 44, 50, 311n9
bribes: Arnold offering, 188; Arnold possibly accepting, 191; Claes offering, 252; for disclosing telegram content, 252; Ericsson offering, 4, 14, 217, 219, 220, 253, 263; for local officials, 14, 131, 219; Veitch asking about, 131
Britain: and Brazil, 29, 31, 32, 70, 87–88, 307n10, 308n18, 324n5; colonialism of, 6; and national-imperial infrastructure supporting plant hunters, 90; and Portugal, 29; Swainson not receiving funding from, 25, 29, 36; trade with Russia, 62–63, 318n2
British Empire, 6–7
British Gardening, 142
British Guiana (Guyana), 88
British Museum: Hooker offered position at, 50, 314–315n16; orchid drawing at, 242; Swainson applying for position at, 48, 50
Broderip, William John, 47
bromeliads, 249
Bromelia fastuosa, 80
Brontë, Charlotte, 319n20
Brown, Robert: on classification systems, 59, 77; Fox engraving painting of, 73; as librarian for Banks, 48, 55–56, 60; and Lindley, 57, 60, 66, 77
Bruges (Belgium), Sander business in, 2, 256, 280, 284–285
Brunner, Jerome, 301n33
Brussels: Ericsson in, 216; L'Horticulture Internationale and Linden family in, 4, 150, 202, 204, 207, 212
Bull, William, 150, 194
Bulletin des séances et bulletin bibliographique de la Sociétié Entomologique de France, 233
Bull's nursery, 116
Bungeroth, Erich, 3–5; card playing by, 223, 224, 263–264; and *Cattleya labiata*, 222, 240, 266, 360n12; and *Cattleya warocqueana*, 208, 210, 216, 217, 221, 244, 254, 264, 359n20; and Claes, 251, 358n8; in Colombia, 5, 217, 218, 219, 220; Ericsson as competitor of, 3–4, 14, 183, 218–224, 236, 251, 254, 263–264; Linden family as employer of, 4, 206, 208, 236, 238, 239, 244, 250, 266; local Brazilians working with, 14, 244, 263; and lost orchid, 3–5, 14, 183, 186; Sander collectors following route of, 220, 233, 240, 250, 251; and Silva painting, 239, 240; Wanderley working with, 264, 265f, 266
Burton, Richard, 234
Byng, George (Lord Torrington), 96

Cabral, Pedro Álvares, 18
cacti, Gardner collecting, 90–91

Index

Caeté tribe in Brazil, 18
caiman, Waterton encounter with, 88–89
Cain, P. J., 297n15, 324n5, 354n8
Calanthe ×dominii, 144, 145
Calanthe furcate, 144
Calanthe masuca, 144
calceolarias, 149
Caldas, Francisco José de, 24, 305n23
Camerini, Jane R., 325n12
Caminha, Pero Vaz de, 18–19, 20, 22, 302n5
Canadian Rockies, 130
Canhotinho, 252, 254, 264, 266
Cañizares-Esguerra, Jorge, 24, 305n23
cannibalism, 18, 19, 20, 302n3
Canning, George, 87–88
Capeira, 249
Capibaribe River, 35
capitalism, 112, 197, 198, 289
Capunga neighborhood of Recife, 29, 35
Caracas, 275
Carmilla (Le Fanu), 75
Carnival celebration, 32
Caruaru, 250
Castle, Lewis, vii, 14, 234, 293, 301n35
Castro, Barbosa de, 33
Catalogue of Plants Contained in the Royal Botanic Garden of Glasgow (Hooker), 68
Catasetum, 64, 162, 206
Catasetum arnoldi, 220
Catasetum bungerothi, 206, 207, 210, 220, 272
Catasetum hookeri, 313n24
Catasetum pileatum, 206–207, 352nn6-8
Catinga, 249
Cattley, John, 62
Cattley, Stephen, 62, 63
Cattley, William: confusion about orchid sent to, 298n17; financial concerns of, 64–66; financial support of *Collectanea*, 64–65, 66, 68, 79; Hooker sending orchid to, 49; and Lindley correspondence, 65; Lindley naming orchid in honor of, 67, 68, 226; Lindley taking position at Barnet with, 60–61, 62–66; offset of orchid blooming in Barnet, 52, 61, 73; in Russian trade, 62–63; showing orchid at Horticultural Society, 69–70
Cattleya, 145, 148, 160–170
Cattleya amethystina, 145
Cattleya autumnalis, 183
Cattleya citrina, 67
Cattleya crispa, 74, 99, 327n1
Cattleya diganceana, 258
Cattleya dolosa, 235

Cattleya ×dominiana, 145, 146–147
Cattleya forbesii, 67, 74, 109, 320n2
Cattleya gaskelliana, 212, 220
Cattleya guttata, 140, 327n1
Cattleya intermedia, 99, 327n6
Cattleya labiata, 310n46; advertisements for, 110, 193, 226–229; alternative names for, 289–290; Arnold hunting for, 186–192; *autumnalis*, 211, 212, 213, 226–229, 290, 305n24, 310n43; *autumnalis vera*, 226–229, 236; botanical illustrations of, 73, 74, 77, 81–82, 83–84, 85, 185, 320n1; and Bungeroth, 222, 240, 266, 36on12; *Cattleya* showing at Horticultural Society, 69–70, 320n28; in *Collectanea Botanica*, 66, 73, 74; color of flowers, 64, 66, 69, 81–82, 185, 291–292; cultivation requirements, 84, 109, 174; and *C. warocqueana*, 211, 245, 290, 292; deforestation affecting, 195, 199; differentiated from other orchids, 97–102, 182–186, 187, 211–212, 213, 231–234, 245, 292; domestication affecting characteristics of, 292; *Floral Cabinet* description of, 81–82; flowering season of, 98, 174, 190, 245, 262, 305n25, 310n47; and Gardner, 92–97, 100, 132–133, 182, 326n23; growth patterns of, 305nn24-25; habitat of, 183, 187, 195, 240, 248–250, 251, 305n24, 310n43; and Hooker, 83, 182, 313n23, 313n25; hybrids of, 140, 145; of Lawrence, 115, 332n1; and Lindley, 64, 67, 68, 69, 84, 183, 226, 323–324n38; Lobb hunting for, 132–133; at Loddiges nursery auction, 122–123; as lost orchid, 67, 68, 69, 182–184, 194, 212, 291–292; naming and classification of, 185, 211, 231–234, 245, 289–290, 292; nursery conflicts over, 131; as old, 147, 149, 150, 245; propagation of, 123; publications increasing interest in, 74, 81–85, 226–231; purchase price of, 73–74, 109–110, 122, 246; Rio Pinto as collection site, 233–234; Sander hunting for, 242; Sander photographs of, 229, 231, 232f; sexual implications of name, 67, 319n18; Silva painting of, 239–244; Swainson collecting, 182, 183, 248; in United States, 175, 282, 290; and Veitch, 123, 131, 132–133; *vera*, 183, 184, 185, 199, 225, 226, 231, 234, 248, 290
Cattleya labiata atropurpurea, 122
Cattleya labiata candida, 100
Cattleya labiata pieta, 100
Cattleya labiata var. *atropurpurea*, 99, 327n7, 328n15
Cattleya labiata var. *mossiae*, 98, 99

Cattleya labiata var. *warocqueana*, 290. *See also Cattleya labiata; Cattleya warocqueana*
Cattleya lobata, 101
Cattleya loddigessii, 67, 69
Cattleya lueddemanniana, 349n22
Cattleya mendelii, 260
Cattleya mossiae, 109, 121, 173, 187; classification and naming of, 98, 99, 100, 327n3
Cattleya percivaliana, 192
Cattleya rex, 254
Cattleya skinneri, 103, 183
Cattleya speciosissima, 187, 349n22
Cattleya walkeriana, 296n3, 354n11
Cattleya warneri, 101, 182, 187
Cattleya warocqueana, 208–213, 240, 290. (*See also Cattleya labiata*); advertisements for, 209f; and Bungeroth, 208, 210, 216, 217, 221, 244, 254, 264, 359n20; and *C. labiata* compared, 245, 292; Digance hunting for, 354n11; and Ericsson, 217, 220, 221, 224–225, 241f, 251–252; of Linden, 229, 231–233; location site of collection, 216, 217, 218, 221; as lost orchid, 211, 244; naming and classification of, 211–212, 231–233, 245, 280, 290, 292; Sander interest in, 214, 216, 217, 220, 221, 225, 245
Cattley family, as merchants in Russian trade, 62–63, 318n3
Catton, Lindley nursery business in, 53, 54, 68
Cavendish, William Spencer (Duke of Devonshire), 103, 120, 129–130
Ceará State (Brazil), 17, 291–292
Cerrado, 21, 303–304n14
Ceylon (Sri Lanka), 96
Chang, Elizabeth, 289
Chapman, Henry ("H.G.C."), 211
charcoal, 96, 133, 199
"Checkmate! The Romance of an Orchid," 189f, 190, 193
Chelsea (London): Bull nursery in, 150; Veitch nursery in, 141, 145–146, 165
Chiswick, Horticultural Society garden at, 333n3
cholera, 26
Christmas Carol (Dickens), 289
chromolithography, 322n21
Cintra (German steamer), 222
Civil War (US), 10, 174, 198
Claes, Florent, 251–253; and Bungeroth, 251, 358n8; disappearance of, 253, 358n17; and Ericsson, 240, 241f, 251, 254, 358n8, 358n10; on flowering time of *Cattleya*, 262; as Linden employee, 240, 244, 251–253; at Paris Exhibition, 240; Wanderley assisting, 264, 265f

classification and naming systems, 57–60. *See also* naming and classification systems
climate change, 291, 293, 363n14
Clyde (Royal Mail steamer), 224
Coddington microscope, 139, 143
Coelogyne, 67
coffee production, 23, 88, 196
Collectanea Botanica (Lindley): botanical Latin in, 77–78; *Bromelia fastuosa* in, 80; *Catasetum* in, 64; *Cattleya labiata* in, 66, 73, 74, 84; Cattley providing financial support to, 64–65, 66, 68, 79; color illustrations in, 79, 80, 282; Curtis illustrations in, 73, 320n1; exotic plants in, 74–75, 76, 80; *Oncidium* in, 64; parts and publication dates of, 319n13; and *Reichenbachia* compared, 207; textual descriptions in, 80
collecting: age of, 111; culture of, 13, 111, 331n22; of plants, 110–114, 115–118; sense of community in, 111, 112
collective authorship, 298n22
Collins, Wilkie, 147–148
Colombia, 4; Arnold in, 186, 188, 262; Bungeroth in, 5, 217, 218, 219, 220; Caldas in, 24; Millican in, 260; Rio Pinto in, 234
Colonial Bank (Trinidad), 192
colonialism, 6, 17–20; adventure stories based on, 271, 276; local people as invisible in, 325n12; marginalization and oppression in, 12; natural resources as limitless in, 195; recovery of Indigenous history of, 302n4
colors: in botanical illustrations, 78–80, 81–82; of *Cattleya labiata*, 64, 66, 69, 81–82, 185, 291–292; classification of plants based on, 97–98, 99, 347n13; Ericsson on, 223; of hybrid orchids, 142, 145; in Silva painting of orchids, 239–240; and symbol of blue flower, 200, 351n10
Columbia River, death of plant hunters on, 130
Commissariat, Swainson job in, 25
commodities: culture of, 289; exploitable, 90; Marx on, 70, 289, 320n30; orchids as, 10, 39, 70, 85, 110, 118, 273, 289, 362n2; in urban world, 199
Companion to the Botanical Magazine, 93
competitions, horticultural: of Horticultural Society, 115–116, 118–119, 120, 121, 129, 334n17; Rollisson's nursery in, 129; Veitch in, 133; women in, 115–116, 118–119, 120, 121, 122
Conrad, Peter, 331n19
conservation movement, 197
consumers: environmental impact of, 197; in middle class, 288; protection laws for, 185

contraception, orchid species used for, 23, 304n21
Cooke, George, 83
correspondence. *See* letter writing
Cottage Gardener, 121
cotton production, 32, 308n23
Cowan, John, 193
crafting, 117, 118; and Arts and Crafts movement, 274
Creation, religious view of, 135, 136, 137, 168–169
Cresswell, Tim, 331n21
Crimean War, 231
Critical Review, 30
Crowned Lion, de Barros Lima known as, 33
cryptogamia, 87, 94
Crystal Palace, 122, 145
Cuba, Roezl accident in, 2, 150
cubes (mahogany boxes), specimens of Banks kept in, 55
cultivation practices: historical, evidence of, 21; for orchids (*see* orchid cultivation)
Curtis, John, 73, 83, 320n1
Curtis, Samuel, 78–79
Curtis's Botanical Magazine, 78–79; on *Catasetum bungerothi*, 206; on *Cattleya labiata*, 84; and Hooker, 93, 98, 327n3; on Lawrence gardens, 117; on Veitch nursery, 141
Customs House (Lower Thames Street), 56

Darwin, Anne Elizabeth (Annie), 154–155
Darwin, Charles, 151–159, 160–170, 184; "Abstract" of, 157, 158; children of, 151–152, 154–155; on evolution, 158, 167, 168, 170, 293, 364n16; family life of, 152, 339n4; and Herbert, 136, 137, 139; and Hooker, 139, 152–159, 160–167, 184, 339n5; and Lindley correspondence, 161; literary interests of, 9, 152, 299n23, 339n4; and Lyell correspondence, 153, 155, 156; Manda comments on, 176; on natural selection (*see* natural selection); *On the Origin of Species*, 157, 163, 164, 168, 171, 184, 293; "On the Tendency of Species to Form Varieties," 342n29; *On the Various Contrivances by Which British and Foreign Orchids are Fertilised by Insects*, 162, 164, 165, 168; on pollination of orchids, 162, 163–164, 165–168, 169–170; *Variation of Animals and Plants under Domestication*, 292; and Veitch (Junior), 165; and Veitch (Senior), 130–131; and Wallace theory, 153, 155, 156–157, 158

Darwin, Charles Waring, 152, 154, 155, 156
Darwin, Elizabeth, 151
Darwin, Emma, 151, 162
Darwin, Francis, 151
Darwin, George, 151
Darwin, Henrietta (Etty), 151, 152, 154, 155, 341n25
Darwin, Horace, 152
Darwin, Leonard, 152
Darwin, Mary Eleanor, 154
Darwin, William Erasmus, 152
Darwin and the Memory of the Human (Schmitt), 300n26
Davallia, 249
Davidson, Kathleen, 117
Davison, Fiona, 115, 332n1
Davy, Stephen, 131
de Barros Lima, José, 33
de Candolle, Augustin, 55, 66, 321n16
deforestation, 14, 195–199; in charcoal production, 96, 133, 199; Ericsson on, 256; Forget on, 248–249; Gardner on, 95–96; and loss of orchids, 182, 195, 273–274; by Portuguese settlers, 22, 23, 304n17
degeneration concern in modern life, 197–198, 199
democratic language on orchids in US, 175–176
Dendrobium phalaenopsis var. *schroderianum*, 268–269, 272
depression and melancholy of Swainson, 37
depression of orchid flowers, 163, 165
Devon: Darwin family in, 342n1; Veitch nursery in, 127, 128, 141, 165
Devonshire, Duke of (Charles Spencer Cavendish), 103, 120, 129–130
diamond districts in Brazil, Gardner on, 90, 324n6
diaries: as genre, 308–309n27; of Swainson, 32–33, 37–38, 309n31, 309nn33–36
Diario de Pernambuco newspaper, 266
Dickens, Charles, 148, 242, 289
Digance, William: in competition with other collectors, 188; death of, 2, 218, 255, 350n30; on difficulties experienced, 218, 247, 248; as discoverer of "new" plants, 17; environmental concerns of, 274; father of, 296n3; hunting for lost orchid, 2–3, 182; letter writing style of, 94; on Linden collectors, 262; local Brazilians working with, 262–263; orchid named after, 258
Digitalium Monographia (Lindley), 68
diphtheria, 154
Disraeli, Benjamin, 151

Dodman (Charles Bellenden Ker), 104–105, 106, 123, 141, 196, 330n5, 330n8, 331n13
domestication syndrome, 292
Dominy, John, 140, 142–146
Douglas, David, 130
"Dover Beach" (Arnold), 170
Downe village (Kent): Darwin home in, 151, 152, 154, 157; epidemic in, 154, 157
Dracula, 199
Drake, Sarah, 84
Drayton, Richard, 306n6
Drayton Green, Lawrence home in, 116
drought: in Brazil, 32, 240, 255; in England, 151
Duggins, Molly, 117
Dürer, Albrecht, 79
Durrell, Gerald, 25
dysentery, 26, 37, 133

East Anglia, 41, 53, 54, 60; Norfolk, 40, 53; Suffolk, 35, 38, 40, 45, 49, 50, 61, 69, 123, 182, 313n22
Easterby-Smith, Sarah, 53, 78, 315n1
East India Warehouses, 56
East Indies, 57
ecology, historical, 21, 303n12
Edinburgh Philosophical Journal, 35
Edward VII, King, 276
Edwardian period, 276, 290
Edwards, Ambra, 298n17
Edwards's Botanical Register, 84
Eiffel Tower, 203
Elephant Moth Dendrobe, 268–269, 272
Eliot, George, 67, 288
Elliott, William, 179
Ellis, Sarah, 120
Ellis, William, 160, 167
Endeavour (ship), Banks on, 59, 310n48
Endersby, Jim: on cultural history of orchids, 306n6; on fertilization of orchids, 169; on lost orchid, 8; on naming and classification systems, 76, 100, 328n11; on orchids in literature, 302n1; on Orchis myth, 75, 321n9; on prices paid for orchids, 296n10; on rostellum as floral chaperone, 163; on social and scientific acceptance, 94, 310–311n3
Enquiry into Plants (Theophrastus), 17
enslavement: abolition in Brazil, 32, 243, 308n23; Africans in, 5–6, 23, 31–32, 34, 37, 38, 308nn22–23; Albius in, 138; Indigenous people in, 20, 23, 302n4; Manoël in, 47; and Moss as slave owner, 327n4; Peace and Friendship Treaty (1810) on, 32; Swainson on, 37

entomology, 30
environment: Caminha on, 20, 22–23; colonial view of, 195, 303n12; deforestation of (*see* deforestation); Digance concerns about, 274; diversity of flora and fauna in, 23, 195; Ericsson on climate change and, 256; Gardner concerns about, 95–96; and historical ecology, 303n12; Indigenous people on features of, 21–22, 303–304n14; natural resources in (*see* natural resources); orchid habitat in, 247–250, 290 (*see also* habitats preferred by orchids); plant hunting affecting, 3, 10, 112, 195–199, 250, 273–274, 290–291, 296–297n14
Epidendrum, 36, 67, 185
Epidendrum cuspidatum, 129
Epidendrum violaceum, 67, 69
epiphytes: Banks cultivation of, 44, 312n19; Gardner collection of, 94–95; growth habit of, 24; species recorded at Kew, 42; viewed as parasitic plants, 36
Erasmus, 79
erasure narratives, 12, 88
Erickson, Clark L., 303n12
Ericsson, Claes, 251–257, 263–264; anger with Sander, 254; arrival in Brazil (1891), 1, 207; Bungeroth as competitor of, 3–4, 14, 183, 200, 218–224, 236, 251, 253, 254, 263–264; and *Cattleya warocqueana*, 217, 220, 221, 224–225, 241f, 251–252; and Claes (Florent), 240, 241f, 251, 254, 354n8, 358n10; on codes in messages, 253; competitive approach of, 3, 188; as discoverer of "new plants," 17; health issues of, 2; and Kromer, 253, 358–359n17; legal problems of, 264; letter writing style of, 9, 11, 94, 208, 250; Linden comments on, 236; and local Brazilians, 221–222, 250, 263–264, 359–360n27; number of plants collected by, 221, 222, 224; and Oversluys, 250; and Palmer, 218; postcard to Sander, 216, 217f; in pursuit of lost orchid, 1–5, 14, 183; requesting job from Sander (1886), 214; Rio Pinto plants collected by, 234; shipping concerns of, 222, 223; at St. Albans nursery of Sander, 216, 218, 354n12; in Sweden, 214, 216; Swinson reports on, 359–360n27; Wanderley mentioned by, 264, 265f
Ernst, A., 190
"Erreurs géographiques concernant les orchidées" (Forget), 247
Etiquette for Gentlemen, 93
evolution, Darwin on, 158, 167, 168, 170, 293, 364n16

Index   377

Exeter: Harris in, 140, 142; Veitch nursery in, 128, 129, 133, 140, 146, 165
exotic, use of term, 74, 75
Exotic Conchology (Swainson), 47
Exotic Flora (Hooker), 69, 83
exotic locale, Orient as, 74–75
Exotic Nursery, 161
exotic plants, 287; benefits of viewing, 82, 83; in *Collectanea Botanica*, 74–75, 76, 80; orchids as, 289–290; Veitch offering for sale, 128; and xenophobia, 274
exotic women, 75, 320–321n6
extinction threats, 197, 291

Farber, Paul, 30, 307n15
Feira de Santana, 37
ferns, 105, 274
fertilization of orchids, 162, 163–164, 165–168, 169–170
fish, Swainson descriptions of, 38
Fitch, Walter Hood, 99
Fitzwilliam, Earl, 84, 103
Flint, Kate, 81, 323n31
Floral Cabinet, 81–82, 322n28
Flor de Guaresma, 262, 290
Floricultural Cabinet, 73, 133
Florist's Exchange, 229
Flor Natal, 262, 290
Flower Garden, 38
"Flowering of the Strange Orchid, The" (Wells), 71, 76
flowering season: of *Cattleya labiata*, 98, 174, 190, 245, 262, 305n25, 310n47; Ericsson on, 222, 223; local knowledge on, 262; of lost orchid, 24, 36, 39, 64
"Flower of Paradox, The" (McCulloch-Williams), 177
"Flower of the Air" (Waddel), 177
forests: classification systems of Indigenous people on, 21–22, 303–304n14; deforestation of (*see* deforestation)
Forget, Louis, 247–251, 255; arrival in Brazil, 246, 250; and *Cattleya labiata*, 247, 249, 256; and *Cattleya warocqueana*, 252; and Claes, 251, 253; death of, 284; disappearance of, 256; on lobed cattleya, 101; on orchid habitat, 247–250, 290; rivals of, 280; on scarcity of orchids, 274; Wanderley mentioned by, 264, 266–267
"Fortune from Orchids, A," 176
fossils, orchid, 23, 304n19
Fösterman, J., 180
Fox, Charles, 73
Frankenstein (Shelley), 125, 134, 142

freemasons lodges in Recife, 32, 33
French Revolution, 110, 203
Friends' Intelligencer and Journal, 177
Fulham, 149
fungi, symbiotic relationship of orchids with, 290, 291

Galahad, 14, 200, 357n11
Galapagos, Darwin in, 156
Garanhuns (Brazil), 221, 223
Garden, The, 183, 211, 212, 283f
Garden and Forest, 183
Garden and Home Builder, 282
Gardeners' Chronicle: advertisements in, 108f, 109, 161, 226, 227f, 230f; on Arnold orchid, 190; on *Catasetum*, 206, 207; on *Cattleya warocqueana*, 208, 211; Darwin communicating with readers of, 164; Dodman writing in, 104–106, 141; on hybrid orchids, 144–145, 146, 147; Linden writing in, 231; Lindley writing in, 99; on lobed cattleya, 99; Moore writing in, 139, 143; on Moreau, 233; orchid identification in, 192; on orchid seeds, 138, 139–140; "Orchids for the Million" articles in, 104, 106, 140; on Parker and Williams nursery, 161; Rolfe writing in, 244–245; Sander narratives in, 233, 234, 269, 270, 271, 272; on Veitch nursery, 141
Gardener's Magazine, 78, 121, 211, 334n17
Gardener's Monthly, 175, 177, 193
gardening: colorful plants and patterns in, 142, 149; increasing accessibility of information on, 30, 107, 307n16; middle class interest in, 107, 109; native plants in, 149, 274; in suburbs, 107, 109, 148, 149, 330n12; in United States, 174–175; by women, 115–118, 121, 332n1
Gardening Illustrated, 197
Gardening World, 211
Garden Oracle (Hibberd), 173
Gardner, George, 86–96; on Brazil as empty and undiscovered, 88, 89, 90; and *Cattleya labiata*, 92–97, 100, 132–133, 182, 326n23; in Ceylon, 96; childhood of, 86; death of, 96; environmental concerns of, 95–96; and Hooker, 86, 87, 90, 91, 92, 93–95, 132, 202; medical training of, 86–87, 91, 326n18; on number of specimens collected, 94; *Pocket Herbarium of British Mosses*, 87; sense of entitlement, 91, 325–326n17; *Travels in the Interior of Brazil*, 90, 97; and Veitch, 132–133
gatekeeping practices, 112–113, 118, 121
General Plantarum (Jussieu), 60

gentlemen: letter writing skills needed by, 93, 326n26; plant collecting by, 110; as scientists, 41, 311n5
George III, King, 41
geraniums, 142, 149
germination of orchid seeds, 144
Ghent, L'Horticulture Internationale and Linden family in, 150
Gilded Age, 281, 282
Gladstone, William, 151
Glasgow, Gardner living in, 86–87
Glasgow Botanic Garden: Gardner sending plants to, 90, 92; Hooker catalog of plants in, 68; offset of orchid blooming in, 52
Glasgow Medical Society, 86
Glasgow University, 48, 49–50, 87
glass for greenhouses, 42, 105
globalization, 5–6, 7
Godey's Lady's Book, 177
Godey's Magazine, 177, 178f
Godseff, Joseph: business approach of, 246; and Digance, 3, 218; and Ericsson, 2, 222, 223, 250, 252; honesty of, 280; and Moreau, 234; narratives on orchid hunting, 235, 268, 270, 272, 275, 355–356n5
Goethe, Johann Wolfgang von, 79
Gold Banksian award, 118, 129
Golden Jubilee, 1, 174
Gold Knightian award, 119
gold mining, 196
Gold of Kinabalu orchid, 287, 362–363n3
Goodno, Anna Nichols, 177
Gould, Helen, 177
Gould, Jay, 177, 281, 362n15
Gower, William Hugh, 183, 212
Graphic, 273
Gravatá, 36, 39
Gray, Asa, 170
Great Exhibition of London (1851), 288
Great North Road, Cattley home on, 63
Great Reform Act (1832), 127
Great Stink in London, 151, 156
Greece, Swainson in, 25, 316n9
greenhouses: glass for, 42, 105; heating of, 42–43, 44, 69, 105, 312n12, 312–313n22; of Hooker, 44, 69; and hothouses, 44, 312–313n22; of Lawrence, 116; for orchids, 42–43, 44, 105, 107, 109, 331n13; portable, Wardian case as, 105; of Sander, 180f; in United States, 175
Grey, Thomas (Lord Grey of Groby), 84, 324n41
Grey, Zane, 198
Griffiths, Devin, 164, 165, 166, 298n22, 299n23

Groby, Lord Grey of (Thomas Grey), 84, 324n41
Grove, Richard, 306n6
growing orchids. *See* orchid cultivation
growth habits: epiphytic and lithophytic, 24; Ericsson on, 223; of lost orchid, 24, 38; of parasitic plants, 42; Swainson lack of interest in, 248
Guatemala, Bateman on orchids of Mexico and, 103, 207, 282
Guyana (British Guiana), Waterton travel in, 88
Guzmán, Tracy, 21, 303n13

habitats preferred by orchids, 18; for *Cattleya labiata*, 183, 187, 195, 240, 248–250, 251, 305n24, 310n43; damage to, 195–199, 248–249, 273–274, 290–291; Ericsson on, 223; Forget on, 247–250, 251, 290; for lost orchid, 24, 199, 244, 249; and relocation of orchids by Indigenous people, 23, 304n21
Hackney, Loddiges & Sons nursery in, 67, 122, 129
Hager, Christopher, 10, 11, 300n28
Haggard, H. Rider, 199, 234
Halesworth: Hooker living in, 54, 62, 68, 69; and Suffolk, 35, 38, 40, 45, 49, 50, 61, 69, 123, 182, 313n22
Hall, G. Stanley, 317n20
Hancock, Michael W., 331n18, 331n20
Handy, Ellen, 231
Hanne, Michael, 9, 299n24
Harris, John, 140, 142–143
Harris, John Delpratt, 338n8
Harris, John William, 142–143, 338n8
Hartweg, Theodore, 131
Harvard Botanic Gardens, 175
Haughton (Professor from Dublin), 157
Henderson, Peter, 179
Henderson & Co., 179
Hendrick, Vincent F., 215
Heraclitus, vii
herbaria: of academic botanists, 100; of Banks, 55, 63, 66, 316nn10–11; classification and collection practices in, 332n24; of Hooker, 45, 292, 319n13, 320n24; of Kew, 45, 68, 292; of orchid collectors, 112; of Reichenbach, 207; shipping of specimens to, 7; of Swainson, 38, 45
Herbert, William, 135–137, 139; and Darwin, 136, 137, 139, 153; death of, 135, 139; on diversity of orchids, 162; on plasticity of species, 152, 153, 162–163
Hey, Jody, 185
Hibberd, Shirley, 173

Index

Himalayas, 144
historical ecology, 21, 303n12
"History of Second Discovery" (Linden), 236, 239
History of the Orchid, A (Reinikka), 7, 298n17
Holloway, Victoria and Paradise nurseries in, 150
Hood, Samuel, 91
Hooker, Joseph Dalton: and Darwin, 139, 152–159, 160–167, 184, 339n5; descriptions of father, 49, 50, 310n1, 314n14; at Kew Royal Botanic Gardens, 152; and Veitch family, 342n2
Hooker, William Jackson, 35, 40–45; Banks support of, 41, 48, 49, 63; botany interests of, 40, 42, 87; brewery business of, 39, 40, 41, 42, 44, 50, 311n9; *Catalogue of Plants Contained in the Royal Botanic Garden of Glasgow*, 68; and *Cattleya labiata*, 83, 182, 313n23, 313n25; Cattley receiving orchid from, 49; as *Curtis's Botanical Magazine* editor, 93; *Exotic Flora*, 69, 83; financial concerns of, 41, 42, 50, 311n9; and Gardner, 86, 87, 90, 91, 92, 93–95, 132, 202; at Glasgow University, 48, 49–50, 87; greenhouse of, 44, 69; in Halesworth, 54, 62, 68, 69; in Iceland plant-hunting expedition, 41, 311n7; identifying plants collected, 130; at Kew Royal Botanic Gardens, 130; on Lawrence gardens, 117; and Lindley correspondence, 54–55, 56, 57, 60, 62, 63, 65–66; and lost orchid, 43–44, 50, 54–55, 68; on naming and classification of orchids, 68–69, 98; offered position at British Museum, 50, 314–315n16; orchids cultivated by, 45; physique and appearance of, 40, 310n1; requesting Brazilian duplicates from Swainson, 50–51; Smith as supporter of, 41, 311n4; and Smith correspondence, 50, 312n20, 314n14; and Swainson correspondence, 50–51, 314n16; on Swainson emigration decision, 51–52; Swainson sending specimens to, 38, 40, 42, 43–44, 45, 182; and Veitch correspondence, 130, 131–132, 336n12; Veitch nursery visit, 141
Hopkins, A. G., 297n15, 324n5, 354n8
Horticultural Advertiser, 193
Horticultural Society, 112, 333n3; admission of women to, 118, 333n8; Cattley showing *C. labiata* at, 69–70, 320n28; Cavendish in, 129; complaints on judging of exhibitions, 118–119, 334n17; Hartweg as collector for, 131; Lawrence gardens compared to gardens of, 117; Lawrence in, 115–116, 118–119, 120, 121, 332n1; Lindley as secretary of, 119; plant competitions of, 115–116, 118–119, 120, 121, 129, 334n17; prices in orchid sale of, 122; and Royal Horticultural Society, 117, 147, 208, 210, 282–283; Sander awarded by, 272; on Veitch specimen of *C. labiata*, 133
Horticulture Internationale. *See* L'Horticulture Internationale
Hortus Veitchii (Veitch), 131, 137, 142, 143, 335n12
Horwood, Catherine, 115, 332n1
hothouses, 44, 312–313n22. *See also* greenhouses
Humboldt, Alexander von, 24, 305n23
Humildes, 37
hybridization of orchids, 123, 134, 135–140; Bateman on, 147; Herbert on, 135–137, 139, 336–337n4; price of plants in, 146–147; religious issues in, 135, 136, 137, 336–337n4; secrecy about, 146; sterility concerns in, 137, 142, 184; by Veitch nursery, 142–150, 160

Iceland, plant hunting expedition in, 41, 311n7
Illustrated London News, 122
illustrations, botanical. *See* botanical illustrations
immigration, 5–6, 296n13
Imperial Academy of Fine Arts (Brazil), 242, 243
In a Glass Darkly (Le Fanu), 272
India, orchid cultivation in, 174
Indian Gardener, 174
Indigenous peoples, 17–23; Caminha reports on, 18–19; classification systems on natural world, 21–22, 303–304n14; death rate in colonial times, 23; devotional practices of, 18–19; enslavement of, 20, 23, 302n4; and erasure narrative, 88; Gardner on, 89, 90–91; interactions with land, 20–22, 303n12; as invisible, 325n12; Micholitz encounters with, 269, 270; modern movements and self-representation, 302n4; orchid species used and grown by, 23, 304n21; population at time of first European contact, 303n11; recovery of history, 302n4; resistance to plant hunters, 91; response to colonialism, 18, 19, 302n4, 304n20; rock art of, 17, 302n2; seizures of lands of, 20; stereotypes on, 20; supposed cannibalism by, 18, 19, 20, 302n3; Vespucci reports on, 19–20, 302–303nn6–7
Indonesia, Micholitz in, 270
inductive method, deviation of Darwin from, 159, 342n29

industrial growth and mass production, 287–288
"informal empire," 6, 297n15
infrastructure, national-imperial, supporting plant collecting, 90
Ingalls' Home and Art Magazine, 177
insects: as pests of orchids, 173; in pollination of orchids, 162, 163–164, 165–167, 290; specimens collected by Swainson, 38
intellectuals, 23, 24, 305n23
Introduction to the Natural System of Botany, An (Lindley), 66

Jameson, Robert, 35
Jane Eyre (Brontë), 319n20
Java, 133
Jekyll, Gertrude, 121
Joao VI, King of Portugal, 32
Josephine, Empress, 122
Journal de Recife, 266
Journal des orchidées, 202, 208, 209f, 210, 212, 218, 231, 239
Journal of Horticulture and Cottage Gardener, 129, 211, 234
Journal of the Horticultural Society, 92, 96, 97
Journal of the Linnean Society, 157
Jussieu, Antoine-Laurent de, 60, 66

Ka'apor language, 21, 304n14
Kameba, Marcia Wayna, 15
Kayapó tribe, 21, 23, 304n14, 304n21
Keats, John, 82
Ker, Charles Bellenden (Dodman), 104–105, 106, 123, 141, 196, 330n5, 330n8, 331n13
Kew, Royal Botanic Gardens at: archive of documents in, 8, 186; *Cattleya labiata* in, 212; decline in government support of, 131, 335n10; epiphyte species in, 42; herbarium specimens in, 292; Hooker (Joseph) at, 152; Hooker (William) at, 130; Rolfe at, 208, 244
Killerton (Devon), Veitch & Sons nursery in, 127
King, Amy M., 58, 78, 317n19
Kipling, Rudyard, 199, 234, 271, 272
Kircher, Athanasius, 138
Kirton, W. F., 192
Knightian medals, 115, 119, 133
Knowles & Co, 222, 354–355n23
Koster, Henry, 27–29, 33; and Swainson, 27–29, 33, 88, 307n10, 307n13; on tensions in Recife, 32; *Travels in Brazil*, 27–29
Kromer, Edward, 253–254, 255, 358–359n17, 359n22

labellum of orchids, 166
Ladies Companion, The, 122
Ladies' Home Journal, 177
Lady Audley's Secret (Braddon), 242
Lady's Friend, 177
Laelia, 328n8, 328n13
Laelia lobata, 101
Lang, Andrew, 198–199
Langfur, Hal, 18, 20, 302nn3–4
Langsdorff, Georg Heinrich von, 23–24, 36, 305n22
La Plata (Royal Mail steamer), 246
Larat (Indonesia), Micholitz in, 270
Lathams Dinner Pills, 35
Latin language: botanical, 77–78, 83, 89, 322n17; Hooker speech in, 50
Lawrence, Louisa, 115–121, 122, 333n3; death of, 119; garden grotto of, 116–118
Lawrence, Trevor, 334n10
Lebeuf, Godefroy, 193f, 194
Le Fanu, Sheridan, 75, 199, 272
Le jardin, 247, 250
Lemon, Charles, 131
Lenten Flower, 262
Leopold, King of Belgium, 280
letter writing, 10–12; employer expectations on, 11; by gentlemen, etiquette guides on, 93, 326n26; as literary form, 10; and travel time for letters, 188–189, 348n18
Levine, George, 9, 298n22, 299n23
L'Exposition Universelle. *See* Paris World's Fair
L'Horticulture Internationale, 4, 150, 202; Bungeroth as plant hunter for, 4, 206, 208, 238; *Cattleya warocqueana* shown by, 208, 209f, 210; curation of Brazilian display at Paris World's Fair, 204–206, 239–240; Kew herbarium specimen from, 292; Sander as rival of, 202–213, 229–233, 236–238, 240, 245–246, 251–256, 263
lichens, 38, 249
Lietze, Adolpho, 239, 356n1
Liger, Louis, 75–76
L'illustration horticole, 202, 212, 228, 245
Lima, José de Barros, 33
Limnocharis commersonii, 89
Linden, Jean Jules, 202, 220; and *Cattleya labiata autumnalis*, 212–213; in hunt for lost orchid, 150; at Paris World's Fair, 239, 240; Veitch offering plants collected by, 160
Linden, Lucien, 202, 236–238; and *Catasetum bungerothi*, 206, 210; and *Cattleya labiata autumnalis*, 212–213, 236; and *Cattleya warocqueana*, 208, 210; on Ericsson, 236; "History of Second Discovery," 236, 239;

Index 381

Linden, Lucien *(continued)*
   in hunt for lost orchid, 150; Sander as rival of, 206, 207, 212, 236; as secretary of L'Orchidéenne, 208; on Silva painting of orchid, 239–244
Linden family: and *Cattleya labiata autumnalis*, 212–213; and *Cattleya warocqueana*, 208–213, 216, 229, 231–233; L'Horticulture Internationale business of *(see* L'Horticulture Internationale); Sander as rival of, 202–213, 229, 233, 236–238, 240, 245–246, 251–256, 263; sending postcard to Sander, 206, 352n5
Linden family plant hunters. *See specific plant hunters.*: Bungeroth, 4; Claes, 240, 244, 251–253; Kromer, 253–254; Wanderley, 266
Lindenia journal, 202, 204, 206, 210, 252
Lindley, George, 53, 54, 60
Lindley, John, 53–61; *An Introduction to the Natural System of Botany*, 66; at Banks library and herbarium, 55–57, 60, 63, 66; botanical illustrations of, 63, 65, 185, 282; and Brown, 57, 60, 66, 77; and *Cattleya labiata*, 64, 67, 68, 69, 84, 183, 226, 323–324n38; Cattley position in Barnet for, 60–61, 62–66; *Collectanea Botanica* of *(see Collectanea Botanica*); and Darwin, 161, 166; death of, 190; *Digitalium Monographia*, 68; education of, 54; expertise of, compared to Hooker, 49, 314n10; and Hooker correspondence, 54–55, 56, 57, 60, 62, 63, 65–66; as Horticultural Society secretary, 119; identifying plants collected, 130; language skills of, 54; in Linnean Society, 63; and lost orchid, 54–55, 64, 67, 68, 69, 313n24; on naming and classification systems, 66–69, 77, 98, 99–100; Old Antiquity as nickname of, 54; and Paxton collaboration, 38; Veitch visiting, 144
Lindley, Mary, 53
Lindley, Nathaniel, 316n4
Lindleyana, 328n13
Linnaean classification system, 58–59, 66, 76, 77, 83, 110, 317n19
Linnaeus, Carl, 58–59; influence of, 59, 60, 317n22, 317–318n23; on number of species, 152; *Systema Naturae*, 58
Linnean Society: Brown attending dinner at, 55–56; correspondence in library of, 26, 306n5; Curtis in, 73; Darwin and Wallace theories presented to, 156, 157, 158, 341n20; and growth of professional societies, 112; Harris in, 143; Hooker in, 41; Lindley in, 63; Smith in, 27; Swainson in, 30, 35, 306n7
Lisbon, British use of ports in, 29
Lissochilus speciosus, 319n12
Literary Gazette, 322–323n28
lithophytic growth habit, 24
Liverpool Botanic Garden, 26–27, 306n5
Liverpool Customs Office, 25
Liverpool Mercury, 151
Lloyd's List, 35
Lobb, Thomas, 133, 160
Lobb, William, 131–133, 160, 182, 336n12
lobed cattleya, 99, 101
local Brazilians, 258–267; as artists, 242, 243, 244; Black, 244, 263; botanical knowledge of, 262, 263; bribes to, 4, 14, 131, 188, 217, 219, 252, 263; as guides and assistants for European plant hunters, 4, 14, 188, 221–222, 252, 258–267; as hotel keeper, 262–263; identification of plants by, 262; Indigenous *(see* Indigenous peoples); Wanderley, 264–276
Loddiges, Conrad, 67
Loddiges, George, 83
Loddiges & Sons nursery: auction of orchid collection, 122–123; *Botanical Cabinet* magazine, 67, 83, 323n30; *Cattleya loddigessii* named in honor of, 67; closing of, 122–123, 146, 150; *Epidendrum cuspidatum* imported by, 129; *Epidendrum violaceum* imported by, 67; lobed cattleya imported by, 99, 101; plant hunters funded by, 130; Veitch purchasing collection of, 146
London: air pollution in, 129; Banks in, 55, 56; *Cattleya* offered by nurseries in, 74; Cattley as merchant in, 49; Curtis in, 73; Customs House in, 56; Ericsson in, 214; Golden Jubilee in, 1; Great Exhibition (1851) in, 288; Great Stink in, 151, 156; Horticultural Society in *(see* Horticultural Society); Lawrence in, 116; Lindley in, 55, 56, 57, 66, 68, 73, 98, 144; Linnean Society in, 26, 30; Loddiges auction in, 122; Sander in, 1, 220, 221, 226, 245, 254; stagecoach route from, 63; suburbs of, 52, 129, 150, 151; Veitch nursery in, 141, 145, 150, 280; weather of, 133, 151; Williams in, 106, 150
London, Jack, 198
Longman's, 182
L'Orchidéenne society, 202, 208, 212
L'orchidophile, 183, 239
lost orchid: Boyle on, 69, 182, 268, 273; *Cattleya labiata* as, 67, 68, 69, 182–184, 194, 212, 291–292; *Cattleya labiata autumnalis* as, 226–229; *Cattleya warocqueana* as, 211,

244; claims on disappearance and rediscovery of, 7–8, 298n18; colors of, 64, 66, 69; compared to symbol of blue flower, 351n10; Digance hunting for, 2–3, 182; Ericsson hunting for, 1–5, 14, 183; evidence on rediscovery of, 291–293; flowering season of, 24, 36, 39, 64; great discovery of, 226–238; growth habits of, 24, 38; habitat of, 24, 199, 244, 249; and Hooker, 43–44, 50, 54–55, 68; idea of, in modern life, 199–201; legends and myths on, 69; Linden family advertisements on, 229; and Lindley, 54–55, 64, 67, 68, 69; publications increasing interest in, 7; as "Queen of the Orchids," 4; Sander advertisements on, 226–229; Sander on discovery of, 244–245; second great discovery of, 239–246; in Swainson collection of parasitic plants, 38–39, 43–44; Swainson not commenting on, 45, 46–47, 50, 69; written reports on, 7

Loudon, Jane, 122, 333n3

Loudon, John Claudius, 78, 107, 109, 116, 122, 330n12, 333n3

Low, Hugh, 150, 191

Low's nursery: advertisements of, 193; and *Cattleya labiata*, 150; and *Cattleya speciosissima*, 187; in growing orchid business, 116, 129; plant hunters in Brazil, 187–188, 189–190, 191; purple orchid introduced by, 99; sending orchids to US, 177

Lyceum of Arts and Crafts (Rio de Janeiro), 242

Lyell, Charles, 153, 155, 156, 157, 158

macaws, 38

Mackay, David, 331n23

Macleay, Alexander, 47, 48, 313–314n6

Madagascar, 57, 160, 271, 350n30

Magazine of Exotic Botany, 81, 322n28

Magazine of Natural History and Journal of Zoology, Botany, Mineralogy, Geology, and Meteorology, 329n3

Magazine of Zoology and Botany, 329n3

magazines. *See specific magazines*.: advertisements in, 83, 215, 229; botanical illustrations in, 69, 78–79, 81, 186; on *Cattleya labiata*, 69–70, 74, 83–84, 95, 212; on collecting, 111; community of readers of, 11; on conservation efforts, 197; development and influence of, 329n3; horticultural, 9, 138, 208; in India, 174; of Linden family, 202; list of worldwide publications, 329n3; for middle class, 9, 106, 164; orchid cultivation information in, 138, 174; orchid hunting stories in, 93, 235, 236; on orchid hybrids, 140; in United States, 174, 175, 191; on World War I losses, 284

malaria, 26

Malaysia, Gold of Kinabalu orchid from, 287

Malory, Thomas, 246

Malta, Swainson in, 25, 26

Manda, W. A., 175, 176, 179

mangel wurzel, 53

Manoël (enslaved boy), 47

Manual of Orchidaceous Plants (Veitch), 131, 132–133; on Lobb, 335–336nn12; on logging in Brazil, 182; on orchid hybridization, 143–144; sections and bindings of, 296n8, 336n13

Manuel I, King of Portugal, 18–19

Maracaibo (Venezuela), 219

Margaret and Anne (ship), sinking of, 41, 311n7

"Mark of the Beast, The" (Kipling), 271

Markovits, Daniel, 113

Markovits, Stefanie, 351n10

Marx, Karl, 70, 289, 320n30

Masdevallia tovarensis, 187

Mason, Peter, 75

Masonic lodges in Brazil, 33

Massachusetts Horticultural Society, 174

Matses people, 21

McCulloch-Williams, Martha, 177

McKenzie, Charles, 99

Mēbêngôkre tribe, 22

medals and awards in horticultural competitions, 115, 118, 119, 129, 133

medical education: botany in, 86–87, 143; of Gardner, 86–87, 91, 326n18

medical uses of orchids, 23, 304n21

Medici, Lorenzo di, 20, 303n7

Mellorchis caribea, 304n19

Menezes, L. C., 185, 305nn24–25, 310n43

merit, recognition of, 112–113

Mexico, Bateman on orchids of Guatemala and, 103, 207, 282

Micholitz, Wilhelm, 269–270, 271, 280, 361n2; fictionalized letter of, 270, 271, 272

middle class: anxiety in, 148; as collectors, 104–105, 110–111; as consumers, 13, 81, 95, 102, 288, 322n28; as gardeners and horticulturalists, 106–107, 174, 273, 290, 317n22; as gatekeepers, 112–113; income in, 109–110, 146; as the "million," 9, 104, 105; orchid cultivation by, 104–110; pastimes of women in, 117; print media for, 9, 106–107, 164; social mobility in, 144; suburban expansion, 52, 63, 107, 128–129, 288, 318n5, 335n2; values and belief systems in, 113, 148; as writers and readers, 11, 49, 81

Miers, John, 90
Millican, Albert, 259f, 260, 273–274
"million," the, 288; middle class as, 9, 104, 105; orchids for, 104, 105, 106, 109, 122, 140, 194, 196, 290, 330n8
Mill on the Floss, The (Eliot), 288
Minas Gerais, 93, 218, 235
mining, 23, 196, 304n17
Minnesota State Horticultural Society, 291
modern life: degeneration as concern in, 197–198, 199; environmental concerns in, 197, 199; idea of hunt for lost orchid in, 199–201
monkey-puzzle tree, 133
Monthly Magazine, 122
Montrose, Duke of, 48
Moore, David, 138–139, 140, 143
More, Thomas, 88
Moreau, E., 8, 233, 236
Morte d'Arthur (Malory), 246
Moscow, in British-Russian trade, 62
Moselle (steamship), 188, 189, 348nn15
Moss, Hannah, 98, 121, 327n3
Moss, John, 327n4; as slave owner, 327n4 (*see also* enslavement)
mosses, 87
moths, in pollination of orchids, 167, 344n25
Mount Radford, Veitch purchase of land in, 128
Mount Roraima, 242
Mozambique, 37
Murray, Stewart, 92
Muscologia Britannia (Hooker), 311n9

names in nineteenth-century novels, importance of, 68, 319n20
naming and classification systems, 57–60, 66–69, 306n7; in botany, 58, 100, 317n19, 328n11; *Cattleya* in, 98–102, 185, 211–212, 231–234, 245, 280, 289–290, 292; color of flowers as basis of, 97–98, 99, 347n13; Hooker on, 68–69, 98; of Indigenous people, 21–22, 303–304n14; Latin language in, 77–78, 83, 89, 322n17; Lindley on, 66–69, 77, 98, 99–100; Linnaean, 58–59, 66, 76, 77, 83, 110, 317n19; of local Brazilians, 262; Natural System in (*see* Natural System); in plant collecting, 110–111; in plant hunting, 186, 188, 190; Quinarian method in, 51; Reichenbach expertise in, 206
Napoleon, 29
Napoleonic Wars, 25, 39, 110
narratives and storytelling, 9–10, 270–276; on *Dendrobium* discovery, 268–269; "fake news" in, 355–356n5; fictionalized letter of Micholitz in, 270, 271, 272; Hanne on, 9, 299n24; historical analysis of, 13, 299n24, 300–301nn32–33; on Indigenous tribes, 18; by Linden, 236–238; in "Orchid" play, 276; on role of local people, 258–262, 270; by Sander, 234–236, 245, 268–269, 270, 272, 274–276, 355–356n5; use of terms, 299n24
Nascimento, Luiza Guilhermina do, 264
native plants in wild gardens, 149, 274
natural history as emerging field, 30, 36, 38, 55, 136, 307nn14–15
Natural History Museum, herbarium of Banks in, 55
naturalists, 7, 9; in Brazil, 23–24; British government funding to, 29; Brown as, 55; collecting as business of, 110; in Colombia, 24; Darwin as, 150, 158, 166; development of core principles and practices, 30; European, 24, 131; Koster as, 27; Miers as, 90; naming and classification systems of, 57, 59, 184; number of specimens collected by, 38; orchids as focus of, 23, 24; packing and shipping of collected specimens, 42, 43; Swainson as, 24, 29, 32, 38, 43, 47
natural resources, 14, 20, 22–23, 90; Caminha reports on, 20, 22–23; colonial view of, 195, 303n12; extraction and exploitation of, 6, 7, 14, 22, 23, 195–200, 301n34; Gardner interest in, 90
natural selection, 152–153, 156–159, 293, 339n5; orchids in, 162, 167, 168, 169–170; religious concerns about, 152, 153, 158, 168–170; Wallace on, 153, 155, 156–157, 158
Natural System, 60, 98, 327n2; and Artificial System, 59, 66, 317n21, 327n2; Lindley on, 66, 77, 321n15; Linnaeus on, 59
Nevill, Dorothy, 122
New Guinea, Micholitz plant hunting in, 269, 270
New Jersey, Sander nursery in, 179–181
newness, in Victorian period, 147–148
New Remedies, 174
Newsham, J. N., 346n15
newspapers, 81, 215, 288, 323n31
New York, nursery businesses in, 179
New York Horticultural Society, 174
New Zealand, Swainson in, 46
Nordau, Max, 197
Norfolk, 40, 53, 57, 73
Norwich, 40, 54
"Notes on the Revolution" (Swainson), 34
Novalis, Georg, 351n10

novels: Gothic, 75; of nineteenth-century, importance of names in, 68, 319n20
Nuphar pumila, 86
nursery businesses, 1–2, 4. *See also* specific nursery businesses.; growing interest in, 53, 315n1; Hooker recommending to Darwin, 159, 160–161; importance of location for, 53–54, 110, 127, 128, 129; proximity to railway stations, 110, 129; transportation and technology developments affecting, 5, 6, 288; in United States, 179–181; World War I affecting, 280, 284, 291
Nymphoea ampla, 89

O'Brien, James, 182, 212
"Of Queens' Gardens" (Ruskin sermon), 115, 332n1
Old Antiquity, Lindley known as, 54
Oliver Twist (Dickens), 242
Oncidium, 64
Oncidium barbatum, 313n24
Oncidium pactum, 132
"one culture," 9, 298n22
"On Hybridization Amongst Vegetables" (Herbert), 136
"On Seeing the Elgin Marbles" (Keats), 82
On the Origin of Species (Darwin), 158, 163, 164, 168, 171, 184, 293
"On the Tendency of Species to Form Varieties" (Darwin), 342n29
"On the Tendency of Varieties" (Wallace), 153, 156
On the Various Contrivances by Which British and Foreign Orchids are Fertilised by Insects (Darwin), 162, 164, 165, 169
Ophrys ananifera, 139
"Orchid, The" (musical), 276, 277f–279f
Orchidaceae of Mexico and Guatemala, The (Bateman), 103, 207, 282
Orchid Album (Warner & Williams), 194
orchid-craze in United States, 174, 177
orchid cultivation: Banks success in, 44; democratic language on, 175–176; and domestication syndrome, 292; as elite interest, 43, 103–104, 105; greenhouses for, 42–43, 44, 105, 107, 109, 331n13; by Hooker, 44–45; and hybridization (*see* hybridization of orchids); in India, 174; insect problems in, 173; by Lawrence, 115–121, 122; as middle class interest, 104–110; and pollination, 138, 139 (*see also* pollination of orchids); and propagation (*see* propagation of orchids); Rand manual on, 183; and relocation by Indigenous people, 23, 304n21; from seeds, 137–140, 144, 174, 292; in United States, 174–181; Williams on, 106–109; by women, 115–121, 177
"Orchidées" (Silva painting), 242
orchidelirium, 1, 174, 272
orchid fossils, 23, 304n19
Orchid Grower's Manual, The (Williams), 106–107, 330n11
orchidology, 111; role of women in, 121–122; in United States, 175, 176, 181
orchidomania, 14, 173, 257, 258, 282, 291; Bateman on, 103–104; and Darwin, 164, 165, 168; decline of, 273; environmental impact of, 197; Ericsson in, 1, 14; Forget on, 247; in globalization, 7, 194; lost orchid in, 150; in middle class, 106, 194; narratives of, 276; and orchidelirium, 1, 174, 272; Sander in, 1; technology developments leading to, 288; Veitch in, 129
orchidophiles, 138, 150, 175, 208, 281
Orchid Review, 274, 280, 283, 284
Orchids (Rand), 175
orchid seeds, 137–140, 144, 174, 292
Orchid Seekers, The (Russan & Boyle), 235–236, 237f
"Orchids for the Million," 104, 106, 109, 140, 196, 331n13
Orchid Thief, The (Orlean), 14
Orchis, 75–76, 139
"Orchis Longicornu," 27
Organ Mountains: Lobb exploring, 132, 133, 182, 336n12; Swainson exploring, 36, 182
Orient, as exotic locale, 74–75
Orientalism (Said), 74
Orinoco River (Venezuela), 219
orkhis, 17, 75
Orlean, Susan, 14, 298n17
Osmers, J. D., 186, 187
ovaries of orchids, 138, 139
Oversluys, Cornelius, 252, 254; arrival in Brazil, 246, 250, 358n10; and Claes, 251, 358n10; on codes in messages, 253; death of, 359n26; departure from Brazil, 266; in Pernambuco, 250, 251; Wanderley mentioned by, 264; yellow fever concerns, 255
Owen, Richard, 158, 342n29

packing of specimens for shipping, 26–27, 43, 306n5
Pádua, José Augusto, 195
"Painter Estêvao Silva, The" (Pereira), 243
Palermo, 26, 306n7
Pal-Lapinski, Piya, 320–321n6

Index

Palmer, Charles, 192, 216, 217–218, 349–350n28
palms, 122, 193, 196, 203, 204, 240, 242, 303–304n14, 334n21
Panton, Jane Ellen, 273
papermaking, 298n16
Paphiopedilum rothschildianum, 287, 362–363n3
Paquevira, *Cattleya warocqueana* collected in, 254
Paradise Nursery, 150, 161
Paraíba (Brazil), 17, 292; Bungeroth in, 224; Forget on *Cattleya* habitat in, 249; Gardner in, 93; rebellion and provisional government in, 33; republican idealism in, 33
parasitic plants, orchids viewed as, 42; by Swainson, 36, 37, 38, 39, 40
Parc Léopold, 202, 208, 210
Paris World's Fair (1889), 8, 203–206, 239–240, 252, 352n1; image of orchids in catalog of, 205f; L'Horticulture curation of display at, 204–206, 239–240; Silva painting of orchids at, 239–244, 251
Parker, Robert, 159, 161, 162, 164
Parkes, John, 315n22
parlour plants, 330n6
Parthenon Sculptures, 110
Patmore, Coventry, 120
Pauwels (former Sander collector), 284
Paxton, Joseph, 38, 69, 83, 120, 130
Paxton's Flower Garden, 100
Paxton's Magazine of Botany, 83, 84, 99, 101, 129
Peace and Friendship Treaty (1810), 32
Pearce, Richard, 160
Pedra Bonita mountain, 92, 96
Pedra de Gávea, 91–92
Pedro (assistant of Gardner), 89–90, 91
Pedro II, Emperor, 195, 203, 243
Pedrosa, Pedro da Silva, 33
Peel, Robert, 105
Penang, 196
Penedu (Penedo), 35
Peninsular War, 29
Pennsylvania Horticultural Society, 174
Pereira, Edimilson de Almeida, 243
Perkins, Harold, 112
Pernambuco (Brazil), 17, 292; British presence in, 31; Bungeroth in, 220, 238, 239, 240, 244; enslaved people in, 31–32, 308nn22–23; Ericsson in, 186, 207, 220, 250, 255; Forget in, 247, 249, 250–251, 274; map of, 28f; orchid habitat in, 310n43; Oversluys in, 250, 251; rebellion in, 31–34, 309n29; Swainson in, 25, 31, 88, 220, 305n1; Waterton in, 88
Perthius, Léon, 256

Perthius, Louis, 256–257, 359n25
Peru, 133
Petropolis, 262
Phalaenopsis, 161, 270, 287
phenotypes, 162
Philippines, 144
Philodendron, 249
Piauí (Brazil), 17, 302n2
Picture of Dorian Gray, The (Wilde), 242
Picturing Plants (Saunders), 77, 321n13
pistils, as wives in Linnaean system, 58
plague (1814), 26
plantations: for coffee production, 23, 88, 196; for cotton production, 32; deforestation for, 23; labor of enslaved people in, 37; for sugar production, 23, 32, 88, 196
plant collecting, 110–114; and classification, 110–111, 112; culture of, 13; increase in plant hunting for, 112; tools and equipment in, 110, 331n17; as wealthy pastime, 110; by women, 114, 115–118
plant hunters. *See specific plant hunters*: of British Empire, 6–7; Bungeroth, 3–5; codes in messages of, 253; competition between, 188, 189, 250–256; conditions experienced by, 3, 12; death on Columbia River, 130; Digance, 2–3; on emptiness of Brazil, 88, 89; environmental impact of, 3, 10, 112, 195–199, 250, 273–274, 290–291, 296–297n14; Ericsson, 1–5, 216–224; funding of, 2, 253, 254, 295n2; Gardner, 86–96; letter writing by, 10–12; of Linden family (*see* Linden family plant hunters); Lobb, 131–133; local Brazilians working with, 258–267; of Low's nursery, 187–188, 189–190; Micholitz, 269–270, 271, 272; naming and classification of specimens by, 186, 188, 190; narratives and stories on, 9–10, 299n24 (*see also* narratives and storytelling); national-imperial infrastructure of Britain supporting, 90; published accounts on, 258–262; resistance of Indigenous people to, 91; of Sander (*see* Sander plant hunters); shipping problems experienced by, 132, 133; Swainson, 24, 25–39; from United States, 181, 346n15; of Veitch nursery, 131–133, 160; Wanderley, 264–267
Play-Pictorial, 276, 277f–279f
Pliny the Elder, 77
Pocket Herbarium of British Mosses (Gardner), 87
pollen of orchids, 163–164, 165, 166, 167
pollination of orchids, 138, 139; Darwin on, 162, 163–164, 165–168, 169–170; environmental damage affecting, 290, 291; insects in, 162, 163–164, 165–167

pollinia, 23, 163, 167, 304n19; orchid classification based on number of, 101, 328n13
Port of Spain (Trinidad), 192, 216
Portugal: arrival of colonists in Brazil, 17–20; Brazil declaring independence from, 70, 87; Brazil under rule of, 29, 32, 33; British use of ports in, 29; royal family living in Brazil, 29, 32
Posey, Darrell Addison, 304n21
Pothos, 249
Potiguara tribe in Brazil, 19
Power of the Story, The (Hanne), 9, 299n24
Pratt, Mary Louise, 324–325n7
Priceless Orchid, The (Ainslie), 235, 258, 261f
prices paid for orchids, 4–5, 73–74, 109–110; for *Cattleya labiata*, 73–74, 109–110, 122, 246; Endersby on, 296n10; for hybrids, 146–147; in Loddiges auction, 122; in Stevens auctions, 122, 206; in United States, 177
Prince's nursery (NY), 179
printing press technologies, 81
print media: advertisements in, 215, 288 (*see also* advertisements); *Cattleya labiata* featured in, 7, 74, 81–85, 226–231; on collecting, 111; color illustrations in, 78–80, 81–82, 84; correspondence in, 10–12; development of, 5, 6, 81; for general public, 215; magazines in, 329n3 (*see also* magazines); for middle class, 9, 106–107, 164; narratives and storytelling in, 9–10, 299n24 (*see also* narratives and storytelling); orchid growing books in, 106–107; papermaking for, 298n16; for United States orchid growers, 174, 175
professional groups and societies, 112, 113
propagation of orchids, 84, 123, 292–293; myths on, 138; from seeds, 137–140; by Veitch nursery, 84, 142
Protheroe, William, 280
Protheroe & Morris auctioneers, 226
Provincia newspaper, 266
pseudobulbs in orchids, 67
Pytchley Horticultural Society, 113

"Queen of the Orchids," lost orchid as, 4
Quinarian system, 51

railways, 5, 6, 110, 129
Rand, Edward Sprague, Jr., 175, 176–177, 183, 346n15
Raphael's Almanac, 127
Recife (Brazil): British communities in, 29, 31; drought and famine in (1816), 32; enslaved people in, 31–32, 308nn22–23; independent-minded secret societies in, 32; lodges and republican ideals in, 32, 33, 308n24; racial make-up of population in, 31–32, 308n21; rebellion in (1817), 31–34, 309n29; Swainson in, 25, 31, 35, 309n35
Reddie, John, 96
Rees-Mogg, Jacob, 12–13
Reichenbach, Heinrich Gustav, 145–146, 190, 192, 206, 207, 352n6
Reichenbachia (Sander), 207, 282, 362n17
Reinikka, Merle A., 7, 298n17
religious issues: in hybridization, 135, 136, 137, 336–337n4; in natural selection theory, 152, 153, 158, 168–170
Representation of the People Act (1832), 127
reproduction, Linnaean system based on, 58–59
republicanism in Recife, 32, 33, 308n24
Republic of Pernambuco, 33
research methods: narrative approach in, 13; sources of information in, 10–13
Richards, Thomas, 288
Rio de Janeiro: British communities in, 29; Digance death in, 218; Portuguese royal family in, 32; Silva in, 242; Swainson travel to, 35–39
Rio Formoso, 250–251
Rio Ipojuca, 36
Rio Meta, 5, 219
Rio Parahyba, 93
Rio Pinto, 233–234, 244, 250
Rio Tinto, 234
Rise of Professional Society, The (Perkins), 112
Robert (ship), Swainson cargo on, 35
Robinson, William, 197
rock art, 17, 302n2
Rockefeller, John D., 171
Roezl, Benedict, 2, 150, 176, 235, 237, 350n30, 356n13
Rolfe, Robert Allen, 101, 208, 210, 211, 244–245, 292
Rölker & Sons nursery, 179
Rollisson's nursery, 116, 129, 130, 150, 335n5
rostellum of orchids, 163, 166, 344n25
Rothschild's orchid, 287, 362–363n3
Royal Botanic Gardens: in Glasgow (*see* Glasgow Botanic Garden); in Kew (*see* Kew, Royal Botanic Gardens at)
Royal Botanic Society, 116
Royal Horticultural Society, 117, 147. *See also* Horticultural Society; *Cattleya warocqueana* shown at, 208, 210; in post-World War I period, 282–283
Royal Mail steamers, 222, 224, 246
Royal Mail Steam Packet Company, 252–253
Royal Society, 41, 310n48
rubber tree seeds, 196

Rubery, Matthew, 215
"Rules of the Pytchley Horticultural Society," 113
Ruskin, John, 115, 120, 332n1
Russan, Ashmore, 235–236
Russell, John (Duke of Bedford), 87, 90, 92, 103
Russia, trade with Britain, 62–63, 318n2

Said, Edward, 74
Sala, George Augustus, 148
Sander, Charles Fearnley, 284
Sander, Frederick, 1–5; advertisements on orchids for sale, 226–229, 268; Beaufort compared to, 282; Boyle as supporter of, 190, 193, 228, 242, 244, 272–273; business approach of, 215–216, 246, 256, 274–276; and *Cattleya warocqueana*, 214, 216, 217, 220, 221, 225, 245; death of, 285; on discovery of lost orchid, 244–245; and environmental damage in Brazil, 197; establishing St. Albans business, 150; "Fortune from Orchids," 176; Godseff as business manager for (*see* Godseff, Joseph); health of, 280, 284–285; honors awarded to, 272; identity of orchids sold by, 185; international locations of nurseries, 1–2, 176, 179–181; Kew archive of correspondence, 186; Linden postcard to, 206, 352n5; marketing stories on orchids, 234–236, 245–246, 268–269, 270, 271, 274–276, 355–356n5; Moreau visited by, 8; personality of, 190–191, 236; photographs of *Cattleya labiata*, 229, 231, 232f; promoting British greatness, 216; and Queen Victoria, 1, 174, 272; *Reichenbachia*, 207, 282, 362n17; rival businesses of, 2, 4, 202–213, 229–238, 240, 245–246, 263; United States business of, 2, 176, 179–181
Sander plant hunters, 250–257. *See also specific plant hunters*.; Arnold, 186–193; correspondence of, 11, 216, 250; Digance, 2–3; Ericsson, 1–5; Forget, 247–251; local people working with, 258, 263–266; marketing stories on, 234–236, 245–246, 268–269, 270, 271, 274–276; Micholitz, 269–270, 271, 272, 280; at Mount Roraima, 242; Osmers, 186, 187; Oversluys, 250, 251, 252, 254; payment of, 2, 253, 254, 295n2; Perthius, 256–257; spying on competitors, 216; Swinson on, 347n11
Santos, Cecília Resende, 203
São Francisco River, 35, 187
Sapucaya (Brazil), Gardner at, 93, 101
Saturday Review, 228

Saul, John, 179
Saunders, Gill, 77, 321n13
scarlet fever, 26, 154
Schaffer, Talia, 118
Schiebinger, Londa, 306n6
Schmitt, Cannon, 90, 300n26, 325n13
Schroder, Johann [John] Heinrich, 206
science: as delightful, 81, 82, 83; gentlemen in, 41, 311n5; local people as invisible in writing on, 325n12; rise of profession in, 30, 310–311n3
Scottish Highlands, 87
sedan chairs, 31, 308n21
Seden, John, 146
seeds: of orchids, 137–140, 144, 174, 292; of rubber trees, export of, 196; Swainson collection and shipping of, 38, 43
Selborne Society, 197, 351n6
Select Orchidaceous Plants (Warner), 173
self-fertilization, 162, 163, 291
sequoia, 133
"Serapius," 147
Serra da Capivara National Park, 17
Serra das Russas, 36
Serton (Certon), 249
sexuality: of exotic women, 75, 320–321n6; Linnaean system based on, 58–59, 76, 77; orchids associated with, 75–76, 137–138
Sheffield Independent, 273
Shelley, Mary, 125, 134
Shephard, Sue, 133, 146, 298n17
Shepherd, J., 26–27, 306n5
shipping: cost of, 254; development of networks for, 5, 6; Ericsson concerns about, 222, 223; Lobb loss of plants in, 132, 133; Shepherd advice to Swainson on, 26–27; survival of specimens during, 26, 42; Swainson packing of specimens for, 26–27, 43, 306n5; to wrong importers, 191–192, 349n26
Shore, Marci, 13–14, 301n34
Shtier, Ann, 115, 332n1
Sicily, 25, 26, 27, 306n7
Siebrecht & Wadley nursery (NY), 179, 179f, 193, 197
Silva, Estêvao: death of, 244; orchid painting of, 239–244, 251
Silva Pedrosa, Pedro da, 33
Silver Knightian medal, 115
Singapore, 133
"Sketch of a Journey Through Brazil" (Swainson), 29, 31, 34, 35–38, 324n6
Skinner, George Ure, 103
Skirving, Mr. (in Liverpool), 92

slavery. *See* enslavement
Smith (Low's plant hunter), 187–188
Smith, James Edward, 27; and Banks, 59; and Hooker correspondence, 50, 312n20, 314n14; as Hooker supporter, 41, 311n4; purchasing Linnaeus collection, 59; and Swainson correspondence, 27, 306n7
Smith, Jonathan, 321n12
Soho Square (London), Banks library and herbarium in, 55–57, 60, 316n10
soil erosion in agriculture, 196
Solander, David, 59
South America: Bungeroth in, 239; Waterton in, 88–89
Southwark (London), 57
Spathoclottis ericssoni, 356n13
species: Darwin on, 152–153, 184; debates on varieties and, 137, 184, 186, 215, 270, 347n6; extinction threats, 197, 291; Herbert on plasticity of, 152, 153, 162–163; Linnaeus on number of, 152; Wallace on, 184
"Splendid-flowered Catleya," 83
Spruce, Richard, 207
Sri Lanka (Ceylon), 96
Staffordshire, Bateman at, 103–104, 147, 328–329n1
St. Albans (England), Sander business in, 1–2; Boyle description of, 193; Ericsson sending plants to, 224, 255; Ericsson visit to, 216, 218, 354n12; establishment of, 150; international expansion of, 179–180; and marketing stories, 236; photograph of, 180f; plant hunters reporting to, 186, 252
stamens, as husbands in Linnaean system, 58
Stamp Duty, 81
Standard (newspaper), 46
Stanley, Henry Morton, 234–235
"Stanley's Adventures in Africa," 235
sterility of hybrids, 137, 142, 184
Stevens, J. C., auction business of, 122, 187, 206
Stewart, Susan, 200
St. James's Gazette, 228, 268
Stockholm, Ericsson in, 3, 353n3, 359n23
Stoker, Bram, 199
storytelling and narratives. *See* narratives and storytelling
stoves for greenhouse heating, 42–43, 44, 312n12, 312–313n22
St. Petersburg, 7, 62, 63, 239
Strand (magazine), 81
Strunz, August, 252, 295n2
St. Salvador, 35
St. Thomas, 255

Suburban Gardener and Villa Companion (Loudon), 107, 109, 116, 333n3
suburbs: gardening in, 107, 109, 149, 330n12; growth and development of, 52, 63, 107, 128–129, 151, 288, 318n5, 335n2; house plants in, 273; middle class in, 52, 63, 107, 128–129, 288, 318n5, 335n2; new homes and villas in, 147, 273; orchids in, 148
Suffolk, 35, 38, 40, 45, 49, 50, 61, 69, 123, 182, 313n22
sugar production, 23, 32, 88, 196, 308n23
Sumatra, 56, 63
Summit, NJ, Sander nursery in, 2, 179–181
Swainson, William, 25–39; accused of unpaid debts, 47–48; on agriculture in Brazil, 196; application for natural history position, 48; arrival in Brazil (1816), 24, 25, 31, 305n1; and Banks correspondence, 39, 310n48; and Broderip, 47; *Cattleya labiata* collected by, 182, 183, 248; collections and letters sent to England (1817), 35; departure from Recife (1817), 35, 309n35; diaries of, 32–33, 37–38, 309n31, 309nn33–36; diversity of specimens collected by, 38; emigration decision of, 51–52; Ericsson following route of, 220; *Exotic Conchology*, 47; family of, 51, 315n20; health of, 32–33, 34, 35, 48; and Hooker correspondence, 50–51, 314–315n16; Hooker receiving specimens from, 38, 40, 42, 43–44, 45, 182; Hooker requesting Brazilian duplicates from, 50–51; hunting ground of, 36, 226, 233, 240, 244, 248; and Koster, 27–29, 33, 88, 307n10, 307n13; lack of interest in lost orchid, 45, 46–47, 50, 69; in Linnean Society, 30, 306n7; and Macleay correspondence, 47, 48, 313–314n6; in New Zealand, 46; "Notes on the Revolution," 34; number of plants sent from, 55; "Orchis Longicornu" introduced by, 27, 306n7; in Organ Mountains, 36, 182; in Pernambuco, 25, 31, 88, 220, 305n1; Quinarian system used by, 51; and rebellion in Recife (1817), 31–34; self-depiction by, 31; and Shepherd correspondence, 26–27, 306n5; "Sketch of a Journey Through Brazil," 29, 31, 34, 35–38, 324n6; and Smith correspondence, 27, 306n7; specimen packing and shipping skills of, 26–27, 43, 306n5; speech impediment of, 25, 47, 48, 313n4; spelling problems of, 47, 48; traveling from Recife to Rio de Janeiro, 35–39; whistling and humming by, 25–26; wide-angle approach of, 38, 47; zoology interests of, 47

Swainson, William J., 315n20
Swan, Joseph, 83
Sweden, Ericsson in, 214, 216
Swinson, Arthur, 188, 347n11, 348n14
Systema Naturae (Linnaeus), 58

Tabajara tribe in Brazil, 18
taxes: of Britain in Ceylon, 96; on glass for greenhouses, 42, 105
telegraph network: development of, 5, 6, 189, 348n18; interception of cables in, 252–253
Tennyson, Alfred, 174, 289
Thames (Royal Mail steamer), 222
Thames River, drought affecting (1858), 151, 156
Thelwall, John, 47, 313n4
Theophrastus, x, 17
Through Forest and Plain (Russan & Boyle), 236
Tijuca mountains, 91–92, 96
Times (of London), 156, 165, 274
Tooting area of London, 128–129, 161
Topsail Mountain, 91–92
Torquay, Darwin residence in, 165
Torrington, Lady (Mary Anne), 96
Torrington, Lord (George Byng), 96
trade, 5–6; annual global sales of orchids in, 287, 362n11; of Brazil and Britain, 29; orchids as commodity in, 70; of Russia and Britain, 62–63, 318n2
transportation: new networks in, 196, 216, 288; shipping networks in, 5, 6; by ships and boats, 36, 39, 89, 90, 130, 132, 173, 188, 207, 219, 222, 255, 260, 355n23; in suburbs, 128–129; by trains and railways, 3, 5, 6, 23, 109, 110, 129, 194, 196, 215, 253, 289, 327n4
Travels and Adventures of an Orchid Hunter (Millican), 259f, 260
Travels in Brazil (Koster), 27–29
Travels in the Interior of Brazil (Gardner), 90, 97
Trinidad: Arnold death in, 192; Ericsson in, 218; orchid from, 99; Palmer in, 216
Triunfo, 264, 266
tuberculosis, 26, 154–155
tulip fever, 103, 174
Tupi-Guarani peoples, 304n17
Tupinambá people, Caminha reports on, 18–19
Turner, Dawson, 41, 42, 50, 63, 314n12
turnips, 53

unifoliate plants, 67
United States: Civil War in, 10, 174, 198; orchids in, 174–181, 282, 290; plant hunters from, 181, 346n15; and republican idealism in Brazil, 33, 309n29; urbanization and modern life as concerns in, 198
Upper Xingu region, 21
Uppsala University, 59
urbanization, 198
Ure-Skinner, George, 160
Utricularai, 89

Valpy, Edward, 54
vampires, 75
Vanda caerulea, 160
van den Berg, Cássio, 292, 310n43, 310n47, 328n13, 363n13
Van Houtte, Louis, 228, 355n2
vanilla, 75, 138
Variation of Animals and Plants under Domestication, The (Darwin), 292
Veitch, Harry, 132, 133, 182, 280
Veitch, James (Junior), 128–129, 133–134, 141; business approach of, 148–149, 160; and Darwin correspondence, 165; Lindley visited by, 144
Veitch, James (Senior), 123, 127–128, 130–133; and Darwin relationship, 130–131; displaying *C. labiata* in competitions, 133; health problems of, 141; and Hooker correspondence, 130, 131–132, 336n12; and Lobb, 131–133
Veitch, James Herbert, 142, 280
Veitch, John, 127, 280
Veitch, Thomas, 127, 335n1
Veitch & Sons nursery, 141–150; and Darwin, 159, 160, 164, 165, 166; Dominy as horticulturalist in, 140, 142–146; on epiphyte species at Kew, 42; growth of business, 116; Harris visit to, 140, 143; and Hooker, 141, 159, 342n2; *Hortus Veitchii*, 131, 137, 142, 143, 335n12; locations of, 127, 128, 129, 141, 149; on lost orchid, 4; *Manual of Orchidaceous Plants* (see *Manual of Orchidaceous Plants*); orchid propagation and hybridization by, 84, 142–150, 160; orchid varieties at, 183; plant hunters employed by, 131–133, 160; Reichenbach visit to, 145–146; as Sander rival, 4; sending orchids to US, 177
Venezuela, 4; Arnold in, 186, 188, 262, 275; Bungeroth in, 219; Ericsson in, 219; orchids collected in, 98, 99, 206, 217
Verschaffelt, Ambroise, 228, 355n2
Vespucci, Amerigo, 19–20, 302–303nn6–7
Victoria, Queen, 1, 121, 174, 272
Victorian Medal of Honor, 272

Victorian period, 12–13; anxiety in, 288, 289; collecting culture in, 13, 110–111; colorful gardens in, 142, 149; crafting in, 117, 118; environmental concerns in, 197–198, 199; gardening books in, 107; middle class as the "million" in, 9; narratives on orchid hunting in, 9; orchid collections in, 112; orchidomania in, 14 (*see also* orchidomania); print advertisements in, 215, 353n5; role of women in, 115–121, 332n1; science in, 93, 123, 131, 134, 142, 152, 157, 174, 248; "Sketch" of Swainson in, 35, 36; suburbs in, 109, 128, 129, 148, 333n3, 335n2; values on new and old in, 147–150; wild garden aesthetic in, 149, 274
Victoria Nursery, 150, 161
Vila, Pablo, 24
Vincente (Swainson assistant), 37
von Humboldt, Alexander, 24, 305n23
von Langsdorff, Georg Heinrich, 23–24, 305n22
Voskuil, Lynn, 10, 299n25, 312n16, 358n5
voting rights, 118, 127, 333n8
Vratz (plant collector), 219–220

Wachter, J. K., 138
Waddel, Nancy Mann, 177
Wallace, Alfred Russel, 153, 155, 156–157, 158, 184
Wallace, Robert, 130
Wallis, Gustav, 2, 350n30
Wanderings in South America (Waterton), 88
Wanderley, Ovidio Bruno do Nascimento, 264–267
Ward, Nathaniel Bagshaw, 105, 330n6
Wardian cases, 105, 330n6
Warner, Charles, 106
Warner, Robert, 173, 183, 194
Warocqué, Georges, 208, 280
Waterton, Charles, 88–89, 90
Watson & Scull agents, 229
Ways of Seeing (Berger), 321n12
"Wellingtonia" sequoia, 133
Wells, H. G., 71, 76, 199, 234, 271
Wells, Mrs. (of Cowley House), 121–122
Wharton, Edith, 280–282

White (Low's plant hunter), 188; and Arnold, 188–190, 348–349n19
White, Hayden, 13, 300n32
White, Robert Gostling, 314n13
Wickham, Henry, 196
Wide World Magazine, 275
Wiener, Martin, 113
Wilde, Oscar, 242
wild gardens, 149, 274
Williams, Benjamin Samuel, 106–109; on *Cattleya labiata*, 183; and Darwin, 159, 161, 162; *Orchid Album*, 194; on orchid culture in US, 175; orchid exhibition of, 108f, 109; *Orchid Grower's Manual*, 106–107, 330n11; Victoria and Paradise nurseries, 150, 161
Withering, William, 30
Wittman, Emily O., 49, 314n11
Wollaston, Thomas, 152
women: as gardeners, 115, 121, 332n1; in horticultural competitions, 115–116, 118–119, 120, 121, 122; as orchid collectors, 114, 290; orchid cultivation by, 115–121, 177; as orchidologists, 121–122; in professional organizations and societies, 113; in United States, 177; voting rights of, 118, 127, 333n8
World's Fair of 1889 (Paris). *See* Paris World's Fair
World War I, 14, 280, 282, 291; orchids in years following, 282–284
Wray, Martha, 122
Wrench, Jacob, 54, 65, 316n6
Wrench & Sons seed business, 54
Wulf, Andrea, 42, 311n10, 316n10

Xanthopan morganii praedicta, 167
xenophobia, 274

yellow fever, 26, 255
Youell, H., 109
Young, I. J., 179
Young & Elliott Seedsmen, 179, 191, 192

Zoological Illustrations (Swainson), 47
zoology, 29, 47
Zurich Botanic Garden, 196